An Introduction to Mechanics and Modelling

An Introduction to Mechanics and Modelling

D. G. Medley
*Department of Applied Mathematical Studies,
University of Leeds*

 Heinemann Educational Books

Heinemann Educational Books Ltd
22 Bedford Square, London WC1B 3HH

LONDON EDINBURGH MELBOURNE AUCKLAND
HONG KONG SINGAPORE KUALA LUMPUR NEW DELHI
IBADAN NAIROBI JOHANNESBURG
EXETER (NH) KINGSTON PORT OF SPAIN

© D. G. Medley 1982
First published 1982

British Library Cataloguing in Publication Data
Medley, D. G.
   An introduction to mechanics and modelling.
   1. Mechanics
   I. Title
   531   QC125.2
   ISBN 0-435-52560-3

Printed by J. W. Arrowsmith Ltd, Bristol BS3 2NT

Contents

	Page
Preface	ix
Acknowledgements	xii
Introduction: Space, time and modelling	1
0.1 Some remarks about modelling	1
0.2 Modelling of time	3
0.3 Modelling of distance	4
0.4 Space	4
0.5 Symbols, definitions and laws of nature	6
0.6 Conclusions	10

Part I   Modelling of space by vectors, and an introduction to modelling in the real world

1 Vector algebra   15

   1.1 Introducing vector algebra
   1.2 Meaning conveyed by a vector symbol **a** and the algebra which follows   17
   1.3 Geometrical results demonstrated by use of vector algebra   24
   1.4 Uses of base vectors **i**, **j**, **k**; components   27
   1.5 Examples   30
   1.6 Magnitude and direction of three-dimensional vectors   34

2 Differentiation of vectors; accelerations; scalar and vector products and geometry in three dimensions   38

   2.1 Sequences of vectors   38
   2.2 Velocity and acceleration   39
   2.3 Components of velocity and acceleration   46
   2.4 Products of vectors   52
   2.5 Use of scalar and vector product notations   55
   2.6 Algebraic manipulation of scalar and vector products   62
   2.7 Applications of vector algebra to geometry in three dimensions   65
   2.8 Linear dependence and independence of vectors   76

3 Modelling in the real world: a particular problem and some general comments   79

   3.1 A problem in plumbing and geometry   79
   3.2 General comments   84

4 Rudimentary ideas of force and mass; conventional modelling in elementary statics and dynamics; Newton's definitions and laws of motion — 86

- 4.1 Chosen mathematical attributes of force — 86
- 4.2 Physical attributes of force; forces in relation to physical objects — 87
- 4.3 Modelling the downward force of gravitation; 'weight' and 'weightlessness' — 93
- 4.4 Modelling of strings and tension; Hooke's law — 94
- 4.5 Newton's second law: the nature of mass — 100
- 4.6 Examples of the use of Newton's second law — 103
- 4.7 Identifying forces — 113
- 4.8 Newton's definitions, his laws of motion, and a useful consequence — 116

Part II Understanding the motion of real bodies

5 Mass and force; momentum; forces of nature — 121

- 5.1 Mass and force — 121
- 5.2 Momentum and impulse — 125
- 5.3 Forces of nature — 132
- 5.4 Empirical laws, accuracy and approximation — 147

6 A system-of-particles model; centre of mass; motion of a rigid body in two dimensions — 150

- 6.1 Modelling a body as a system of particles — 150
- 6.2 Interpretation of $\Sigma m_i \mathbf{r}_i$; centre of mass — 153
- 6.3 Summary of the principal algebraic properties of the centre of mass — 155
- 6.4 Identification of additive sets of forces — 156
- 6.5 Examples using centre of mass relations — 156
- 6.6 Methods of finding the position of the centre of mass of simple rigid bodies (without integration) — 159
- 6.7 Motion of rotation of a rigid body — 163
- 6.8 Rigid body rotation about a fixed axis — 176
- 6.9 Summary of formulae valid for any body of matter — 182

7 Motion and differential equations — 186

- 7.1 Some basic tools of calculus — 186
- 7.2 Special significance of the exponential function in discussions of linear differential equations — 188
- 7.3 Forced oscillations and inhomogeneous linear differential equations — 203
- 7.4 Models of fall in a resisting medium — 210
- 7.5 Linear equations with variable coefficients — 214
- 7.6 Fall of a raindrop – an exercise in practical modelling — 217

8 Equivalent systems of forces; centre of gravity; forces between rigid bodies — 225

- 8.1 'Equivalent' systems of force in two dimensions — 225
- 8.2 Equivalent systems of force in three dimensions — 229
- 8.3 Centre of gravity — 235
- 8.4 Dynamically equivalent bodies in two dimensions — 240
- 8.5 Representation of the forces of contact between rigid bodies (in two dimensions) — 243

9 Mechanics applied to the real world; frames of reference in the real universe — 251

- 9.1 A 'real' problem (loaves of bread down a chute), and the application of mechanics to it — 251
- 9.2 The loaf of bread problem (stage 2), and the idea of 'impulse' — 252
- 9.3 Laboratory, earth-centred and sun-centred frames of reference — 258
- 9.4 'Pot Black': a discussion of the motion of billiard/snooker balls — 265
- 9.5 Appendix: a first model of loaves of bread down a chute — 274

Part III Energy modelling in elementary mechanics

10 The energy model — 285

- 10.1 You pays your money and you takes your choice — 285
- 10.2 Particle mechanics and energy — 286
- 10.3 Power — 302
- 10.4 Work done by forces acting on extended systems, and consequences for the use of potential energy — 303
- 10.5 Kinetic energy of a system of particles, or rigid body; applications of energy modelling — 309
- 10.6 Discussion of the equilibrium of bodies acted upon only by conservative forces — 320

Answers to exercises — 327

Abbreviations and symbols — 336

Index — 338

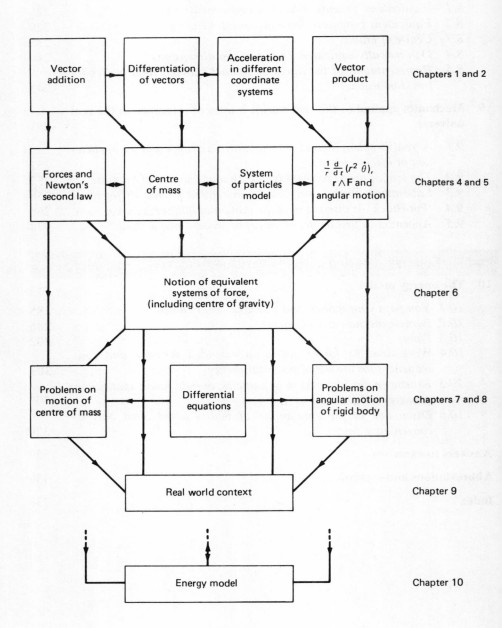

Flow-diagram model of mechanics

Preface

To the student

Your lecturer will not just work his way through this book. He will probably look at individual problems, explain the mechanisms of 'solving' them, and expect you to solve other problems of similar type by imitation.

Sometimes even though you grasp the main lines of the routines to be followed, you will be uncertain about *why* they are applicable or *when* they may be used. At other times, your lecturer will raise questions which are too deep to discuss fully during lecture hours. This book is concerned to enable you to see *why* and to recognize *when* and to penetrate to the root of elemental mechanical ideas.

If time is short (as it usually is!) you need not attempt to read this book from beginning to end; there is a good index – use it. Reference data may also be tracked down by use of the index.

Some of the questions set in the exercises are easy, others difficult. Make sure you can do the easy ones. When in difficulty, a look at the answers in the back of the book may help. If the question still defeats you, don't struggle with it hopelessly, but press on to the next topic. As your knowledge builds up, earlier difficulties often drop away.

After you have grasped the method of tackling several problem types, consecutive reading of the book should help you to integrate your ideas, and thereby fortify your understanding of mechanics over a broad field.

Numbering of equations, diagrams and paragraphs

The **equations** within each chapter are labelled by consecutive natural numbers prefixed by the chapter number. Thus the equations of Chapter 5 are labelled 5–1, 5–2, 5–3 and so on. **Diagrams** are labelled according to a similar scheme, but use a point rather than a hyphen; thus the diagrams of Chapter 2 are named FIG. 2.1, 2.2, 2.3 etc. The **paragraphs** run in decimal order throughout the book. Since the decimal number 5.33 is less than 5.4, paragraph 5.33 (§ 5.33) comes before § 5.4 in the book. This system has the advantage of indicating that all the paragraphs between § 5.31 and § 5.399 are subsections of § 5.3.

Short lists of abbreviations and of those symbols which are used in a standardized way throughout the book can be found immediately before the Index at the end of the book.

Important phrases or words are shown in **bold** type; and

> important longer statements are displayed, like this.

To the lecturer

It has been said (a) that the difficulty of mechanics lies in its many implicit assumptions; (b) that mathematics students should receive greater encouragement to read, and to be able to communicate in non-mathematical language – to use ideas as well as symbols; (c) that mechanics without the ability to model is like an egg without a yolk. This book offers – in addition to the standard content of a first course in mechanics – a body of reading matter relevant to the alleviation of (a), the pursuance of (b), and nourishment of the yolk of (c).

The three objectives are more closely related than might at first sight be supposed. Modelling is a process of identifying and representing significant characteristics, and the ability to model depends on a capacity to analyse. When ideas are indistinct, it is necessary to reformulate them in clear and explicit terms. In discussing ideas and making assumptions explicit we are taking the first steps along the road to modelling.

The stage of modelling which, more than any other, gives trouble to students of mechanics, is 'putting in the forces' on a diagram. The modelling of forces and the steps which lead to the concept of equivalent systems of force are therefore given priority in the early chapters of this book. It makes sense to find out how to set up mathematical equations of motion before discussing methods available for their solution: an early acquaintance with centres of mass is more important for the understanding of mechanics than is an early introduction to special techniques for integrating differential equations.

Such views lead to the order of treatment adopted in this book. It is a logical order closely related to the theoretical structure of mechanics as illustrated on page viii. It covers in a systematic way all the essentials of elementary mechanics, and it provides the reader with the tools which will help him when he embarks on original modelling.

In the 1980s most universities will have to cater for students who have little or no prior acquaintance with mechanics, as well as students who have taken mechanics as a subject at G.C.E. Advanced Level or equivalent. Because the approach to mechanics presented in this book differs substantially from that usually chosen in schools, it opens up the possibility of catering in an appropriate way both for those with some knowledge of mechanics and for those without – if necessary without splitting the class. It will provide new insight into the material which some students have already met: they will be helped to see mechanics as a coherent structure rather than the sequence of disconnected topics – 'projectiles', 'impulses', 'pendulums' and the like – with which they are familiar. The needs of the novice should be equally well met by the orderly introduction of ideas in Parts I and II. It is recommended that students in the novice category should attempt the majority of the questions set in the exercises of Chapter 4.

The author hopes that no mechanics student who reads this book will emerge from his first/second year mechanics course feeling that he has been required to perform tricks whose underlying logic remains a mystery to him.

In courses for first-year honours mathematics students, Part I is suitable for the support of a short induction course. As elsewhere in the book, omission of the more technically demanding sections, while retaining the line of development of mechanics, is recommended for lower-level courses. Where it is teaching policy to emphasize the use of 'cartesian vectors', i.e. linear combinations of **i**, **j**, **k**, a subsequent re-reading of the *early* part of Chapter 1 should be a salutary reminder of the essential vector properties which are appealed to in mechanics.

Part II contains the key to mechanics in the sense that, if the student has a good understanding of this material, he should be able to make creative use of his knowledge. At the most prestigious universities a mechanics course could start with the Introduction, followed immediately by Part II. Part III deals with energy modelling.

A proposed sequel volume, called *More Mechanics and Modelling*, will cover topics of mechanics which some university departments may think are more suitably treated in a second rather than a first university mechanics course, namely elementary continuum modelling (including area and volume integrals and rockets), 3D rigid body mechanics, orbits, Lagrange's equations, introductory relativity, and the relationship between mechanics and quantum mechanics.

The locations of some popular first-year topics in the present volume are as follows:

Particle in a resisting medium: §§ 7.312, 7.4;
Damped and forced harmonic motion: § 7.32;
Conservation laws: momentum § 5.22, energy §§ 10.1–5;
Centre of mass and centre of gravity: §§ 6.2, 8.4;
Rolling: §§ 9.24, 9.52.

The exercises are mainly straightforward, especially those in the early chapters. They are not meant as tests of ingenuity, but as reinforcements of understanding. Nevertheless, they are of varying standards of difficulty, and the lecturer can be sure that they are suitable for his students only if he vets them himself. SI units are mainly used, but, so that the student may become a practising applied mathematician, examples involving conversion of units are also included.

It is intended that book and lecture course should be complementary. The techniques of solution are best demonstrated by the lecturer in action. He quite properly skates over a certain amount of the thin ice of presumed knowledge in order to point the way ahead. The student requires a textbook *first* to consolidate the ice and provide a firm plank on which he (she) may follow the lecturer's course without the uneasy feeling that crucial supporting elements may be missing; *then* to promote confidence by demonstrating how the twin underpinnings of common sense and standard skills allow successful excursions to be made into the territory of real-world problems. It is desirable that the lecturer should guide the student to appropriate sections of the recommended text.

It is hoped that the grounding given by this book will be found appropriate for those who at some future date may find themselves teaching mechanics, as well as those who may use it. It may also be of use to high flyers in the sixth form.

Acknowledgements

Many friends have read and commented on parts of the manuscript of this book, or assisted in other ways. The author would like to thank them all. She would like to give special thanks to Professor Sir Hermann Bondi and Mr C. Ormell (who gave encouragement at a very early stage of the enterprise), to my colleague Dr Graham Ferris for reading the entire manuscript, and above all to Mr A. B. Civil whose meticulous care in proof-reading and correction has made a very large contribution to the production of the book. Mr Civil and my colleagues Mr Knapp, Dr Falle and Dr Garlick provided or checked many of the answers given at the back of the book, Dr A. West helped with FIG. 10.25 and Miss H. Diaper did the drawings for Chapter 4.

The author would welcome any comments or suggestions which could be used to improve the book.

Introduction: space, time and modelling

Parts of this introduction may be useful for tutorial, discussion and essay work. A quick look through may suffice for students who have no opportunity for group discussion.

0.1 Some remarks about modelling

The modern phrase describing the application of mathematics to real-life problems is 'mathematical modelling'. The activity itself is not a new one. The Chinese (before 2300 BC), the ancient Egyptians, Babylonians and Greeks all attempted to explain and predict natural phenomena by use of mathematical ideas and/or notation, i.e. they inaugurated mathematical modelling.

A mathematical model is a simplified representation of reality which employs mathematical concepts and/or symbols. Being a simplified representation, it mirrors only some aspects of reality and overlooks others.

The usefulness of a model is not necessarily increased by making the model more faithful, i.e. by overlooking fewer characteristics – though it is certainly a good thing if we can be conscious of what we are ignoring. We know full well that real matter consists neither of structureless 'point particles' nor of structureless 'continuous material', yet these are the useful Newtonian models which we retain in mechanics in preference to any 'more correct' picture provided by twentieth-century views about the wave nature of matter. We know that the earth and indeed the solar system are hurtling through space, yet the useful starting point is to conceive the room in which we sit as providing a stationary frame of reference. It is a matter for great satisfaction that it can be shown later (by the principles of Newtonian mechanics, see § 9.3) that only small corrections of value would result if, not the room, but the solar system or even the galaxy were regarded as providing the basic 'stationary' frame of reference. (Newton himself found it difficult to reconcile the necessity for abstracting selected characteristics from nature with his recognition of the fallibility of the abstraction process. So, although he at one point tried to distinguish the 'absolute time' used in his theoretical work from the 'relative or common time' of clocks, he elsewhere resorted to the statement 'I do not define time, space and motion as being well-known to all'.)

The theoretical development of mechanics requires the representation of real-life entities by mathematical symbols obeying mathematical rules. The representation of an interval of time by an algebraic symbol, t say, will be useful only if the symbol t can be assumed to obey certain rules of algebra, e.g. if the

2 An introduction to mechanics and modelling

sum $t_1 + t_2$ of two intervals of time has a sensible meaning in some contexts. The representation of time by an algebraic symbol is a modelling step. Definitions of physical quantities, and relationships between them are also necessary parts of a mathematical theory. Such steps may be referred to collectively as modelling postulates of the theory.

In describing accepted models, whether of wide areas of theory or of minor applications, we shall try to make the choice of one model rather than another seem reasonable, but it would be a mistake to regard the final choice as logically inevitable. Hindsight as well as rationality often influences the final choice.

In particular applications of established theory, the investigator himself must make decisions about how he is going to attack the problem: in other words, he must select his own modelling postulates. If they prove unsatisfactory, he can modify them and try again.

The mathematical achievements of the Greeks were preserved in the writings of Roman and medieval scholars. Their mathematical models of spatial relationships (geometry), musical relationships and the motion of the heavenly bodies came to be regarded as crystallizations of eternal wisdom rather than as approximations or pragmatic tools. Only after Newton did scholarly philosophers begin to recognize explicitly that the connection between reality and theory is an investigative one rather than a logically or theologically necessary one.

After a major new theory (i.e. a new model of nature itself) has been proposed, it is the task of the scientist to subject its predictions to stringent testing. Each test which the theory passes adds to its credibility. No theory can be proved beyond doubt to have universal validity, but well-tested models (such as those of Newtonian mechanics) may reasonably be regarded as reliable when applied within the range of circumstances for which, though searching tests have been carried out, no failure has been detected. The theory may be regarded as validated in this range. The success of Newtonian mechanics has placed it in some danger of being regarded – as was Greek science for so many centuries – as a crystallisation of eternal wisdom!

In the last few decades there has been increased effort among teachers of applied mathematics to break away from the more fossilized treatments found in traditional textbooks; the emphasis on 'modelling' is part of this attempt. We may distinguish two extreme varieties of modelling: on the one hand, the construction of an underlying theoretical framework; on the other hand, the application of those theoretical ideas to particular problems. The formulators of models of the first kind are the giants of mathematics – Euclid, Newton, Einstein and their peers. Most of us cannot aspire to that kind of achievement. We can only hope to benefit from the illumination provided by the great modelling schemes, and to apply those ideas in small ways and particular cases. These lesser applications constitute the sphere of modelling to which readers of this book may aspire.

Sciences less securely founded than geometry, mechanics and physics provide a less favourable field for creative modelling by beginners. Every new modelling is adventurous, but it is best to take off from relatively secure ground.

Readers may take heart and acquire resolve from the following assessment of Newton by one of his biographers (N. W. Chittenden, *c.* 1860): 'Quickness of apprehension or intellectual nimbleness did not belong to him. He dwelt fully, cautiously upon the least subject; while to the consideration of the greatest he brought ... a matchless clearness that bore with unerring sagacity upon the prominencies of the subject, and, grappling with its difficulties, rarely failed to surmount them.'

0.11 From basic theory to modelling

The student's first need is for a framework of ideas, so the first step in this book is:

> (1) To describe the elements of Newtonian modelling. Since the overall validity of Newtonian mechanics may be regarded as well-corroborated, an ordered – even didactic – account of its elements seems appropriate.

Mathematical arguments (sometimes called rational mechanics) within the exposition serve the purpose of making the subject hang together as a whole, and so assist understanding and creativity. Following (1) the next steps are:

> (2) To observe how the standard tools provided by Newtonian mechanics are used in problems of orthodox type. At this stage interest is concentrated on how the methods work, and what kind of 'predictions' can be made.
>
> (3) To provide similar artificial problems for the reader to 'solve' using the standard tools in the orthodox way. This phase of learning is of great value in spite of the artificiality of the problems.
>
> (4) To encourage the reader to take his courage in both hands, and himself choose idealizations of real-life situations. At this point instruction is self-defeating, but suggestions, guidelines and encouragement are provided (as well as caveats about conclusions).

Examples of real-life modelling are presented in Chapters 3, 7 and 9 and elsewhere. Unfortunately, once a real situation has been modelled in a textbook it ceases to provide an unexplored field in which the student can try out his independent skills, for the initiatory steps have already been taken. Modelling examples show how disciplined common sense guides the application of basic mathematical knowledge and technique.

We now direct our attention to the first step enumerated above, namely a description of the elements of Newtonian modelling.

0.2 Modelling of time

In our day, as in Newton's, philosophical probing seems to obscure rather than illuminate the nature of time. Our practical purposes are served by adopting the following modelling postulates.

(1) Any interval of time can be modelled by an algebraic variable t ($-\infty < t < \infty$).

4 An introduction to mechanics and modelling

(2) For the purpose of validating a theory the numerical values of t must be derived from readings of suitable clocks. (Two readings t_1 and t_2 are required, and $t = t_2 - t_1$.) The criteria of 'suitability', for purposes of testing, are:

- (i) that there are grounds for supposing that the time measured by the clock is not peculiar to itself; consistency of the chosen clock with other clocks of either the same or different designs supplies a good starting point;
- (ii) that the accuracy of the instrument can be estimated – usually by checking with the clock provided by the rotation of the earth against the background of the stars, or other standard clock;
- (iii) that in the context of the proposed experiment, the likely inaccuracy of the instrument does not affect the conclusions drawn.

The General Conference of Weights and Measures, 1964, proposed as standard an atomic clock using electrons of the ground state of caesium as the basic oscillators. This was confirmed as the international standard in 1967.

The specification of a relationship with a clock associates units with the numbers t used in the model.

Having defined how time is to be quantitatively measured, every subsequent physical quantity used in mechanics, including distance (see § 0.5), can be quantified by use of a postulated law of nature relating it to previously defined quantities. Time, being the first quantity to be introduced, is necessarily the most elusive – the one which must be modelled *a priori* instead of by a definition rooted in a law of nature.

0.3 Modelling of distance

The Newtonian concept of distance (or length) corresponds to the following modelling decision: distance is modelled mathematically by a positive algebraic variable d ($0 \leqslant d < \infty$). The traditional way of assigning a numerical value to the distance d between two points is by use of a rigid measuring rod (e.g. a metre rule) or an inextensible measuring chain.

If the constancy of the speed of light is accepted as a law of nature, a more elegant, though less immediate, definition of distance becomes available, see § 0.5 below. This alternative method of measurement eliminates the need for measuring rods – the clocks of § 0.2 measure the time taken for light to travel (*in vacuo*) between two points, and that time is a measure of the distance between the points.

0.4 Space

The ordinary man's idea of 'space' is much less closely related to numbers than are time and distance. From a mathematical point of view our intuitions about the nature of space are both complex and vague. Some of the vagueness (as well as some of the intuitiveness!) is dispelled by the process of mathematical modelling. As for the complexity, it may either be built into the structure of the

Introduction: space, time and modelling

chosen model – this is the case in vector geometry – or, if the basic model is simple, the complexity must be recognized in added 'laws of nature', i.e. supplementary modelling postulates, which are often called axioms of geometry.

Euclidean geometry starts from the concept of a fixed point. Space is modelled as a collection of points (FIG. 0.1); and this basic model is supplemented by further concepts, such as distance and direction, as well as chosen axioms, for example that space is three-dimensional, and that only one straight line can be drawn through a given point parallel to a given direction. The axioms are additional to the basic model and could be modified or abandoned without jettisoning the underlying model. Theoretical deductions from the chosen axioms, in the context of the model, result in Euclidean geometry.

Fig. 0.1

An alternative approach to the description of space can be made by postulating a more complex initial model, one which has a larger number of the essential characteristics of space (i.e. 'laws of nature') written in. The coordinate method, wherein space is described by use of numerical triads (x, y, z) will probably be familiar to the reader. The coordinate model of space automatically features its three-dimensionality, its infiniteness (provided x, y and z are allowed to take all real values) and, with the same proviso, its continuity in the x-, y-, and z-'directions'. It has the disadvantage of demanding a fixed origin and fixed directions for the axes, i.e. there is an in-built suggestion of preferred directions which has no counterpart in real space.

Another alternative to the Euclidean model is equally useful in mechanics. This is the vector model presented in Chapter 1; while largely avoiding false in-built suggestions in its notation, this model incorporates within itself even more than does the coordinate model of the essential characteristics of space. Thus there is an in-built mirroring of the property of space which is illustrated in FIG. 0.2, namely 'if the magnitude and direction of the displacement AB are known, and the magnitude and direction of the displacement BC are known, then the magnitude and direction of AC are uniquely determined'. The advantage of having so essential and so far-reaching a property written into the symbols and their postulated law of combination has to be paid for by a conscious labour

6 An introduction to mechanics and modelling

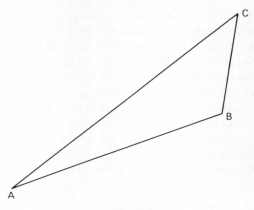

Fig. 0.2

of 'understanding' the model and how to derive mathematical results from it (Chapter 1).

0.5 Symbols, units, definitions and laws of nature

In what follows, we shall neglect the distinction between vectors and number triads (because either is easily converted into terms of the other by the methods of § 1.4) and we shall refer to all the mathematical models of space and time as quantitative, i.e. numerical models. Newtonian mechanics (Chapter 4 onwards) is firmly based on numerical modelling. As well as space and time, masses and forces and all the other entities which are to be discussed in this book are essentially numerical quantities, with units attached. They are represented by algebraic symbols s, t, m, **F** etc.

0.51 Algebraic symbols

The reader will readily appreciate that the indication of units, explicitly, throughout intricate pieces of mathematical evaluation is often impracticable. This is especially true when particular cases are being considered, in which numerical values and algebraic variables are intermingled within a single equation. In such circumstances, **for purposes of mathematical manipulation**, the algebraic symbols are regarded as representing mere numerical values. The onus of correctly specifying consistent units is placed on the modelling procedures which precede and follow the purely mathematical steps. **In the modelling steps** it is usually desirable to regard an algebraic symbol as standing for a number with units attached. Thus, in the modelling stages it would be proper to specify $t = 3$ seconds, while in the interludes of mathematical manipulation we might find the statement

$$y = 4 \exp(-0.2t) = 4 \exp(-0.6) = 2.195$$

which might in turn be followed by the interpretative statement

the distance, $y = 2.195$ metres

Introduction: space, time and modelling   7

This not-altogether-happy ambivalence of usage of algebraic symbols could be avoided by excluding from consideration all mathematics other than the development of general theories in general algebraic notation; the expression of the results predominantly in terms of ratios of similar quantities; and only finally the substitution of numerical values for the ratios occurring in the general formula and numbers-with-units for the remaining variables, to end up with a result in which correct units are automatically incorporated. The practical engineer may be prepared to accept such a restriction; so also may the ivory-tower theoretician. The investigative scientist or applied mathematician, however, cannot so severely restrict himself; for such fetters would deprive him of the option of considering particular cases, in which numbers and algebraic variables are inevitably mingled. And particular cases are sometimes the only soluble cases.

In this book the ambivalent use of algebraic symbols is accepted as the lesser of two evils. Statements such as '$t = 3$ seconds' will appear in modelling contexts, and this will often be interpreted as '$t = 3$' in sections of pure manipulation. As the student's manipulative skills improve, more sophisticated presentations will be recommended.

0.52 Definitions
Because the entities of mechanics are essentially numerical quantities, it follows that the definition of every new quantity that is introduced must be put on a numerical basis. By and large there are two ways of doing this – by specifying a measuring device/procedure *or* by using a definition based on a formula.

As examples of measuring devices, we might instance wrist-watches and speedometers for the measurement, respectively, of time and velocity. Such devices are very much of the 'black-box' variety, in that the user does not concern himself with the details of their mechanisms. To define a physical entity as 'that which is measured by a specified black-box device (in black-box units)' provides a viable, if not very attractive, starting-point for scientific use of the concept. Rather more attractive is the specification of a measuring procedure by comparison, or ratio. The length of a room is measured by ascertaining how many times the length of a standard measuring rod can be laid down end to end. The room's length is *defined* as the number of rod-lengths, with the unit 'rod-length'. (The name *metre* is given to the standard rod-length of the SI system.) The same kind of definition can be provided for any physical quantity, provided that (i) we equip ourselves with a physical instance of the quantity, which will be taken as the unit, and (ii) we specify exactly what we mean by the number of times the unit of the quantity 'goes into' the sample we are attempting to measure, i.e. how ratios are to be determined.

For all save the most directly appreciated kinds of physical quantity, there is a considerable 'black-box' element in step (ii) of any quantitative definition by measurement. Before physical quantities are defined it is not possible to establish the body of laws which would 'justify' the apparatus. 'Black-box' measurement is the fundamental starting-point in the exploratory stages of a science: it allows definitions to be made of any number of logically unrelated types of quantity. Comparison of the magnitudes of different measured quantities opens up the

8   An introduction to mechanics and modelling

opportunity of conjecturing relationships between them, and of testing the conjectures.

As the science develops, some confirmed conjectures take on the status of laws of nature, embodied in equations and formulae; at this point the opportunity arises of replacing the less satisfactory black-box definitions by definitions-by-formula. Before seeing how this works out, a comment about the phrase 'laws of nature' is appropriate.

0.53   Laws of nature
Each of the phrases 'law of nature' and 'empirical law' signifies an experimentally-founded modelling hypothesis, typically expressible in equations or inequalities. The distinction between the two appellations is not one of essence, but one of usage – the choice is made according to the degree of faith which it seems appropriate to repose in the modelling referred to. The description 'empirical law' is given to an experimentally-founded hypothesis which provides only an admittedly crude approximation to observation, or a model of strictly limited applicability. By contrast, the term 'law of nature' carries the suggestion that the relation in question closely mirrors a fundamental characteristic of nature. Although the subjective character of the distinction will not endear it to the logician, the author thinks the usage on the whole assists rather than obstructs physical understanding.

0.54   Definitions based on laws of nature
Consider the following way of approaching the idea of speed: speed v is (initially) defined as the reading on a car's speedometer. Let us suppose that this remains more or less constant over the periods of time with which we are concerned. An experimentalist observes that the distance s travelled by the car is proportional to the reading v as well as time t. He formulates the hypothesis

$$s = kvt \qquad \qquad 0\text{--}1$$

where the value of the constant k depends on the units in which he has chosen to measure distance and time. (There is no guarantee that these will 'match' the units in which the speedometer is calibrated; k is a *scale-dependent* constant.)

Using equation 0–1 he may now redefine speed in a way free from all reference to the speedometer, and as part of the redefinition he postulates a specific relationship between the unit of v and those of s and t. For simplicity he usually expresses his definition in units which make the constant $k = 1$; thus he defines

$$\text{speed } v = s/t \qquad \qquad 0\text{--}2$$

with the proviso that whenever the definition 0–2 is used, the unit of $v = $ (unit of s)/(unit of t). (The quantity s/t is more correctly called 'average speed', but the distinction is immaterial to our present discussion.)

A second example shows the same procedure in use. When the radius a, arc length s, and angle θ of a sector of a circle are measured in independent units, the observer notes that $s = ka\theta$. In particular, this form arises when the angle θ is defined as a multiple of the arbitrary unit 1° (one degree). The mathematician chooses to abandon measurement as a multiple of the degree, and to redefine angles on the basis of the formula $s = a\theta$, i.e. he/she redefines angle $\theta = s/a$ (with the further postulate that the lengths s and a are both measured in the same units). The unit angle, corresponding to $s = a$, is called one radian.

In the same way, any mathematical relation between variables $u, v, w, \ldots$ which can be rearranged or manipulated to give an equation of the form $u = f(v, w, \ldots)$ may be used to define the quantity u in terms of the quantities v, w, etc.

As a final example, consider the law of nature referred to in § 0.3, viz. the constancy of the velocity of light (*in vacuum*). This law is expressible by the equation

$$s/t = \text{constant } (c) \qquad \qquad 0\text{--}3$$

where $t =$ time taken for light to travel the distance s in vacuum.

The significance of c as a velocity may now be disregarded, and its constancy be regarded as the crucial characteristic. Equation 0–3 is then seen as a possible basis for the redefinition of distance, viz. the distance between two points (measured in light-time units) is the time taken for light to travel (through vacuum) between those points; i.e.

$s = t$ light-secs if t is measured in secs,

or $\quad s = ct$ if the constant c is given a non-unit value in order to maintain conformity with units of length defined traditionally.

The foregoing definition corresponds to that implicit in the measurement of distance by radar.

0.55 Definitions, fundamental units and derived units

Since any number of 'black-box' definitions-by-ratio could be introduced, there is no compulsion to define physical quantities in one order rather than another. There are several different rational bases of choice. Quantities can be defined in the historical order in which they were recognized; or in that order which allows the highest degree of experimental accuracy in their measurement. A third alternative would be to define first the seven quantities length, time, electric current, temperature, mass, amount of substance (i.e., moles), and luminous intensity, which are by international convention (rather than necessity) regarded as independent, and then develop definitions of coherent derived units. Or finally quantities may be defined in the order which makes it easiest to understand their nature and significance.

In a book whose object is to make mechanics clear and usable, the last of these four choices is the best one. This does not preclude other choices of order being made by other workers for other purposes.

The SI system of units is based on the seven chosen quantities enumerated above. This choice determines the nominal classification of units into 'fundamental units' – the chosen seven – and 'derived units'. It also determines the usual form taken by the technique known as dimensional analysis.

10 An introduction to mechanics and modelling

0.6 Conclusions

The structure of mechanics may be described as follows. It is founded on numerical models of time and distance, and either a Euclidean, a coordinate or a vector model of space. The three ways of modelling space are equivalent because and in so far as they mirror the same characteristics of space; whichever of the three is regarded as the primary model, the techniques of the other two can be built in by means of subsidiary postulates.

Additional physical concepts (e.g. velocity, force, mass, etc.) are incorporated into the basic model of time and space. Typically in the physical sciences, quantitative definition is approached indirectly by a sequence of modelling steps culminating in a quantitative formula derived from a 'law of nature' (see § 0.5). This indirect procedure is exemplified in the definition of 'force' in Chapters 1, 4 and 5.

Should two or more laws of nature be available for quantifying a new concept at any stage in the edifice of Newtonian mechanics, either one may be chosen as the basis for defining the unit of the new entity. If the first law is chosen, the constant of proportionality in the second law must usually, to ensure consistency, take a definite non-unit value which is often scale-dependent. The constant G in Newton's law of gravitation, relating mass and force, see Chapter 5, especially § 5.311, is of this kind.

Investigation of the mathematical consequences of both the laws of nature on which definitions are based and all additional laws of nature which it is desired to incorporate, results in a theory. The whole theory may be regarded as a highly sophisticated model of the mechanical world, from which predictions can be made. No model describes nature perfectly. Apparently different models may describe nature equally well; sometimes, as in the case of coordinate geometry and vector geometry, apparently different models can be shown to possess an underlying mathematical equivalence. In modern expositions of mechanics, including this book, some details of modelling differ from Newton's own, but the resulting system is still basically Newtonian mechanics.

Predictions of theory can be tested by insertion of numerical values in the equations modelling a physical situation, and checking against observation. The predictions of Newtonian mechanics have received massive confirmation, and the theory may be applied with confidence to all problems of motion except (a) atomic and nuclear phenomena, and (b) electromagnetic phenomena, including the interpretation of relativistic astronomical observations.

'Understanding' mechanics is partly a matter of familiarity. As Mach said, 'when once we have reached the point where we are everywhere able to detect the **same** few simple elements, combining in the ordinary manner ... the phenomena no longer perplex us, they are **explained**' (Ernst Mach, *Principles of Mechanics*, first German edition, 1883).

Assignment
Either, write your own paraphrase of the contents of this chapter,
Or choose any topic discussed in this chapter, and enlarge on it in essay form.

Further reading

1. Einstein, A. *On the method of theoretical physics* (Herbert Spencer lecture) (Clarendon Press, 1933). This is short, clear and easily read.
2. Magee, B. *Popper* (Fontana/Collins, 1973). Chapters 2 and 3 provide a simple and interesting summary of the important scientific ideas of Karl Popper.
3. Jeffreys, H. *Scientific Inference*, 2nd edn (CUP, 1957). Do not be misled by the dry title: this is a fascinating book. First-year students will probably only wish to read the first few pages of each chapter: even that amount of study is likely to be very thought-provoking. Whereas Popper's ideas apply to innovation, Jeffreys's analysis is relevant to the *application* of theoretical science.
4. Feather, N. *Matter and Motion* (Penguin, 1970). The section on Newtonian relativity (§ 8.1) is strongly recommended. Students interested in the historical development of science will find many other parts of this book interesting.
5. Ravetz, J. R. 'Galileo on the strength of materials', *School Science Review* (14 November 1961). This is a delightful, brief, but thought-provoking account of a modelling attempt which failed.

The remaining books are more appropriate for the mature student than for the first-year undergraduate.

6. Campbell, N. R. *Physics: the elements* (CUP, 1920). This is long, but full of interest to the thoughtful scientist.
7. Mach, E. *The Science of Mechanics* (trans.) (Open Court Publishing Company, 1919). An old-fashioned presentation illuminated by much wisdom.
8. Whitrow, G. J. *The Natural Philosophy of Time* (Nelson, 1961). A book for philosophers of science.
9. Popper, K. R. *The Logic of Scientific Discovery* (Hutchinson, 1959). A key work, but not at all easy to read.
10. Lakatos, I. *Proofs and refutations* (CUP, 1976). This is a book which will delight mathematicians. The reader of this book will the more readily appreciate the unity between mathematics and the other sciences, or between 'pure' and 'applied' mathematics.
11. Lakatos, I. *The methodology of scientific research programmes* (CUP, 1979). It is more difficult to read than *Proofs and refutations*. Pages 31–47 and 159–67 discuss the conflict, and the possibility of reconciliation, of the views of Popper with views such as those expressed by Jeffreys.
12. Dingle, H. *Through science to philosophy* (Clarendon Press, 1937). Although long, it is fairly easy to read.
13. Newton, I. *Principia* in any translation available.

Part I

Modelling of space by vectors, and an introduction to modelling in the real world

1
Vector algebra

1.1 Introducing vector algebra

The study of what causes objects to move, and how they move (or stay still), is called mechanics. If we restrict ourselves to the study of objects – mainly terrestrial objects – visible to the naked eye, a particularly elegant system of mechanics is sufficient for nearly all needs. This is the system rooted in the work of Sir Isaac Newton (1642–1727) and known as Newtonian mechanics. The reader of this book need not concern himself/herself, at this stage, either with relativistic mechanics, which is appropriate when discussing objects moving with speeds near to that of light, or with quantum mechanics, which is needed for understanding the behaviour of atomic and subatomic particles. The central postulates of Newtonian mechanics will be summarized at the end of Chapter 4.

Mathematics deals with abstractions, like points. For the purpose of describing the motion of objects visible to the naked eye, visual marks such as ink spots sufficiently well identify the small regions called points in common parlance. In turn, these regions may be thought of, or **modelled**, as mathematical points.

> By a model we mean a simplified representation.

Mathematicians commonly use points, straight lines, diagrams, symbols and equations as constituents of their models of the real world.

The motion of a body can be modelled by the changes in position of its points with time. To describe changes of position mathematically is the prime function of the geometric vectors to which this chapter is devoted; other uses accrue.

The needs of Newtonian mechanics are best served by the non-localized, three-dimensional (3D) vectors described here. Other types of vector differ in small but significant ways: some are regarded as having a prescribed origin; other vectors are used for the representation of complex numbers and therefore possess a special multiplication law which would be inappropriate to the vectors of mechanics. Other vectors again, namely those discussed in the subject called linear algebra, are a distillation of properties common to all vectors, and therefore lack some of the ingredients which are essential for discussions in real three-dimensional space. For these reasons

> we concern ourselves specifically with spatial, three-dimensional vectors.

Two-dimensional vectors are a specially restricted set of three-dimensional vectors.

16 An introduction to mechanics and modelling

The reader may well be familiar with vectors from school work. The applications of vector algebra in later chapters may nevertheless seem open to doubt unless the student has, in addition to a general familiarity, a clear consciousness of the specific justification of each algebraic law. By careful reading of this chapter, doubt need not arise later about the applicability of techniques of vector manipulation.

1.11 Modelling by use of directed line segments

A change in position, for example the change involved in a journey from Leeds (L) to Newcastle (N) and then to Edinburgh (E), may be represented by a directed line segment $\overrightarrow{LE}$ on a map, as in FIG. 1.1. For reasons which become clear later, the notation **LE** for $\overrightarrow{LE}$ is undesirable.

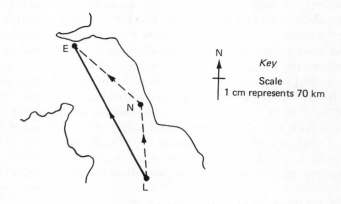

Fig. 1.1

The location of point L on the paper is immaterial so long as the relative positions of N and E are right. The scale and reference directions of the map, i.e. the modelling conventions, are shown in the key.

In mathematical modelling we deal with **selected** characteristics, not **all** the characteristics of the subject under discussion. When discussing an orange in the context of the week's housekeeping bill, only the number representing the cost of that orange, 7 (pence) say, is of importance. In a different context only the weight of the orange, 0.15 (kg) say, might be significant. When taking an orange to a friend in hospital, the best guide might be provided by the quality grading, Class 1 or Class 2. The nature of the appropriate arithmetic or algebra (if any) applicable to the numbers 7, 0.15 or 2, and the nature of the selected characteristics are interdependent.

In discussing journeys in this chapter, the crucial relation to be embodied is the equivalence of consecutive journeys such as $\overrightarrow{LN}$ and $\overrightarrow{NE}$ to the single journey $\overrightarrow{LE}$: we choose to pay attention only to the direction and magnitude of the eventual displacement from the starting-point. The scenery, the duration of the journey and the location of the starting-point are some of the many characteristics with which our present mathematical investigations will not be

Vector algebra 17

concerned. The chosen selection of characteristics is embodied in (modelled by) the notion of a vector.

Any quantity possessing magnitude and direction may be represented by a directed line segment; but

> the term **vector** is reserved for quantities which not only have magnitude and direction, but for which the third side of triangles constructed as in FIG. 1.1 have significance.

An example of a kind of entity with magnitude and direction which is nevertheless not a vector will be found in question 2 of Exercise 1 below.

Only if the same units and reference frame apply to both of the 'added' line segments will the 'third side' be interpretable. Line segments are segregated into sets according to the type of quantity they represent.

1.2 Meaning conveyed by a vector symbol, a, and the algebra which follows

The notation **a** (heavy type) – or, in manuscript **a** or a – is used to indicate that a direction and magnitude (of some entity whose units, frame of reference and physical nature are specified elsewhere) are simultaneously carried in the symbol **a**; and, moreover, that **a** belongs to a set of similar quantities **b**, **c**, etc., which are meaningfully combined according to the triangle law of composition stated formally below (§ 1.21), but already encountered in § 1.11 above. No other characteristics are carried by the vector symbol.

Other symbols, such as **r**, **s**, **A**, **B**, $\mathbf{F}_1$, $\mathbf{F}_2$ etc may be introduced to denote other vectors. No meaning automatically attaches to any of the juxtapositions **ab**, **AB**, **r** · **s** or **A** × **B**. The common but rather undiscerning use of **AB** for $\overrightarrow{AB}$ will not be adopted in this book.

Vector equality refers specifically to the prescribed characteristics, so

> **a** = **b** means (i)   **a** and **b** can be regarded as vectors of the same set: if representing physical quantities, they are measured in the same units;
> (ii)   the direction of **a** is the same as, i.e. is parallel to, that of **b**: this may be written **a**∥**b**; and
> (iii)   the magnitude of **a** equals the magnitude of **b**; this we write $|\mathbf{a}| = |\mathbf{b}|$.   1–1
>
> More generally, we may introduce the notation **a** = k**b** where k represents a positive numerical quantity (for k negative, see § 1.25);
> **a** = k**b** means (i)   the units of $|\mathbf{a}|$ are equal to the units of k multiplied by the units of $|\mathbf{b}|$;
> (ii)   **a** is parallel to **b**;
> (iii) $|\mathbf{a}| = k|\mathbf{b}|$, i.e. the magnitude of **a** is k times the magnitude of **b**.

Conversely, if **a** and k**b** belong to the same set, and **a**∥**b**, and $|\mathbf{a}| = k|\mathbf{b}|$, then **a** = k**b**.

18 An introduction to mechanics and modelling

FIG. 1.2 shows directed line segments representing equal vectors. Note that if **a** = **b**, then **b** = **a**, a property which is very useful in vector manipulation. For, if we have an equation, no matter how complex, between an expression on the left-hand side representing a vector, and another expression on the right-hand side, then the stated property of equal vectors provides the justification for interchanging the left-hand and right-hand sides of that equation.

Fig. 1.2

1.21 Triangle law of composition ('vector law of addition')

Definition of **a** + **b**. Choose a scale and reference directions so that **a** and **b** can be represented by directed line segments (FIG. 1.3(i)). From an arbitrary point P draw $\overrightarrow{PQ}$ to represent **a**. (We commonly write $\overrightarrow{PQ}$ = **a**.) From the point Q which has been arrived at, draw $\overrightarrow{QR}$ to represent **b** (as in FIG. 1.3(ii)). The directed line segment $\overrightarrow{PR}$ represents the combination (or resultant, or sum) of the vectors **a** and **b** on the same scale and reference conventions as were used when representing **a** and **b** by $\overrightarrow{PQ}$, $\overrightarrow{QR}$. It is denoted **a** + **b**.

1–2

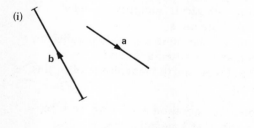
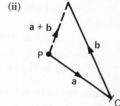

Fig. 1.3

The notation (**a** + **b**) + **c** is used to signify that the combination of **a** with **b**, in that order, (**a** + **b**), is combined with **c**. k**a** is the vector whose direction is parallel to that of **a**, and whose magnitude is k times that of **a**.

Vector algebra 19

1.22 Summary of basic rules of vector algebra

The reader is invited to check the validity of the following rules by reference to the accompanying diagrams (FIG. 1.4(i)–(x)) using the definition of equality from § 1.2 and the definition of addition from § 1.21.

(i)   If $\mathbf{a} = \mathbf{b}$, then $\mathbf{b} = \mathbf{a}$.
(ii) If $\mathbf{a} = \mathbf{b}$ and $\mathbf{b} = \mathbf{c}$, then $\mathbf{a} = \mathbf{c}$.
(iii) If $\mathbf{a} = k\mathbf{b}$ and $k \neq 0$, then $\mathbf{a} \parallel \mathbf{b}$.
(iv) $\mathbf{a} + \mathbf{b} = \mathbf{b} + \mathbf{a}$.
(v) $(\mathbf{a} + \mathbf{b}) + \mathbf{c} = \mathbf{a} + (\mathbf{b} + \mathbf{c})$
       $= \mathbf{a} + \mathbf{b} + \mathbf{c}$ (definition)   1–3
(vi) If $\mathbf{a} = \mathbf{b}$, then $\mathbf{a} + \mathbf{c} = \mathbf{b} + \mathbf{c}$. (When subtraction has been defined, it will be similarly found that $\mathbf{a} - \mathbf{c} = \mathbf{b} - \mathbf{c}$.)
(vii) If $\mathbf{a} = \mathbf{b}$, then $k\mathbf{a} = k\mathbf{b}$.
(viii) $k_1(k_2\mathbf{a}) = (k_1 k_2)\mathbf{a}$.
(ix) $k(\mathbf{a} + \mathbf{b}) = k\mathbf{a} + k\mathbf{b}$.
(x) $(k_1 + k_2)\mathbf{a} = k_1\mathbf{a} + k_2\mathbf{a}$.

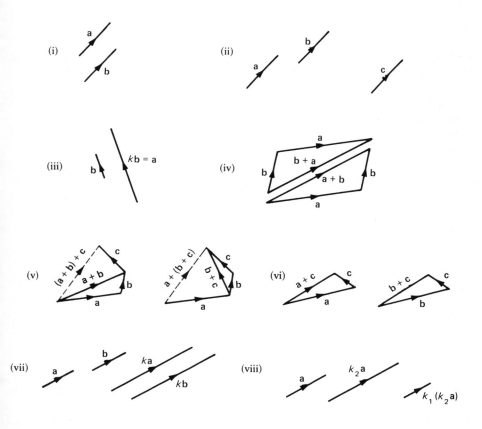

Fig. 1.4

20 An introduction to mechanics and modelling

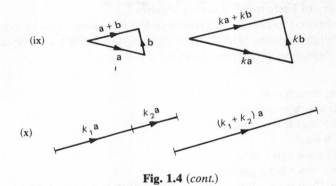

Fig. 1.4 (*cont.*)

It is of interest and importance to recognize the algebraic processes justified by rules (i)–(x). For example, rule (v) justifies the removal of brackets preceded by a + sign. Rule (vi) justifies adding the same vector to both sides of a vector equation, and rule (vii) justifies multiplying both sides of a vector equation by the same real number, and also, since k may be a fraction, division by a real number.

Quite a lot of elementary geometry concerning parallel lines, congruent triangles, similar triangles, etc. is involved in the proof of relations (i)–(x). It will not be altogether surprising to find later (§ 1.3) that geometrical problems can be solved by use of vector algebra, for that algebra is packed with geometrical implications. **Vector algebra automatically models some of the properties of three-dimensional space.**

Note that none of the rules of §§ 1.21 and 1.22 postulate that the vectors concerned must all lie in one specified or unspecified plane. They are general rules, true for any three-dimensional spatial vectors. Moreover, none of the rules requires the symbols **a**, **b** etc to represent immutable quantities: they may equally well represent the instantaneous values of changing quantities. Even functional relationships may be envisaged. Thus, if the three edges of a triangular piece of cardboard are represented by vectors **a**, **b**, **c**,

$$\mathbf{a}+\mathbf{b}+\mathbf{c}=\mathbf{0}$$

If the piece of cardboard is moving, the value of **a** (and likewise **b**, **c**) is time dependent, $\mathbf{a} = \mathbf{a}(t)$ etc; but $\mathbf{a}(t)$, $\mathbf{b}(t)$, $\mathbf{c}(t)$ at every time t satisfy

$$\mathbf{a}(t)+\mathbf{b}(t)+\mathbf{c}(t)=\mathbf{0}$$

1.23 Vectors, scalars, units; the type of work for which vector algebra is appropriate

To distinguish clearly between vectors and the ordinary numbers of arithmetic, or algebraic symbols standing for those ordinary numbers, like k, k_1, k_2 of the previous paragraph, it is customary to refer to the latter as 'scalars'. Scalars are numerical quantities or representative symbols (like k) to which, in the given context, the laws of arithmetic may be applied, the answer being a number. Units of measurement are separately specified.

The real-world vector quantities to which the vector symbols **a** and **b**, etc. relate have units about which the symbols **a** and **b** in themselves carry no information. However, because of the context from which it arises, the magnitude of the quantity represented by $k\mathbf{a}$ typically has units which are the units associated with k multiplied by the units associated with **a**.

The vector notation, the rules (i)–(x) enumerated above, and a few auxiliary relations yet to be discussed, constitute the substance of vector algebra. These are the mathematical tools for discussing **general** laws of geometry and for developing a **general** theory of mechanics.

When **numerical** answers are sought, general formulae must be translated into numbers, i.e. the investigator transfers his interest away from the realm best dealt with by pure vectors. Either a hybrid notation may be used, wherein each vector is represented by a number triad such as $(2, -1, 3)$ or an equivalent algebraic expression $2\mathbf{i} - \mathbf{j} + 3\mathbf{k}$ or matrix form $\begin{pmatrix} 2 \\ -1 \\ 3 \end{pmatrix}$; or else a complete translation of vector equations into arithmetic equivalents is made, three ordinary equations replacing each vector equation.

Thus the arithmetic equivalents of a vector equation $\mathbf{r} = \mathbf{a}$ might be the three equations $x = 2$, $y = -2$, $z = 1$; the process of translating vector equations into numbers will be discussed in § 1.4.

Exercise 1

1. State whether the numbers 3 and 5 are appropriately regarded as scalars in the following contexts. If they are scalars, which of the processes of arithmetic could be applicable?
    (i)   3 oranges and 5 oranges.
    (ii) 3 p.m. (dental appointment) and 5 p.m. (tea-time).
    (iii) No. 3 gas coke and no. 5 gas coke.
    (iv) The number 3 and the length 5 m.
2. I look due north for 8 minutes, and then I look south-east for 6 minutes. Each observation may be represented by a directed line segment. Is the operation of vector addition appropriate?
3. Write the relation 'a cat is an animal' as $c = a$; write the relation 'a dog is an animal' as $d = a$. Can we deduce that $c = d$, i.e. that a cat is a dog? Discuss, referring where appropriate to the steps whereby it could be shown for vectors that $\mathbf{c} = \mathbf{d}$ followed from $\mathbf{c} = \mathbf{a}$ and $\mathbf{d} = \mathbf{a}$.
    (Note: from this example we do not conclude that it is 'wrong' to model the word 'is' by the sign '=', but only that it is a poor modelling, suggesting false conclusions. A completely different mathematical sign, free of such suggestions, would be better.)
4. Two vectors **a** and **b** are initially represented by segments of two **non-intersecting** straight lines. Is there any impediment to finding their vector sum according to the definition of § 1.21?

1.24 Angle between two vectors
In FIG. 1.5, θ is the angle which **b** makes with **a**.

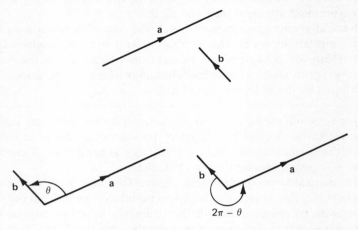

Fig. 1.5

> Note that the angle between two vectors is identified in a figure in which both vectors sprout from the same initial point.

The same angle may be seen when both vectors are drawn pointing inwards to the same point.

The angle which **a** makes with **b** is $-\theta$ or $2\pi - \theta$ (multiples of 2π are not geometrically significant). Two vectors **a** and **b** for which θ is equal to π may be called **antiparallel** when it is wished to stress that the arrows on the vectors point in opposite directions.

When two vectors point in the same direction they are called strictly parallel. It is often convenient to use the word **parallel** in the generalized sense of 'parallel or antiparallel'; **non-parallel** means neither parallel nor antiparallel.

1.25 Negative, zero and unit vectors; subtraction
The reader may check through the following definitions and notations.

Negative. $-\mathbf{a}$ is defined as the vector whose magnitude is equal to the magnitude of **a**; and whose direction is antiparallel to that of **a**.

Zero. The displacement beginning and ending at the same point represents the zero vector **0**. The symbol 0 is often used, rather imprecisely, for **0**.

Unit vectors. A vector whose magnitude is unity is (for brevity) referred to as a unit vector. Its direction must be separately specified. Symbols **i**, **j**, **I** and **e** are commonly reserved for unit vectors, and **â** is sometimes used for the unit vector parallel to **a**.

Subtraction. The difference $\mathbf{a} - \mathbf{b}$ between two vectors is defined as the vector sum of **a** and $(-\mathbf{b})$, which is the same as the sum of $(-\mathbf{b})$ and **a** (FIG. 1.6),

$$\mathbf{a} - \mathbf{b} \equiv \mathbf{a} + (-\mathbf{b}) \equiv (-\mathbf{b}) + \mathbf{a}.$$

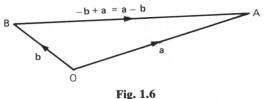

Fig. 1.6

The reader who appreciates that $\overrightarrow{BO}$ in FIG. 1.6 represents $-\mathbf{b}$ will recognize how important it is that

> when line segments are used to represent vectors, (for example $\mathbf{a}, \mathbf{b}, \ldots$ etc.), **arrows must be marked on those line segments** in order to indicate which senses correspond to $\mathbf{a}, \mathbf{b}, \ldots$ rather than $-\mathbf{a}, -\mathbf{b}$, etc.

1.26 An example of algebraic simplification analysed
Whenever there is doubt about whether a certain algebraic process is valid for vectors, recourse must be made to the basic rules stated in § 1.22. In the following example, detailed justification is given for the most common processes.

Example
Prove that $2(\mathbf{F}_1 + 3\mathbf{F}_2) - 5(2\mathbf{F}_1 - \mathbf{F}_2 + \mathbf{F}_3) = -8\mathbf{F}_1 + 11\mathbf{F}_2 - 5\mathbf{F}_3$

$\quad 5(2\mathbf{F}_1 - \mathbf{F}_2 + \mathbf{F}_3) = 5[(2\mathbf{F}_1) + (-\mathbf{F}_2 + \mathbf{F}_3)]$ from (v)

$\qquad\qquad\qquad\qquad = 10\mathbf{F}_1 + 5(-\mathbf{F}_2 + \mathbf{F}_3)$ from (ix) and (viii)

$\qquad\qquad\qquad\qquad = 10\mathbf{F}_1 + (-5\mathbf{F}_2 + 5\mathbf{F}_3)$ using (ix) and (vi)

$\qquad\qquad\qquad\qquad = 10\mathbf{F}_1 - 5\mathbf{F}_2 + 5\mathbf{F}_3$ using (v)

$\therefore \quad 5(2\mathbf{F}_1 - \mathbf{F}_2 + \mathbf{F}_3) = 10\mathbf{F}_1 - 5\mathbf{F}_2 + 5\mathbf{F}_3$ using (ii) three times

$\therefore \quad 2(\mathbf{F}_1 + 3\mathbf{F}_2) - 5(2\mathbf{F}_1 - \mathbf{F}_2 + \mathbf{F}_3)$

$\qquad = 2(\mathbf{F}_1 + 3\mathbf{F}_2) - (10\mathbf{F}_1 - 5\mathbf{F}_2 + 5\mathbf{F}_3)$ using (vi) and (vii)

$\qquad = 2(\mathbf{F}_1 + 3\mathbf{F}_2) - 10\mathbf{F}_1 + 5\mathbf{F}_2 - 5\mathbf{F}_3$ using (ix) with $k = -1$

$\qquad = 2\mathbf{F}_1 + 6\mathbf{F}_2 - 10\mathbf{F}_1 + 5\mathbf{F}_2 - 5\mathbf{F}_3$ using (ix) and (v)

$\qquad = 2\mathbf{F}_1 - 10\mathbf{F}_1 + 6\mathbf{F}_2 + 5\mathbf{F}_2 - 5\mathbf{F}_3$ using (iv) and (v)

$\qquad = -8\mathbf{F}_1 + 11\mathbf{F}_2 - 5\mathbf{F}_3$ using (x) and (i) twice.

whence, using (ii) four times, the result follows.

The reader may not at first recognize why the use of (ii) is required three times in the first part and four times in the second part of the working. The simplification of $5(2\mathbf{F}_1 - \mathbf{F}_2 + \mathbf{F}_3)$ was written down in the form $\mathbf{a} = \mathbf{b} = \mathbf{c} = \mathbf{d} = \mathbf{e}$, but proof was initially given only that $\mathbf{a} = \mathbf{b}$, $\mathbf{b} = \mathbf{c}$, $\mathbf{c} = \mathbf{d}$, and $\mathbf{d} = \mathbf{e}$. The conclusion $\mathbf{a} = \mathbf{e}$ requires three uses of (ii).

To check every piece of algebraic manipulation against rules (i)–(x) would be excessively tedious, and self-defeating too, since the purpose of vector modelling

24 An introduction to mechanics and modelling

is to carry the geometrical properties automatically in the symbols. If the reader has followed the logic of every step in the above example, he may henceforth confidently perform similar operations on vectors, i.e. operations involving +, −, multiplication by scalars and removal of brackets, as though vectors were familiar algebraic quantities. Note that

> vectors are entities whose magnitude and direction are inseparable during the processes of vector algebra.

Exercise 2
1. Simplify $2[3(\mathbf{v}-\mathbf{u})+\tfrac{1}{2}(\mathbf{v}+\mathbf{u})]$.
2. Justify your solution to question 1 step by step as in the example of § 1.26.
3. Given that $m_1, m_2, \ldots m_n$ are the numerical magnitudes of certain physical quantities, that $m_1\mathbf{r}_1 = \mathbf{P}_1$; $m_2\mathbf{r}_2 = \mathbf{P}_2$; $\ldots$; $m_n\mathbf{r}_n = \mathbf{P}_n$, and that the $\mathbf{P}$'s may be added vectorially, verify, by reference to the basic rules of § 1.22, that
$$m_1\mathbf{r}_1 + m_2\mathbf{r}_2 + \cdots + m_n\mathbf{r}_n = \mathbf{P}_1 + \mathbf{P}_2 + \cdots + \mathbf{P}_n.$$
4. Given that $\overrightarrow{OA} = \mathbf{a}$, $\overrightarrow{OB} = \mathbf{b}$, express $\overrightarrow{AB}$ in terms of $\mathbf{a}$ and $\mathbf{b}$. Given that (i) $\overrightarrow{AP} = \tfrac{7}{8}\overrightarrow{AB}$, (ii) that $3\overrightarrow{AP} = 11\overrightarrow{PB}$, express $\overrightarrow{OP}$ in terms of $\mathbf{a}$ and $\mathbf{b}$. [Separate answers for (i) and (ii).]
5. Given that $\overrightarrow{AB} = \mathbf{a}$, $\overrightarrow{CD} = \mathbf{c}$ where lines AB, CD do not intersect, state the meaning of 'the angle between $\mathbf{c}$ and $\mathbf{a}$' (cf. FIG. 1.2 and definition 1–1).

1.3 Geometrical results demonstrated by use of vector algebra

Far from being trivial, the rules (i)–(x) of § 1.22 embody a substantial part of Euclidean geometry. The following examples and exercises show this.

Example 1
To prove that the medians of any triangle are concurrent and that their point of concurrence is a point of trisection of each median. (Note this is a general result: vector algebra will be particularly appropriate.)

Any triangle ABC is specified (apart from its location) by two vectors, e.g. $\overrightarrow{BC} = \mathbf{a}$, $\overrightarrow{CA} = \mathbf{b}$. All geometrical features of the triangle can therefore be expressed in terms of those two vectors alone. Thus $\overrightarrow{BA} = \mathbf{a} + \mathbf{b}$; and, if the medians are AD, BE and CF as shown in FIG. 1.7(i), then

$$\overrightarrow{BD} = \overrightarrow{DC} = \tfrac{1}{2}\mathbf{a}; \quad \overrightarrow{CE} = \overrightarrow{EA} = \tfrac{1}{2}\mathbf{b}; \quad \overrightarrow{BF} = \overrightarrow{FA} = \tfrac{1}{2}(\mathbf{a}+\mathbf{b}).$$

In FIG. 1.7(ii), (iii), (iv), G_1, G_2 and G_3 are points of trisection of medians AD, BE and CF respectively: it is our object to establish whether or not those three points coincide.

From FIG. 1.7(ii), $\overrightarrow{BG_1} = \tfrac{1}{2}\mathbf{a} + \tfrac{1}{3}(\overrightarrow{DA})$

$$= \tfrac{1}{2}\mathbf{a} + \tfrac{1}{3}(\tfrac{1}{2}\mathbf{a} + \mathbf{b})$$

$$= \tfrac{2}{3}\mathbf{a} + \tfrac{1}{3}\mathbf{b}$$

From FIG. 1.7(iii) $\overrightarrow{BG_2} = \tfrac{2}{3}\overrightarrow{BE} = \tfrac{2}{3}(\mathbf{a}+\tfrac{1}{2}\mathbf{b}) = \tfrac{2}{3}\mathbf{a} + \tfrac{1}{3}\mathbf{b}$

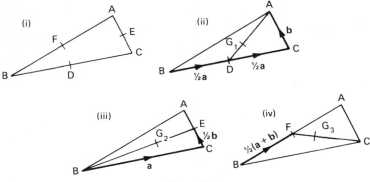

Fig. 1.7

From FIG. 1.7(iv) $\overrightarrow{BG_3} = \frac{1}{2}(\mathbf{a}+\mathbf{b}) + \frac{1}{3}\overrightarrow{FC}$

$= \frac{1}{2}(\mathbf{a}+\mathbf{b}) + \frac{1}{3}[-\frac{1}{2}(\mathbf{a}+\mathbf{b}) + \mathbf{a}]$

$= \frac{1}{2}(\mathbf{a}+\mathbf{b}) - \frac{1}{6}(\mathbf{a}+\mathbf{b}) + \frac{1}{3}\mathbf{a}$

$= \frac{2}{3}\mathbf{a} + \frac{1}{3}\mathbf{b}$

Thus $\overrightarrow{BG_1} = \overrightarrow{BG_2} = \overrightarrow{BG_3}$. The three points of trisection therefore coincide; the required result has been proved.

The application of vector algebra does not – as the above proof might suggest – demand that the answer be known in advance. The median property will be proved below without preselection of the points of trisection for special attention.

The following theorem on vectors is very useful for extending the range of problems which can be tackled:

Theorem
If **a** and **b** are two non-parallel vectors, then: (i) a plane can be identified which contains both **a**, **b**; (ii) any other vector in this plane may be expressed as $k_1\mathbf{a} + k_2\mathbf{b}$; and (iii), given that $\lambda_1\mathbf{a} + \lambda_2\mathbf{b} = k_1\mathbf{a} + k_2\mathbf{b}$, we may deduce both that $\lambda_1 = k_1$ and that $\lambda_2 = k_2$.

1–4

Proof
(i) From an arbitrary point O in space, draw $\overrightarrow{OA} = \mathbf{a}$, $\overrightarrow{OB} = \mathbf{b}$. Then plane OAB contains both **a** and **b**. (All planes parallel to OAB also contain vectors **a** and **b**. Though not geometrically identical, parallel planes are **vectorially** equivalent.)
(ii) Given any vector represented by directed line segment $\overrightarrow{PQ}$ in plane OAB, we require point R such that $\overrightarrow{PR} = k_1\mathbf{a}$, $\overrightarrow{RQ} = k_2\mathbf{b}$. Draw a line through P parallel to **a**, and a line through Q parallel to **b**. Since **a**, **b** are non-parallel, these two lines intersect in a unique point (R), where, since $\overrightarrow{PR} \| \pm\mathbf{a}$, $\overrightarrow{PR} = k_1\mathbf{a}$, and since $\overrightarrow{RQ} \| \pm\mathbf{b}$, $\overrightarrow{RQ} = k_2\mathbf{b}$.
(iii) Consider the construction of $\lambda_1\mathbf{a} + \lambda_2\mathbf{b}$ starting from point P and (because $\lambda_1\mathbf{a} + \lambda_2\mathbf{b} = k_1\mathbf{a} + k_2\mathbf{b}$) ending at Q. The only point R' satisfying both $\overrightarrow{PR'} = \lambda_1\mathbf{a}$ and $\overrightarrow{R'Q} = \lambda_2\mathbf{b}$ is that given by the recipe in (ii) above. Hence $\lambda_1 = k_1$ and $\lambda_2 = k_2$.

26 An introduction to mechanics and modelling

Example 2
The median property of a triangle (alternative method).
   Consider the point of intersection G of medians BE and AD of the triangle shown in FIG. 1.7(ii) and (iii). Because G lies on BE, $\overrightarrow{BG} = k(\mathbf{a} + \tfrac{1}{2}\mathbf{b})$ for some value of k and because G lies on AD, $\overrightarrow{BG} = \tfrac{1}{2}\mathbf{a} + \lambda(\tfrac{1}{2}\mathbf{a} + \mathbf{b})$ for some value of λ.
   So
$$k\mathbf{a} + \tfrac{1}{2}k\mathbf{b} = \tfrac{1}{2}(1+\lambda)\mathbf{a} + \lambda \mathbf{b}$$
and using the theorem,
$$k = \tfrac{1}{2}(1+\lambda) \quad \text{and} \quad \tfrac{1}{2}k = \lambda$$
whence $k = \tfrac{2}{3}$ and $\lambda = \tfrac{1}{3}$. I.e. G is a point of trisection of AD and of BE.

Example 3
In any triangle OAB, $m_1\overrightarrow{OA} + m_2\overrightarrow{OB}$ (where m_1, m_2 are any two positive numbers) is expressible as $(m_1 + m_2)\overrightarrow{OC}$, provided that the point C is suitably chosen. For an arbitrary position of C (FIG. 1.8), $\overrightarrow{OA} = \overrightarrow{OC} + \overrightarrow{CA}$; $\overrightarrow{OB} = \overrightarrow{OC} + \overrightarrow{CB}$.

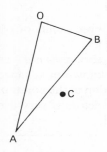

Fig. 1.8

Hence, for any position of C,
$$m_1\overrightarrow{OA} + m_2\overrightarrow{OB} = (m_1 + m_2)\overrightarrow{OC} + m_1\overrightarrow{CA} + m_2\overrightarrow{CB}.$$
In order that $m_1\overrightarrow{CA} + m_2\overrightarrow{CB} = \mathbf{0}$, i.e. that $m_1\overrightarrow{AC} = m_2\overrightarrow{CB}$, we require
(i) that $\overrightarrow{AC} \parallel \overrightarrow{CB}$,
(ii) that $m_1 AC = m_2 CB$ (where AC, CB denote the lengths of the displacements $\overrightarrow{AC}$, $\overrightarrow{CB}$).
These two conditions are satisfied if:
(i) C lies on AB,
(ii) C divides AB in the ratio $AC:CB = m_2:m_1$.

This result will be useful in mechanics. The numbers m_1, m_2 will represent the masses of particles situated at A and B. The quantity $m_1\overrightarrow{OA} + m_2\overrightarrow{OB}$ is then a vector with units mass × distance. The result shows that if the two masses m_1 and m_2 move in any way which maintains the length of AB constant, the point C moves as though it was a fixed point of a rigid rod AB.

Some of the rewards of a study of vectors should now be apparent. First, it is an excellent example of physical modelling: it shows how the expression in mathematical form of one or two postulates about nature enables a whole system of mathematics to be built up. Secondly, that system – vector algebra – is a powerful tool whereby theorems of great generality may be established with comparative ease. In Chapters 6 and 8 we shall make much use of this tool.

Exercise 3

(These exercises will help you, later, to understand 'centre of mass' properties.)
1 ABCD is a parallelogram, with $\overrightarrow{AB}=\mathbf{a}$, $\overrightarrow{BC}=\mathbf{b}$. Find by vector algebra alone, the position vector from A of (i) the mid-point of diagonal AC and (ii) the mid-point of BD. What geometrical theorem have you thereby established?
2 A, B, C and D are any four points in space. P, Q, R and S are the mid-points of AB, BC, CD and DA respectively. Prove (vectorially) that PQRS is a parallelogram. (**Hint:** write $\overrightarrow{AB}=\mathbf{a}$, $\overrightarrow{BC}=\mathbf{b}$, $\overrightarrow{CD}=\mathbf{c}$, and then express $\overrightarrow{PQ}$ and $\overrightarrow{SR}$ in terms of $\mathbf{a}$, $\mathbf{b}$, $\mathbf{c}$.)
3 ABCD is a parallelogram with $\overrightarrow{AB}=\mathbf{a}$, $\overrightarrow{AD}=\mathbf{b}$. Points G, H lie in BC and CD respectively, with $BG=\frac{1}{2}GC$ and $CH=\frac{1}{2}HD$. Point X lies in HG with $HX=\frac{1}{2}XG$.
  (i)   Express the displacement $\overrightarrow{HG}$ in terms of $\mathbf{a}$ and $\mathbf{b}$.
  (ii) Find the displacement $\overrightarrow{AX}$ in terms of $\mathbf{a}$ and $\mathbf{b}$.
  (iii) Hence prove that X lies on the diagonal AC, and find the ratio in which X divides AC.
4 D, E, F are the mid-points of the sides of a triangle ABC. O is any point in space. Prove that

$$\overrightarrow{OA}+\overrightarrow{OB}+\overrightarrow{OC}=\overrightarrow{OD}+\overrightarrow{OE}+\overrightarrow{OF}$$

(**Hint:** write $\overrightarrow{OA}=\mathbf{a}$, $\overrightarrow{OB}=\mathbf{b}$, $\overrightarrow{OC}=\mathbf{c}$, and transform the right-hand side into terms of $\mathbf{a}$, $\mathbf{b}$ and $\mathbf{c}$.)
5 In triangle ABC, P is the point of BC for which $BP=2PC$. Q is the point of CA for which $CQ=2QA$. PQ produced cuts BA (produced) in R. Prove that R divides BA externally in the ratio $4:3$.
6 ABC is a triangle. P is such that $\overrightarrow{BP}=3\overrightarrow{PC}$. Q is such that $\overrightarrow{CQ}=3\overrightarrow{QA}$. Find the ratio in which PQ produced divides BA. (**Hint:** use the method used in the second of the investigations of the median property of a triangle in the text.)

1.4 Uses of base vectors i, j, k; components

By convention, $\mathbf{i}$ denotes a vector of unit magnitude parallel to the x-axis in a fixed rectangular cartesian frame of reference. Vectors of unit magnitude parallel to the y- and z-axes are denoted $\mathbf{j}$, $\mathbf{k}$ respectively. A vector of magnitude p parallel to the x-axis may therefore be written $p\mathbf{i}$, in which form the magnitude as well as the direction is explicitly displayed; thus $6\mathbf{i}$ is of magnitude 6. Similarly a vector of magnitude 4 units in the negative sense of the y-axis is written $-4\mathbf{j}$. A vector $\mathbf{v}=\alpha\mathbf{i}+\beta\mathbf{j}$ lying in the x-y plane is said to have **components** α, β. A vector $\mathbf{v}=\alpha\mathbf{i}+\beta\mathbf{j}+\gamma\mathbf{k}$ is said to have components α, β, γ.

When all the vectors under consideration are expressed as the sum of multiples of $\mathbf{i}$, $\mathbf{j}$, $\mathbf{k}$, those three chosen vectors are called **base vectors.**

To keep the arithmetic simple, the various uses of components are illustrated overleaf by examples in which all the vectors lie in a plane.

28 An introduction to mechanics and modelling

Exercise 4

1. Draw sketches of the four vectors in the x-y plane whose components are, respectively, $(3, -4)$, $(4, 2)$, $(-3, 4)$, $(-1, -5)$. Mark the direction of each vector with an arrow, and describe its direction clearly (preferably using the phrase 'as shown in the diagram'), using tables to find a convenient angle in degrees. Also find the magnitude of each vector.
2. Given that $\mathbf{F}_1 = 2\mathbf{i} + 3\mathbf{j}$; $\mathbf{F}_2 = -6\mathbf{i} - 2\mathbf{j}$; $\mathbf{F}_3 = -\mathbf{i} + 2\mathbf{j}$, find the magnitude and describe the direction of $\mathbf{F}_1 + \mathbf{F}_2 + \mathbf{F}_3$.
3. Find the magnitude and direction of the vector sum of the three vectors whose two-dimensional components are $(2, 3)$, $(-6, -2)$, $(-1, 2)$.
4. Given that $\mathbf{i}, \mathbf{j}$ are unit vectors in two perpendicular directions, draw diagrams representing the following vectors, $\mathbf{a} = 3\mathbf{i} - \mathbf{j}$; $\mathbf{b} = -\mathbf{i} + 2\mathbf{j}$. Calculate the magnitudes and angles specifying the directions of $\mathbf{a}$, $\mathbf{b}$ and $\mathbf{a} + \mathbf{b}$. Calculators may be used or surds retained as convenient. Find the constant k for which $\mathbf{a} + k\mathbf{b}$ makes an angle of $+30°$ with the direction of $\mathbf{i}$.
5. $\mathbf{p}$ and $\mathbf{q}$ are any two vectors (of the same set); the angle between them is θ, which, in the first place, may be assumed to be acute; and the magnitudes of the vectors are P and Q respectively. Find the component of $\mathbf{q}$ in directions parallel and perpendicular to $\mathbf{p}$, and deduce the magnitude of the vector sum of $\mathbf{p}$ and $\mathbf{q}$ in terms of P, Q and θ.

[**Answer:** $\sqrt{(P^2 + Q^2 + 2PQ \cos \theta)}$]   1–5

Note: the recognized definitions of $\sin \theta$, $\cos \theta$ for values of θ outside the range $0 \leq \theta \leq \tfrac{1}{2}\pi$ are those which ensure that the components of a vector of magnitude Q, inclined at an angle θ (measured anticlockwise from the x-axis) are in all cases

$$X = Q \cos \theta; \quad Y = Q \sin \theta. \quad\quad 1\text{–}6$$

It follows that the formula 1–5 derived above for the magnitude of the sum of $\mathbf{p}$ and $\mathbf{q}$ is true in all cases, not merely when θ is acute.

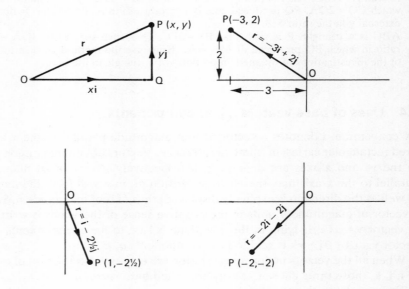

Fig. 1.9

1.41 Position vector of point P whose coordinates are (x, y)

If O is the origin of coordinates, $\overrightarrow{OP}$ is called the **position vector** of point P, $\overrightarrow{OP} = \mathbf{r}$ say. Position vectors are of great importance in the theoretical development of mechanics.

For any point in the x-y plane (cf. FIG. 1.9)

$$\mathbf{r} = \overrightarrow{OQ} + \overrightarrow{QP}$$
$$= x\mathbf{i} + y\mathbf{j} \qquad 1\text{--}7$$

1.42 Position vector in three dimensions; point (x, y, z)

$$\mathbf{r} = x\mathbf{i} + y\mathbf{j} + z\mathbf{k} \qquad 1\text{--}8$$

1.43 Force and components of force

> By a force, we mean a push or a pull or a rub – a quantity with both direction and magnitude.

The concepts which will enable us to give a precise, quantitative definition in Chapter 5 are not yet available to us. It is wholly proper in science to accept a provisional, imprecise definition. As explained in the Introduction, it has to suffice until such a time as a 'law of nature' can be enunciated involving force as the only unknown quantity: from that law of nature, a precise definition of force can be derived. Meanwhile we start from the 'definition' of force as a push or a pull or a rub.

One of the postulates of Newtonian mechanics is that, for some sets of forces, the vector sum is a significant entity, i.e. for some purposes,

> forces 'obey' the vector law of composition, i.e. forces are vector quantities.

The magnitudes of forces are measured in newtons (N); the newton is the SI unit; other (practical) units are the kgf (kilogram-force) and lbf (pound-force, which used to be called the pound-weight).

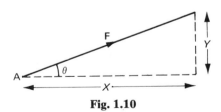

Fig. 1.10

The statement $\mathbf{F} = X\mathbf{i} + Y\mathbf{j}$ (FIG. 1.10), and the statement $\mathbf{F}$ has components (X, Y) are equivalent. If the magnitude and direction of $\mathbf{F}$ are known, e.g. $|\mathbf{F}| = F$, $\angle A = \theta$ in FIG. 1.10, then we find (whether θ is acute, obtuse or reflex)

$$X = F \cos \theta; \quad Y = F \sin \theta \qquad 1\text{--}9$$

30 An introduction to mechanics and modelling

Conversely, if X and Y are known, the magnitude and direction of **F** can be calculated:

$$\left.\begin{array}{l}F = \sqrt{(X^2+Y^2)}; \\ \theta = \text{that value of } \tan^{-1}(Y/X) \text{ for which } \cos\theta \text{ has the same sign as } X.\end{array}\right\} \quad 1\text{–}10$$

Thus, if $X = 6$, $Y = -3$, $\tan\theta = -\tfrac{1}{2}$;

$$\theta = \tan^{-1}(-\tfrac{1}{2})$$
$$= \text{either } 180° - 24.2°$$
$$\text{or } -24.2°$$

Only the latter corresponds to a positive value of X, so $\theta = -24.2°$.

1.44 Vector sum of forces, using components

Given that $\mathbf{F}_1 = X_1\mathbf{i} + Y_1\mathbf{j}$, $\mathbf{F}_2 = X_2\mathbf{i} + Y_2\mathbf{j}$ and so on, then the vector sum

$$\mathbf{F} \equiv \mathbf{F}_1 + \mathbf{F}_2 + \cdots = (X_1 + X_2 + \cdots)\mathbf{i} + (Y_1 + Y_2 + \cdots)\mathbf{j}$$
$$\text{i.e. } \mathbf{F} = X\mathbf{i} + Y\mathbf{j} \qquad\qquad 1\text{–}11$$

where $X = X_1 + X_2 + \cdots$ and $Y = Y_1 + Y_2 + \cdots$. From the values of X and Y, the magnitude and direction of **F** can be calculated explicitly as in the final example of § 1.43.

1.5 Examples

Example 1

ABCD is a square of side 2 units and centre O. Forces of magnitudes 3, 8, 4, 3, √2 and √2 units act in the lines AB, BC, CD, DA, OA and OB respectively, in the senses indicated by the order of the lettering. Find the magnitude of the resultant force, and the angle which it makes with AB. You may assume that

the word **resultant** is synonymous with the phrase **vector sum** of the forces.

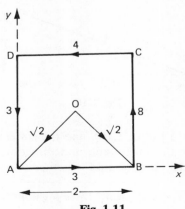

Fig. 1.11

From FIG. 1.11, we see that the vector sum has components

$$X = 3 - 4 - \sqrt{2} \cos 45° + \sqrt{2} \cos 45° = -1$$
$$Y = 8 - 3 - \sqrt{2} \sin 45° - \sqrt{2} \sin 45° = 3$$

The force with components X, Y is represented in FIG. 1.12. Its magnitude is $\sqrt{(3^2 + 1^2)} = \sqrt{10}$, and its direction makes angle θ with the x-axis, where θ is that value of $\tan^{-1}(-3)$ for which $\cos \theta$ is negative, i.e.

$$\theta = 180° - 71.6°$$

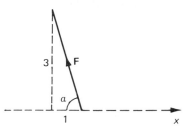

Fig. 1.12

In practice it is often easier to describe the direction by reference to FIG. 1.12, as follows: the force **F** makes angle α with the x-axis as shown, where

$$\tan \alpha = 3/1 = 3$$

i.e. $\alpha = 71.6°$

This style of description recognizes by implication that

> information about direction is often most conveniently communicated by reference to acute angles shown in a diagram.

Example 2
Find the magnitude of (i) the vector sum, (ii) the vector difference of (1, 3) and (5, −2).

Solution (but see the comment at the end).

Vector sum = $(1+5, 3-2) = (6, 1)$

Magnitude of sum = $\sqrt{(36+1)} = \sqrt{37}$

Vector difference = $(1-5, 3+2) = (-4, 5)$

Magnitude of difference = $\sqrt{(16+25)} = \sqrt{41}$

(It should cause the reader no concern that this magnitude is greater than that of the sum.)

32 An introduction to mechanics and modelling

Critical comment. Certain assumptions have been made. Without them the problem would not have a well-defined answer. The implicit assumptions are:

(i) $(1, 3)$ and $(5, -2)$ are components, with the same base vectors, of two compatible vectors: this justifies the addition/subtraction of components.

(ii) the base vectors are mutually perpendicular unit vectors: this justifies the use of Pythagoras' theorem to find the magnitude of the vector sum/difference.

Example 3
Given that $\mathbf{a} = 4\mathbf{i} + \mathbf{j}$; $\mathbf{b} = 3\mathbf{i} - 4\mathbf{j}$, calculate $|\mathbf{a}|$ and $|\mathbf{a} - \mathbf{b}|$. Find the value of λ for which $\mathbf{a} + \lambda \mathbf{b}$ is parallel or antiparallel to the vector $\mathbf{c} = 11\mathbf{i} - 2\mathbf{j}$.

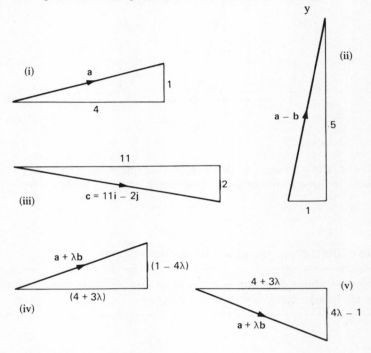

Fig. 1.13

FIG. 1.13(i) and (ii) show

$$\mathbf{a} = 4\mathbf{i} + \mathbf{j}$$

$$\mathbf{a} - \mathbf{b} = \mathbf{i} + 5\mathbf{j}$$

Using Pythagoras' theorem in the relevant right-angled triangles,

$$|\mathbf{a}| = \sqrt{(16 + 1)} = \sqrt{17}$$

$$|\mathbf{a} - \mathbf{b}| = \sqrt{(1 + 25)} = \sqrt{26}$$

We require that the vector $\mathbf{a} + \lambda \mathbf{b} = (4 + 3\lambda)\mathbf{i} + (1 - 4\lambda)\mathbf{j}$ shall be parallel or antiparallel to $\mathbf{c}$ [FIG. 1.13(iii)]. The representation of $\mathbf{a} + \lambda \mathbf{b}$ as in FIG. 1.13(iv) assists us less than the equivalent representation in which the $\mathbf{j}$-component is shown, not as $(1-4\lambda)$ upwards, but as $(4\lambda-1)$ downwards [FIG. 1.13(v)]. To ensure that the triangles of FIG. 1.13(iii) and (v) are similar, we require

$$(4\lambda-1)/(4+3\lambda) = 2/11$$

i.e. $\qquad\qquad 44\lambda - 11 = 8 + 6\lambda,$

whence $\qquad\qquad 38\lambda = 19, \quad\text{i.e.}\quad \lambda = \tfrac{1}{2}$

This value of λ gives $\mathbf{a} + \lambda\mathbf{b} = \tfrac{1}{2}\mathbf{c}$.

It should be noted that if we write $-\mathbf{c} = r\mathbf{i} + s\mathbf{j}$, then $s/r = 2/-11 = -2/11$, i.e. the ratio of the coefficients of $\mathbf{j}$ and $\mathbf{i}$ is unaltered by reversing the direction of the vector.

The condition '$p\mathbf{i} + q\mathbf{j}$ is parallel or antiparallel to $r\mathbf{i} + s\mathbf{j}$' requires that $q/p = s/r$, parallelism occurring when p and r (and therefore q and s) have like signs, and antiparallelism when p and r (and therefore q and s) have unlike signs.

Exercise 5
(For readers with little previous acquaintance with examples of this kind.)
1. Forces of magnitude 2, 4, $\sqrt{2}$ and $4\sqrt{3}$ units act through the origin at angles 60°, 120°, 225° and 330° to the x-axis respectively. Calculate the magnitude and direction of the resultant.
2. What force must be added to the system given in question 1 to produce a resultant of magnitude $\sqrt{3}$ acting in the direction of the y-axis? (**Hint:** we require $X_1 + X_2 + X_3 \cdots = 0$, and $Y_1 + Y_2 + \cdots = \sqrt{3}$.)
3. A regular hexagon ABCDEF is drawn in the x-y plane with A at the origin and AB along the positive x-axis. The vertices are labelled anticlockwise. Forces of magnitude 3, $6\sqrt{3}$, 2, $\sqrt{3}$ and 8 units act along AB, AC, AD, AE and CD respectively. Calculate the magnitude and direction of the resultant force.
4. Find the magnitude and direction of the resultant of the following forces:
   3 newtons at 20° to the x-axis;
   5 newtons at 100° to the x-axis;
   2 newtons at 210° to the x-axis;
   8 newtons at 300° to the x-axis.
   All angles are measured in an anticlockwise sense from the positive direction of x.
5. In FIG. 1.14 inclined, but still orthogonal (i.e. mutually perpendicular), axes of x and y are shown. Lines parallel to the x-axis have been drawn through the starting point

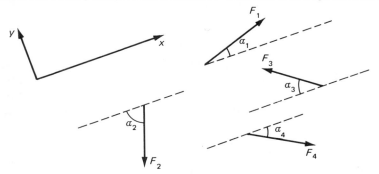

Fig. 1.14

of each of the vectors shown by arrows; their magnitudes are F_1, F_2, F_3, F_4 as indicated, and their directions can be described in terms of the acute angles $\alpha_1, \alpha_2, \alpha_3$ and α_4. Write down the x- and y-components of the resultant of the given forces.

6 In FIG. 1.15 three forces, of magnitudes T, F and W act on a particle at point P. Given that the vector sum of the three forces $= \mathbf{0}$, find T and F in terms of W. (**Hint:** take x-axis parallel to direction of F.)

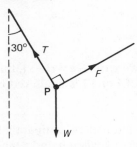

Fig. 1.15

7 In FIG. 1.16 forces of magnitude T_1, T_2 and W act in the directions indicated. Given that their resultant is zero, find the values of T_1 and T_2. (**Hint:** horizontal and vertical axes are convenient. Two simple simultaneous equations must be solved.)

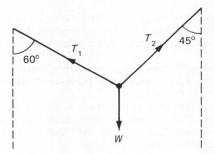

Fig. 1.16

1.6 Magnitude and direction of three-dimensional vectors

By a double use of Pythagoras' theorem (FIG. 1.17), the magnitude of the vector $\mathbf{v} \equiv a_1\mathbf{i} + a_2\mathbf{j} + a_3\mathbf{k}$ is $\sqrt{(a_1^2 + a_2^2 + a_3^2)}$. The direction of $\mathbf{v}$ is not conveniently stated in terms of angles; instead, it is usually specified by describing the unit vector $\mathbf{u}$ parallel to $\mathbf{v}$. If this is

$$\mathbf{u} = l\mathbf{i} + m\mathbf{j} + n\mathbf{k} \qquad 1\text{--}12$$

then we say that $\mathbf{v}$ has direction (l, m, n). Since l is the component of a displacement of unit length, its magnitude is cos [angle between $\mathbf{u}$ (or $\mathbf{v}$) and the x-axis]; m and n are likewise cosines of the angles between $\mathbf{u}$ (or $\mathbf{v}$) and the y- and z-axes. For this reason l, m, n are often referred to as **direction cosines.**

> The direction cosines of a vector $\mathbf{v}$, are the components of the unit vector $\mathbf{u}$ parallel to $\mathbf{v}$.

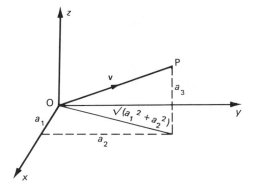

Fig. 1.17

Since **v** is of magnitude $|\mathbf{v}|$, the vector $(1/|\mathbf{v}|)\mathbf{v}$ has unit magnitude. Its direction is parallel to **v**, so $\mathbf{v}/|\mathbf{v}| = \mathbf{u}$. It is often convenient to use a notation which relates the names of the components to that of the vector, i.e. to write $\mathbf{v} = v_1\mathbf{i} + v_2\mathbf{j} + v_3\mathbf{k}$, $\mathbf{b} = b_1\mathbf{i} + b_2\mathbf{j} + b_3\mathbf{k}$. The main exceptions to this are position vectors, which we write as $\mathbf{r} = x\mathbf{i} + y\mathbf{j} + z\mathbf{k}$, and force vectors which we commonly write $\mathbf{F} = X\mathbf{i} + Y\mathbf{j} + Z\mathbf{k}$.

From a simple extension of theorem 1–4 of § 1.3, if

$$\mathbf{a} = a_1\mathbf{i} + a_2\mathbf{j} + a_3\mathbf{k} \quad \text{and} \quad \mathbf{p} = p_1\mathbf{i} + p_2\mathbf{j} + p_3\mathbf{k},$$

the equation $\mathbf{p} = \mathbf{a}$ implies

$$p_1 = a_1; \quad p_2 = a_2; \quad \text{and} \quad p_3 = a_3 \qquad 1\text{–}13$$

and, conversely, the three equations 1–13 imply $\mathbf{p} = \mathbf{a}$.

1.61 Comment on notation

If, in a given (cartesian) coordinate frame the vector **v** has components v_1, v_2, v_3, it is common to write

$$\mathbf{v} = (v_1, v_2, v_3)$$

This means just the same as $\mathbf{v} = v_1\mathbf{i} + v_2\mathbf{j} + v_3\mathbf{k}$. An alternative notation

$$\mathbf{v} = \begin{pmatrix} v_1 \\ v_2 \\ v_3 \end{pmatrix}$$

has some mathematical advantages, but is less used because of its uneconomical use of page-space.

Example

Find the coordinates x, y and z of the point defined by the position vector

$$6(3, 2, 1) + 3(-1, 6, 2) - 2(2, 0, -5)$$

36 An introduction to mechanics and modelling

The given vector $= (18, 12, 6) + (-3, 18, 6) + (-4, 0, 10)$

$\qquad\qquad\qquad = (11, 30, 22)$

The required coordinates are $x = 11$; $y = 30$; $z = 22$.

Comment: the same work could be written

$$\mathbf{r} = 6(3\mathbf{i} + 2\mathbf{j} + \mathbf{k}) + 3(-\mathbf{i} + 6\mathbf{j} + 2\mathbf{k}) - 2(2\mathbf{i} - 5\mathbf{k})$$
$$= 11\mathbf{i} + 30\mathbf{j} + 22\mathbf{k}$$

whence $x = 11$, $y = 30$, $z = 22$.

Exercise 6

1. In a triangle OAB, $\overrightarrow{OA} = 2\mathbf{i} - 5\mathbf{j}$; $\overrightarrow{OB} = -3\mathbf{i} - 4\mathbf{j}$ and P is a point of AB such that AP = 2PB. Express $\overrightarrow{AB}$ and $\overrightarrow{OP}$ in terms of $\mathbf{i}$ and $\mathbf{j}$ and hence find the length of AB and the coordinates of P (taking O as origin).
2. $\mathbf{I}$ is a unit vector making angle α with the x-axis, so $\mathbf{I} = (\cos \alpha)\mathbf{i} + (\sin \alpha)\mathbf{j}$; and $\mathbf{J}$ is unit vector at right angles to $\mathbf{I}$, making angle $\pi/2 + \alpha$ with the x-axis. Express $\mathbf{J}$ in terms of $\mathbf{i}$, $\mathbf{j}$ and α. $\overrightarrow{OB}$ is a vector of unit length making angle β with $\mathbf{I}$, i.e. angle $\alpha + \beta$ with the x-axis.
   (i) Express $\overrightarrow{OB}$ in terms of $\mathbf{I}$ and $\mathbf{J}$.
   (ii) Substitute the expressions for $\mathbf{I}$ and $\mathbf{J}$ in terms of $\mathbf{i}$ and $\mathbf{j}$ into (i), and deduce an expression for $\overrightarrow{OB}$ in terms of $\mathbf{i}$ and $\mathbf{j}$.
   (iii) By comparing the expression of (ii) for $\overrightarrow{OB}$ with the value of $\overrightarrow{OB}$ expressed directly in terms of $\mathbf{i}$ and $\mathbf{j}$ and $(\alpha + \beta)$, prove that
   $$\cos(\alpha + \beta) = \cos \alpha \cos \beta - \sin \alpha \sin \beta,$$
   and $\qquad\sin(\alpha + \beta) = \sin \alpha \cos \beta + \cos \alpha \sin \beta.$
3. A is the point whose position vector $= a_1\mathbf{i} + a_2\mathbf{j}$. P is a point such that $\overrightarrow{AP}$ is parallel to $\mathbf{b}$, where $\mathbf{b} = b_1\mathbf{i} + b_2\mathbf{j}$ (write $\overrightarrow{AP} = \lambda \mathbf{b}$). Write down the position vector of point P and express it in terms of $\mathbf{i}$, $\mathbf{j}$, a_1, a_2, b_1, b_2, and λ; hence find the coordinates (x, y) of P in terms of a_1, a_2, b_1, b_2 and λ.
   By eliminating λ and writing $b_2/b_1 = m$, show that all points on the line through A parallel to $\mathbf{b}$ satisfy the equation $y - a_2 = m(x - a_1)$.

> **Note:** the equation $\mathbf{r} = \mathbf{a} + \lambda \mathbf{b}$ may be regarded as the equation of the straight line, in vector form, λ taking all real values $-\infty < \lambda < \infty$.

4. (i) Find the vector sum of $2\mathbf{i} - \mathbf{j} + \mathbf{k}$, $-3\mathbf{i} - 4\mathbf{k}$, $2\mathbf{j} - \mathbf{k}$, $-\mathbf{i} + 3\mathbf{j}$.
   (ii) Find the vector sum of forces with components $(2, 2, -1)$, $(-3, 0, 4)$, $(1, -5, 3)$. (Consistency of the force units is assumed.)
5. Find the magnitude of the resultant of the vectors given in question 4(ii) and hence find the unit vector parallel to the direction of the resultant. Describe the direction of the resultant in terms of its direction cosines.
6. Three forces are parallel, respectively, to $\mathbf{i} - 2\mathbf{j} + \mathbf{k}$, $\mathbf{j}$, and $2\mathbf{i} - \mathbf{k}$. The vector sum of the three forces is $\mathbf{k}$. Find the magnitude of each. [**Hint:** write the first $c_1(\mathbf{i} - 2\mathbf{j} + \mathbf{k})$, the second $c_2\mathbf{j}$, etc.]
7. Find the angle made by $\mathbf{i} - 2\mathbf{j} + \mathbf{k}$ with each of the x-, y- and z-axes, and hence find the angle it makes with each of the coordinate planes.

Modelling exercises outside mechanics.

8. Attempt to formulate a simple algebra relevant to the palatability of food containing salt, e.g. let mass m_i of foodstuff i contain mass s_i of salt, and its palatability be denoted by $p(s_i, m_i)$. Propose laws of addition $p(s_1, m_1) + p(s_2, m_2) = ?$, and of multiplication by a scalar, $kp(s, m) = ?$ Go further in any direction which seems possible.

Vector algebra 37

9 In what ways is your attack on question 8 satisfactory, and what shortcomings do you think should be mentioned?
10 Attempt to construct an algebra of nouns or verbs, or language (you may not get very far, but you will see the kind of difficulty such attempts usually encounter; textbooks usually show you only exceptional, successful models).
11 (Suitable for group discussions.) Potatoes in the shop cost 20p per kg. A 25 kg sack of potatoes sold at a farm gate costs £3. To assess whether the sack is good value for money, what calculations/measurements could be made (either before buying or subsequently)? Illustrate with calculations, using hypothetical data (e.g. mean diameter of potatoes in sack = 10 cm, proportion rotten = 0.5%, ...). After each calculation, summarize the conclusions so far, and list the factors which you know you have ignored; state whether, *to you*, what you have calculated is more/less important than what you have ignored at that stage.

2

Differentiation of vectors; accelerations; scalar and vector products and geometry in three dimensions

The subject matter of this chapter falls into three main compartments.

§§ 2.1–2.3 Differentiation of vectors; velocity, acceleration and their components.
§§ 2.4–2.6 Scalar and vector products and their manipulation.
§§ 2.7–2.84 Application of vector algebra to three-dimensional geometry.

The reader is advised to master these component parts one at a time.

2.1 Sequences of vectors

Consider an (infinite) sequence of vectors $\mathbf{a}_1, \mathbf{a}_2, \mathbf{a}_3, \ldots$ as shown in FIG. 2.1.

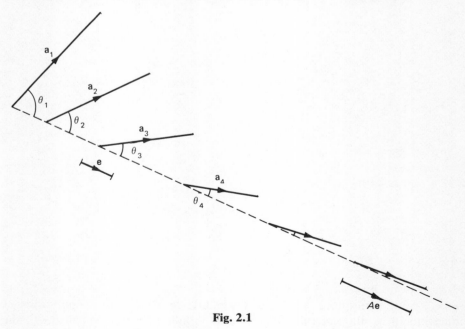

Fig. 2.1

Differentiation of vectors; accelerations; scalar and vector products 39

If a direction (specified by unit vector **e**) exists such that the angles $\theta_1, \theta_2, \theta_3, \ldots$ made by $\mathbf{a}_1, \mathbf{a}_2, \mathbf{a}_3 \ldots$ with **e** form a sequence with limit zero (see FIG. 2.1) then the vectors have a limiting direction, viz. **e**. If also the sequence of magnitudes $|\mathbf{a}_1|, |\mathbf{a}_2|, |\mathbf{a}_3|, \ldots$ has a limiting value, A say, then the sequence of vectors is said to have limit $A\mathbf{e}$.

The values of **e** and A can be determined only when the sequence $\mathbf{a}_1, \mathbf{a}_2, \ldots$ is known; just as in ordinary algebra, the procedures are best illustrated by particular cases, for which see § 2.21.

2.2 Velocity and acceleration

If P is a moving point, and O is a fixed origin, the position vector of P at time t_1 will, in general, be different from its position vector at time t_2. We may use the notations $\mathbf{r}(t_1)$ and $\mathbf{r}(t_2)$ to distinguish the two position vectors. The further generalization wherein $\mathbf{r}(t)$ is used to signify the position vector of the moving point P at any time t, follows naturally.

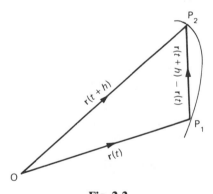

Fig. 2.2

FIG. 2.2 shows the path of a moving point, and the position vectors of the point at times t and $t+h$. The displacement $\overrightarrow{P_1P_2} = \overrightarrow{P_1O} + \overrightarrow{OP_2} = -\mathbf{r}(t) + \mathbf{r}(t+h)$. Thus $\mathbf{r}(t+h) - \mathbf{r}(t) =$ change of position achieved in time interval h. The left-hand side can be multiplied by the number $1/h$.

$$(1/h)[\mathbf{r}(t+h) - \mathbf{r}(t)] = \text{mean rate of change of position}$$
$$= \bar{\mathbf{v}}(t, h) \text{ say.}$$

Construction of the vector $\bar{\mathbf{v}}(t, h)$ has involved only the standard processes of subtraction of two vectors, and multiplication by a scalar. Keeping the same t, we may now consider a sequence of values of h, viz. $h_1, h_2, h_3, \ldots$ tending to zero. If the corresponding sequence of values of $\bar{\mathbf{v}}(t, h)$ has a limit which is independent of the particular sequence of h values through which $h \to 0$, the limit $\mathbf{v}(t)$ $[\equiv \bar{\mathbf{v}}(t, 0)]$ is called the derivative (with respect to time) of the position vector $\mathbf{r}(t)$.

$$\frac{d\mathbf{r}}{dt} \equiv \lim_{h \to 0} \frac{1}{h}[\mathbf{r}(t+h) - \mathbf{r}(t)]$$
$$= \mathbf{v}(t) \quad \text{say} \qquad 2\text{-}1$$

The expression 2-1 constitutes the **algebraic** definition of the derivative $d\mathbf{r}/dt$. Its physical significance becomes clear in § 2.21. Only the algebraic definition is needed for Exercise 1.

Exercise 1

The methods required are the exact counterpart of methods of the calculus with which the student should already be familiar.

1. Given that $\mathbf{V} = \mathbf{p} + \mathbf{q}$, prove that $d\mathbf{V}/dt = d\mathbf{p}/dt + d\mathbf{q}/dt$.
2. Given that $\mathbf{V} = u_1 \mathbf{a} + u_2 \mathbf{b}$, where u_1 and u_2 are (scalar) functions of time, and $\mathbf{a}$ and $\mathbf{b}$ are constant vectors, prove that

$$\frac{d\mathbf{V}}{dt} = \frac{du_1}{dt}\mathbf{a} + \frac{du_2}{dt}\mathbf{b}.$$

3. Given that $\mathbf{V} = u\mathbf{p}$, where $\mathbf{p}$ is a time-dependent vector, and u is a scalar function of time, prove that

$$\frac{d\mathbf{V}}{dt} = u\frac{d\mathbf{p}}{dt} + \frac{du}{dt}\mathbf{p}$$

Hint: first show that $\mathbf{V}(t+h) - \mathbf{V}(t) = u(t+h)[\mathbf{p}(t+h) - \mathbf{p}(t)] + [u(t+h) - u(t)]\mathbf{p}(t)$. Subsequently you may assume the result (which follows directly from the corresponding result for scalar functions) $\lim_{h \to 0}[u(t+h)\mathbf{v}(t,h)] = \lim_{h \to 0} u(t+h) \lim_{h \to 0} \mathbf{v}(t,h)$ provided the limits on the right-hand side exist.

4. Given that $\mathbf{a}$ and $\mathbf{b}$ are constant vectors, find the derivatives with respect to t of each of the following vectors: $3t^2\mathbf{a} + \mathbf{b}$; $(\mathbf{a} - \mathbf{b})t$; $\mathbf{a}/t - \mathbf{b}/t^2$.

2.21 Magnitude and direction of velocity

In FIG. 2.2, the vector $\mathbf{v}(t, h) = (1/h)\overrightarrow{P_1 P_2}$ is parallel to $\overrightarrow{P_1 P_2}$. The limit of the sequence of **directions** of the vector $\mathbf{v}(t, h)$ is therefore the limiting direction of $\overrightarrow{P_1 P_2}$, i.e. is the direction of the tangent at P_1.

The limit of the sequence of **magnitudes**,

$$\lim_{h \to 0} \frac{1}{h}(\text{chord } P_1 P_2) = \lim_{h \to 0} \frac{1}{h}\left(\frac{\text{chord } P_1 P_2}{\text{arc } P_1 P_2}\right) \text{arc } P_1 P_2$$

For a smooth curve,

$$\lim_{P_2 \to P_1} \frac{\text{chord } P_1 P_2}{\text{arc } P_1 P_2} = 1$$

and the limit of the other factor,

$$\lim_{h \to 0} \frac{\text{arc } P_1 P_2}{h}$$

is the quantity ds/dt whose magnitude is the speed of point P along the curve.

Differentiation of vectors; accelerations; scalar and vector products 41

Hence the velocity is

> $\mathbf{v}(t)$ = vector in the direction of the tangent at P_1, of magnitude ds/dt
> = $(ds/dt)\mathbf{e}$ where $\mathbf{e}$ is unit vector along the tangent 2–2
> = $d\mathbf{r}/dt$

Example 1
A point P moves along a circular arc, starting from Q. At any time t the arc distance $QP = s$ satisfies the relation $s = 3t^3 - t$. Find the velocity vector of P at time t.

We have $ds/dt = 9t^2 - 1$, so, if $\mathbf{e}_t$ represents unit vector in the direction of the tangent at P, the required velocity

$$\mathbf{v} = (9t^2 - 1)\mathbf{e}_t$$

Example 2
A point moves in the y-axis in such a way that $y = \sin kt$. Its velocity at time t is therefore

$$\mathbf{v} = (k \cos kt)\mathbf{j}$$

where $\mathbf{j}$ is unit vector in the y-direction. The speed is $|k \cos kt|$.

Less straightforward velocity calculations are discussed in § 2.4.

2.22 Vectorial definition of acceleration
The time derivative of the velocity vector of a point is called the acceleration of the point,

> we define acceleration
> $$\mathbf{a} = \frac{d\mathbf{v}}{dt}$$ 2–3

2.23 Relative velocity and relative acceleration
Given two moving points Q_1, Q_2, whose position vectors are $\mathbf{r}_1$, $\mathbf{r}_2$, the position of Q_2 relative to Q_1 is characterized by

$$\overrightarrow{Q_1Q_2} = \mathbf{r}_2 - \mathbf{r}_1 \qquad 2\text{–}4$$

The velocity of Q_2 relative to Q_1 is defined as

$$\frac{d}{dt}(\overrightarrow{Q_1Q_2}) = \frac{d}{dt}(\mathbf{r}_2 - \mathbf{r}_1) = \frac{d\mathbf{r}_2}{dt} - \frac{d\mathbf{r}_1}{dt}$$

42 An introduction to mechanics and modelling

i.e. the relative velocity, **V** say,	$= \mathbf{v}_2 - \mathbf{v}_1$	2–5(a)
and, the (true) velocity of Q_2	$= \mathbf{v}_1 + \mathbf{V}$	2–5(b)

Similarly, the acceleration of Q_2 relative to Q_1 is defined as

$$\frac{d^2}{dt^2}(\overrightarrow{Q_1 Q_2}) = \frac{d^2}{dt^2}(\mathbf{r}_2 - \mathbf{r}_1) = \frac{d^2 \mathbf{r}_2}{dt^2} - \frac{d^2 \mathbf{r}_1}{dt^2}$$

i.e. the relative acceleration, **A** say,	$= \mathbf{a}_2 - \mathbf{a}_1$	2–6(a)
and, the (true) acceleration of Q_2	$= \mathbf{a}_1 + \mathbf{A}$	2–6(b)

which is equal to the acceleration of Q_1 + the acceleration of Q_2 relative to Q_1.

2.24 Differentiation with respect to variables other than time

Given that a vector **V** is a function of an arbitrary scalar variable w say, representations of **V** can be drawn from a chosen origin O, so that, for example, $\overrightarrow{OP_1}$ represents $\mathbf{V}(w)$ and $\overrightarrow{OP_2}$ represents $\mathbf{V}(w + h)$. Apart from the change of nomenclature, FIG. 2.2 and the definition of the derivative given in § 2.2, may be used as before with **r** replaced by **V**.

2.25 Example: motion in a circle (velocity)

A point P moves in a circle, centre O, radius a (constant). Find the velocity of P in terms of the angle θ which OP makes with a fixed direction in its plane of motion (and derivatives of θ). The arc length $P_1 P_2$ (FIG. 2.3) $= a\delta\theta$, provided that $\delta\theta$ is measured in radians.

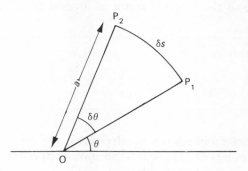

Fig. 2.3

$$ds/dt = \lim_{\delta t \to 0} \frac{a\delta\theta}{\delta t} = a\, d\theta/dt \qquad 2\text{–}7$$

Hence the velocity at any point P is $a(d\theta/dt)\mathbf{e}$ where **e** is unit vector in the direction of the tangent at P.

Differentiation of vectors; accelerations; scalar and vector products 43

2.26 Dot notation for time derivatives

Quantities like $d\theta/dt$, dx/dt, dy/dt, etc. are in constant use in mechanics because they arise in the calculation of velocities. Like velocities, they can vary with time. The values of, say, $d\theta/dt$ at times t_1 and t_2 could be distinguished by writing $(d\theta/dt)_{t_1}$ and $(d\theta/dt)_{t_2}$, but the notation is not a very happy one. The dot notation is introduced in mechanics to denote derivatives with respect to the time variable t; thus

$$d\theta/dt = \dot{\theta}; \quad dx/dt = \dot{x}; \quad dy/dt = \dot{y}: \quad \text{etc.}$$

The values of $\dot{\theta}$ at times t_1 and t_2 are then neatly written as $\dot{\theta}(t_1)$ and $\dot{\theta}(t_2)$.

2.27 Speed, velocity and time derivatives

The derivative dy/dt is neither exactly a speed nor a velocity. It is not precisely a speed because dy/dt may take negative as well as positive values. It is not precisely a velocity because it is a scalar not a vector quantity. However, **if a diagram is available which clearly shows the direction of measurement** of y, then the symbol dy/dt carries information, albeit not in explicit vector form, about the direction as well as the magnitude of the velocity. Where there is no possibility of misunderstanding, dy/dt or $\dot{y}$ will sometimes be referred to as a (linear) velocity; if s is a curvilinear path length in a prescribed sense, ds/dt or $\dot{s}$ may be called a (curvilinear) velocity; and if θ is an angle of rotation about a specified axis, or a polar angle in a specified plane, $d\theta/dt$ or $\dot{\theta}$ may be called an angular velocity; $\ddot{y}$, $\ddot{s}$ may be loosely referred to as accelerations, $\ddot{\theta}$ as an angular acceleration.

The preferred notation for velocity or acceleration remains the one in which its vectorial character is explicitly shown, e.g.

$$\text{if} \quad \mathbf{r} = y\mathbf{j}; \quad \mathbf{v} = \dot{y}\mathbf{j}; \quad \mathbf{a} = \ddot{y}\mathbf{j}$$

The axis of rotation, not the direction of motion of any particular point, is chosen as the direction associated with an angular speed $\dot{\theta}$, where θ is a polar angle like that used in the example of § 2.25. By definition, angular velocity is equal to $\dot{\theta}\mathbf{k}$ where $\mathbf{k}$ is unit vector parallel to the axis of rotation, the sense of $\mathbf{k}$ corresponding to the direction of motion of a right-hand screw subjected to turning in the positive direction of θ.

Note that $\theta\mathbf{k}$ does not represent a linear displacement, and cannot be added vectorially to ordinary displacements. It remains to be discussed in a later chapter whether the vector law of addition has any significance for quantities such as $\theta\mathbf{k}$, $\dot{\theta}\mathbf{k}$, $\ddot{\theta}\mathbf{k}$. In the outcome, angular rotations are **not** vectorial in character, but angular velocities and angular accelerations **are** appropriately added by the vector law. Consequently the notation $\theta\mathbf{k}$ is rejected, but angular velocity and acceleration are usefully represented by the notations $\dot{\theta}\mathbf{k}$, $\ddot{\theta}\mathbf{k}$ respectively.

2.28 Example: motion of a jointed system

In FIG. 2.4, the two rods AB and BC of lengths $2a$, $2b$ respectively, are swinging in a plane. The angles which the rods make with the vertical at any time t are written θ, ϕ respectively. Calculate the speed of the mid-point G of rod BC.

44 An introduction to mechanics and modelling

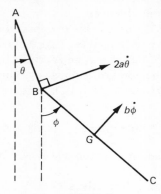

Fig. 2.4

Write the velocity of $G = \mathbf{v}$; the velocity of $B = \mathbf{v}_1$; the velocity of G relative to $B = \mathbf{v}_2$. Then

$$\mathbf{v}_2 = \mathbf{v} - \mathbf{v}_1, \quad \text{i.e.} \quad \mathbf{v} = \mathbf{v}_1 + \mathbf{v}_2 \qquad 2\text{-}8$$

Now BA is constant in length, so B moves with a velocity $2a\dot{\theta}$ in a direction perpendicular to AB (see FIG. 2.4). Since BG is constant in length, G's motion relative to B is a circle; its velocity relative to B is $b\dot{\phi}$ in the direction perpendicular to BG as shown in FIG. 2.4. Note that the positive sense of $\dot{\theta}$ must correspond to the direction of θ increasing, and similarly for $\dot{\phi}$.

Equation 2-8 tells us that the true velocity of G is the vector sum of the two velocities we have just described and shown in FIG. 2.4. The angle between the two velocities to be added is the same as the angle between the rods, i.e. $\phi - \theta$. By use of formula 1-6 of the last chapter, or from first principles, the required speed is therefore

$$\sqrt{[4a^2\dot{\theta}^2 + b^2\dot{\phi}^2 + 4ab\dot{\theta}\dot{\phi}\cos(\phi - \theta)]} \qquad 2\text{-}9$$

Exercise 2
A light aircraft A is flying due east at 150 km h^{-1} on a straight course which will take it directly along the line of the east–west runway of an airport, at a height of 1 km above the runway. At the instant when the aircraft's horizontal distance from the end of the runway is 1 km, a jet plane J takes off from the airport in the east–west direction, and climbs at a constant angle α to the horizontal at constant speed 600 km h^{-1}. To avoid accident, the two aircraft must at no time be closer than 0.366 km. Impose suitable restrictions on the take-off point and on the angle α of the climb of the jet.

Hints for solution
(i) Identify vectors representing the velocities of the two planes.
(ii) Write down the components of the vector $\mathbf{V}$ which represents the velocity of the jet relative to the light aircraft.
(iii) Establish that, from the viewpoint of either pilot, the line of approach of the craft makes an angle $\gamma = |45° - \beta|$ with their initial relative displacement $\overline{AJ}$, where

$$\tan \beta = \frac{4 \sin \alpha}{4 \cos \alpha + 1}$$

(iv) Show that, to avoid accident, either $\beta > 60°$ or $\beta < 30°$.

(v) Find the corresponding restrictions on α. This may be done **either** by drawing intersecting graphs of $y = 4 \sin x$, $y = \sqrt{3}(4 \cos x + 1)$ and $y = \sqrt{3}(4 \cos x + 1)/3$, **or** by the substitution $t = \tan(\alpha/2)$ and solution of a quadratic equation for t, accepting only roots corresponding to $0 < \alpha < \pi/2$.

2.29 Differentiation of unit vectors rotating in a plane

Consider a unit vector **I** whose direction varies with time but which can always be represented in a constant plane (i.e. its variation is in two dimensions, not three). Let us write $\mathbf{I} = \mathbf{I}(t)$, and evaluate the derivative $d\mathbf{I}/dt$ in terms of the angle θ which **I** makes with a specified constant line in the plane (the x-axis say), and two perpendicular base vectors of which one is **I**, and the other is unit vector **J** in the same plane, but always 90° ($\pi/2$ radians) ahead of **I** in the same, usually anticlockwise, sense as the angle θ is measured.

No matter what points in space are suggested by the physical context as the 'natural' starting points of **I** at different times, we choose to represent $\mathbf{I}(t)$ at all times starting from the same fixed origin O, FIG. 2.5a and b. Because of the given restriction on **I**, all these displacements from O lie in a plane.

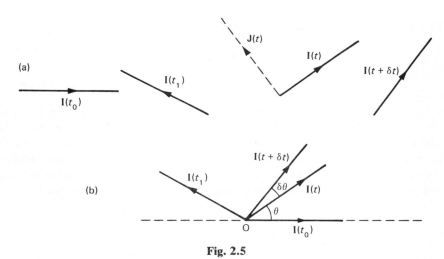

Fig. 2.5

In this representation, the end-point P of the vector $\mathbf{I} = \overrightarrow{OP}$ is always at unit distance from O, i.e. point P moves on a circle, centre O, radius = 1, as the inclination of **I** varies. From § 2.25, equation 2–7, it follows that the magnitude of $d\mathbf{I}/dt$ is $d\theta/dt = \dot{\theta}$, and the direction of $d\mathbf{I}/dt$ is that of the tangent at P to the unit circle, the sense being that of θ increasing. That is, the direction of $d\mathbf{I}/dt$ is that of the unit vector **J** previously introduced. Hence,

$$d\mathbf{I}/dt = \dot{\theta}\mathbf{J} \qquad 2\text{–}10$$

Similarly,

$$\frac{d\mathbf{J}}{dt} = \frac{d(\theta + \pi/2)}{dt}\mathbf{L} = \dot{\theta}\mathbf{L}$$

where **L** is unit vector $\pi/2$ radians in advance of **J**, i.e. is unit vector antiparallel to **I** (FIG. 2.6), $\mathbf{L} = -\mathbf{I}$.

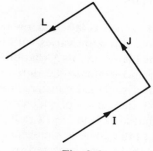

Fig. 2.6

So,
$$\mathrm{d}\mathbf{J}/\mathrm{d}t = -\dot{\theta}\mathbf{I} \qquad 2\text{-}11$$

From equations 2-10 and 2-11, it further follows that
$$\frac{\mathrm{d}^2}{\mathrm{d}t^2}\mathbf{I} = -\dot{\theta}^2\mathbf{I}$$

Example

An automatic pen moves over a piece of graph paper in such a way that its coordinates at time t are $(3t^2, t-2)$. Find the velocity of the pen point (a) assuming the graph paper to be stationary, (b) assuming that the paper is rotating in its own plane with constant angular velocity Ω about the origin of coordinates.

[In (b), the coordinates are defined by reference to marked axes on the rotating graph paper, but the required velocity is a true velocity in 'fixed' space.]

(a)
$$\mathbf{r} = 3t^2\mathbf{i} + (t-2)\mathbf{j}$$
$$\mathbf{v} = 6t\mathbf{i} + \mathbf{j}$$

(b)
$$\mathbf{r} = 3t^2\mathbf{I} + (t-2)\mathbf{J}$$
$$\mathbf{v} = 6t\mathbf{I} + 3t^2\,\mathrm{d}\mathbf{I}/\mathrm{d}t + \mathbf{J} + (t-2)\,\mathrm{d}\mathbf{J}/\mathrm{d}t$$

so, using equations 2-10 and 2-11 with $\dot{\theta} = \Omega$,
$$\mathbf{v} = 6t\mathbf{I} + 3t^2\Omega\mathbf{J} + \mathbf{J} - (t-2)\Omega\mathbf{I}$$
$$= (6t - \Omega t + 2\Omega)\mathbf{I} + (1 + 3t^2\Omega)\mathbf{J}$$

2.3 Components of velocity and acceleration

Given two (or three) mutually perpendicular unit vectors, **I**, **J**, (**K**), called base vectors, any vector **V** in the plane of **I**, **J** may be expressed as a multiple of **I** plus a multiple of **J**. In three dimensions any vector is expressible as a sum of multiples of **I**, **J**, **K**. The coefficients of **I**, **J**, (**K**) are called the components of **V**

Differentiation of vectors; accelerations; scalar and vector products 47

in the **I**-, **J**-, (**K**)-directions; this terminology is used whether or not the directions of **I**, **J**, (**K**) vary with time.

2.31 Rectangular cartesian components of velocity and acceleration
The base vectors are the constant vectors **i**, **j**, **k**.

$$\mathbf{r} = x\mathbf{i} + y\mathbf{j} + z\mathbf{k}$$

$\mathbf{v} \equiv \dot{\mathbf{r}} = \dot{x}\mathbf{i} + \dot{y}\mathbf{j} + \dot{z}\mathbf{k}$	2–12
$\mathbf{a} \equiv \ddot{\mathbf{r}} = \ddot{x}\mathbf{i} + \ddot{y}\mathbf{j} + \ddot{z}\mathbf{k}$	2–13

Thus, the cartesian components of velocity are $(\dot{x}, \dot{y}, \dot{z})$; and the components of acceleration are $(\ddot{x}, \ddot{y}, \ddot{z})$.

2.32 Tangential and normal components of acceleration
Sometimes it is convenient to describe a particle's position through an arc length OP, where O is a fixed point on, and $s = \mathrm{OP}$ is measured along, the particle's path (FIG. 2.7). When the path is a plane curve, we further specify a base line, XX' say, from which to measure angles. Let ψ denote the angle between XX' and the tangent at P to the particle's path. Let **I'** denote unit vector along this tangent, in the sense of s increasing. From equation 2–10, expressed in the present variables, $\mathrm{d}\mathbf{I}'/\mathrm{d}t = (\mathrm{d}\psi/\mathrm{d}t)\mathbf{J}'$, i.e. $\dot{\mathbf{I}}' = \dot{\psi}\mathbf{J}'$ where **J'** is unit vector along the normal PC which is $\pi/2$ ahead of **I'**.

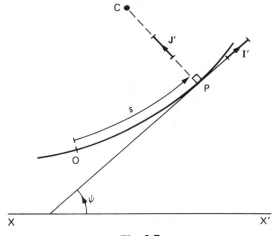

Fig. 2.7

The velocity of the particle at P

$$\dot{\mathbf{r}} = \dot{s}\mathbf{I}' \qquad\qquad 2\text{–}14$$

i.e.   the tangential component of velocity $= \dot{s}$
the normal component of velocity $= 0$

48 An introduction to mechanics and modelling

From 2–14,
$$\ddot{\mathbf{r}} = \ddot{s}\mathbf{I}' + \dot{s}\dot{\mathbf{I}}'$$
$$= \ddot{s}\mathbf{I}' + \dot{s}\dot{\psi}\mathbf{J}' \qquad 2\text{–}15$$

The normal component
$$\dot{s}\dot{\psi} = \dot{s}\frac{d\psi}{dt} = \dot{s}\frac{d\psi}{ds}\frac{ds}{dt} = \frac{\dot{s}^2}{\rho} \qquad 2\text{–}15(a)$$

where $\rho = ds/d\psi$. When $ds/d\psi$ is positive, as in FIG. 2.9(i), the normal acceleration has the same sense as $\mathbf{J}'$, i.e. is pointed along the **inward** normal to the curve. When $ds/d\psi$ is negative, the curve bends the other way, but the sign of $\dot{s}\dot{\psi}\mathbf{J}'$ is also reversed, so the sense of the normal component of acceleration is still **along the inward normal to the path** FIG. 2.9(ii).

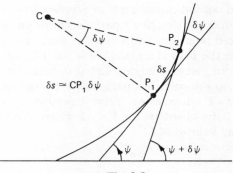

Fig. 2.8

The quantity $\rho = |ds/d\psi|$ is called the radius of curvature at point P; the reason for this name can be seen from FIG. 2.8. In that figure, P_1 and P_2 are the end-points of arc δs, and the normals at P_1 and P_2 to the given curve intersect at C. For δs sufficiently small, the approximation $CP_2 = CP_1$ is acceptable, and C is the centre of an elementary circular arc which nearly coincides with δs. It follows that, in FIG. 2.8, $\delta s \simeq CP_1 \delta\psi$, and, in all cases,

$$CP_1 \simeq \left|\frac{\delta s}{\delta\psi}\right|$$

whence
$$\rho = \lim_{\delta s \to 0} CP_1 = \left|\frac{ds}{d\psi}\right|$$

The statement (equation 2–15) that the component of acceleration in the $\mathbf{J}'$ direction is $\dot{s}\dot{\psi}$ can therefore be re-expressed.

The normal component of acceleration $= \dot{s}^2/\rho$ along the **inward** normal.
$$\qquad 2\text{–}16$$

We have also, from equation 2–15, the tangential component of acceleration $= \ddot{s}$, as in FIG. 2.9.

Differentiation of vectors; accelerations; scalar and vector products   49

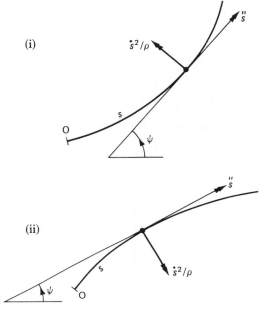

Fig. 2.9

2.33 Radial and transverse components of velocity and acceleration

We take $\mathbf{I}$ = unit vector parallel to $\mathbf{r}$, and $\mathbf{J}$ = unit vector $\pi/2$ ahead of $\mathbf{I}$ (in the sense of the polar coordinate θ increasing) (see FIG. 2.10a). These will be our

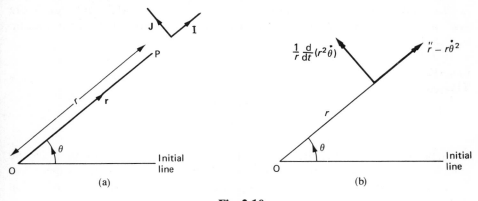

Fig. 2.10

base vectors. The direction of $\mathbf{I}$ is called the radial direction. The direction of $\mathbf{J}$ is called the transverse direction.

$$\mathbf{r} = r\mathbf{I}$$

so,
$$\dot{\mathbf{r}} = \dot{r}\mathbf{I} + r\dot{\mathbf{I}}$$
$$= \dot{r}\mathbf{I} + r\dot{\theta}\mathbf{J} \text{ (using equation 2–10)} \qquad 2\text{–}17$$

> i.e. the velocity of the point has radial and transverse components $\dot{r}$ and $r\dot{\theta}$ respectively.

A second differentiation gives,

$$\ddot{\mathbf{r}} = \ddot{r}\mathbf{I} + \dot{r}\dot{\mathbf{I}} + (\dot{r}\dot{\theta} + r\ddot{\theta})\mathbf{J} + r\dot{\theta}\dot{\mathbf{J}},$$

whence, using equations 2–10 and 2–11,

$$\ddot{\mathbf{r}} = (\ddot{r} - r\dot{\theta}^2)\mathbf{I} + (2\dot{r}\dot{\theta} + r\ddot{\theta})\mathbf{J}$$

$$= (\ddot{r} - r\dot{\theta}^2)\mathbf{I} + \frac{1}{r}\frac{\mathrm{d}}{\mathrm{d}t}(r^2\dot{\theta})\mathbf{J} \qquad 2\text{–}18$$

> i.e. the acceleration of a point moving in a plane has radial and transverse components
>
> $$\ddot{r} - r\dot{\theta}^2 \quad \text{and} \quad \frac{1}{r}\frac{\mathrm{d}}{\mathrm{d}t}(r^2\dot{\theta}) \quad \text{respectively} \qquad 2\text{–}18(a)$$
>
> (see FIG. 2.10b).

Note: the re-writing of the **J**-component $2\dot{r}\dot{\theta} + r\ddot{\theta}$ as $(1/r)(\mathrm{d}/\mathrm{d}t)(r^2\dot{\theta})$ proves to be very rewarding: the reader should verify that the latter form is equivalent to the other, and thereafter regard it as the standard form.

2.34 Example: motion in a circle (acceleration)

As an exercise, we shall compare the convenience of the different coordinate systems discussed in §§ 2.31–2.33 when applied to the particular case of motion in a circle.

(a) Polar coordinates
The path has equation

$$r = a \quad (= \text{constant})$$
$$\dot{r} = 0, \quad \text{and} \quad \ddot{r} = 0$$

So, from equation 2–17, the velocity has components 0, $a\dot{\theta}$. From 2–18 the components of acceleration are $-a\dot{\theta}^2$ radially (i.e. $a\dot{\theta}^2$ towards the centre) and $(1/a)(\mathrm{d}/\mathrm{d}t)(a^2\dot{\theta})$, i.e. $a\ddot{\theta}$ directed along the tangent to the circle, in the direction of θ increasing.

(b) s, ψ coordinates
The path has equation

$$s = a\psi$$

$$\frac{\mathrm{d}s}{\mathrm{d}\psi} = a \quad \text{whence } \rho = a.$$

Differentiation of vectors; accelerations; scalar and vector products

From equation 2–14, the velocity is $\dot{s}$ along the tangent (which for comparison with method (a) may be expressed as $a\dot{\psi}$), and the acceleration components are

$$\ddot{s} \; (= a\ddot{\psi})$$

tangentially, and

$$\dot{s}^2/a \; (= a^2\dot{\psi}^2/a = a\dot{\psi}^2)$$

towards the centre of the circle, in agreement with (a).

(c) Cartesian coordinates

In (a) and (b) above, knowledge of the path traced out by the moving point enabled the velocity and acceleration of the moving point to be expressed in terms of the time derivatives of a single parameter, viz. θ in (a) and s in (b). The equation of a circle in cartesian coordinates, viz. $x^2 + y^2 = a^2$, by contrast does not directly lead to neat expressions for $\dot{x}$, $\dot{y}$, $\ddot{x}$ and $\ddot{y}$ in terms of a single parameter x or y.

From $x^2 + y^2 = a^2$, differentiation gives $x\dot{x} + y\dot{y} = 0$. So,

$$y = \pm\sqrt{(a^2 - x^2)} \quad \text{and}$$

$$\dot{y} = \frac{-x}{y}\dot{x}$$

$$= \mp \frac{x\dot{x}}{\sqrt{(a^2 - x^2)}}$$

In terms of x and $\dot{x}$, the velocity components are therefore $(\dot{x}, \mp x\dot{x}/\sqrt{(a^2 - x^2)})$. The equivalence of this to previous results may be established by writing $x = a \cos \theta$. The moral to be drawn is that the use of coordinates other than the familiar x, y and z sometimes makes calculations much easier and more direct.

From (a), (b) and (c), the reader may draw three conclusions, viz.

> (1) The choice of one coordinate system rather than another will not affect the validity of the conclusions (which are equivalent).
> (2) The choice of one coordinate system rather than another may well affect the amount of labour involved in a calculation.
> (3) When a point moves in a circle of radius a with (tangential) velocity $v = \dot{s} = a\dot{\theta}$, its acceleration has components
>
> $$\dot{v} \; (= \ddot{s} = a\ddot{\theta}) \text{ tangentially}$$
>
> and
>
> $$v^2/a \; (= \dot{s}^2/a = a\dot{\theta}^2) \text{ towards the centre of the circle.}$$
>
> 2–19

Exercise 3

1 Derive the results of equations 2–19 by direct differentiation of the vector $\mathbf{r} = a\mathbf{I}$.

52 An introduction to mechanics and modelling

2 Given that $\mathbf{r} = 3t^2\mathbf{i} - (2t-3)\mathbf{j}$, find the cartesian components of acceleration of the point whose position vector is $\mathbf{r}$.
3 Given that $y = r \sin \theta$, find $\dot{y}$ in terms of $\dot{r}$, $\dot{\theta}$ and θ.
4 Given that $y = r \sin \theta$, find $\ddot{y}$ in terms of time derivatives of r and θ. Evaluate $\ddot{y}$ when $\theta = 90°$, i.e. when $y = r$ (momentarily). Explain why $\ddot{y} \neq \ddot{r}$.

2.35 Acceleration in mechanics

The reader may be puzzled by our apparent pre-occupation with the evaluation of $\ddot{\mathbf{r}}$ in various coordinate systems. The importance of the acceleration $\ddot{\mathbf{r}}$ arises as follows. Newton's second law tells us that **the acceleration of the centre of a rigid body is a constant multiple of the resultant force acting on the body**; this is the central equation of mechanics. From it, when the forces acting on a body are known – whether as constant values, as functions of time, or as functions of position coordinates – the acceleration of the central point can be deduced. By writing the components of acceleration as expressions involving derivatives of the coordinates and equating these to the corresponding components of force (functions of time/coordinates), the mathematical essence of the body's motion is encapsulated. From the equations involving derivatives, standard mathematical techniques of integration etc. allow the positions and velocities of the body at all times to be deduced. Sometimes the mathematical processes are intransigent, but in many cases, by choice of an appropriate coordinate system, mathematical difficulties melt away. Familiarity with the most commonly useful mathematical methods, and the problem types to which they are applicable, will be gained as the reader works through this book.

2.4 Products of vectors

If, at this point, we wrote two vector symbols in juxtaposition in any of the forms (i) $\mathbf{a}\mathbf{b}$ (ii) $\mathbf{a} \cdot \mathbf{b}$ (iii) $\mathbf{a} \times \mathbf{b}$ (iv) $\mathbf{a} \div \mathbf{b}$, it would be a nonsense; for no meaning of these combinations follows from the only properties embodied in vector symbols. On the other hand, there is no impediment to **ascribing** a meaning to these or any other so far undefined combination of symbols; the only practical desideratum is that the notation shall be helpful rather than misleading – that considerably more of the expectations aroused by the notation shall prove true than false. The combinations (ii) and (iii) will be defined in this section; the combination (iv) will be regarded as utterly abhorrent – the obliquely implied analogy with division in arithmetic could only be misleading in the context of the chosen definitions of 'products' (ii) and (iii). The notation (i) will not be needed.

The student will need the vector product ($\mathbf{a} \times \mathbf{b}$ or $\mathbf{a} \wedge \mathbf{b}$) for mechanics. His/her most urgent needs are met in §§ 2.53–2.56 and § 2.62.

2.41 Component of a vector in an arbitrary direction

Consider a vector $\mathbf{a} = a_1\mathbf{i} + a_2\mathbf{j} + a_3\mathbf{k}$, of magnitude $|\mathbf{a}| = a$. The component of $\mathbf{a}$ parallel to $\mathbf{k}$ (i.e. to the z-axis) is equal in magnitude to the side parallel to $\mathbf{k}$ of a right-angled triangle whose hypotenuse is $\mathbf{a}$ (FIG. 2.11). Thus

$$a_3 = a \cos (\text{angle between } \mathbf{a} \text{ and } \mathbf{k})$$

Differentiation of vectors; accelerations; scalar and vector products 53

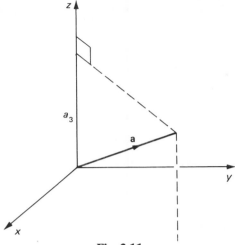

Fig. 2.11

It further follows that if l_1, m_1 and n_1 are the direction cosines of **a**, so that n_1 is the cosine of the angle between **a** and **k**, $a_3 = n_1 a$, and in the same way

which, with
$$\left.\begin{array}{l} a_1 = l_1 a \\ a_2 = m_1 a \\ a_3 = n_1 a \end{array}\right\} \qquad 2\text{-}20$$

completes a useful triad of relationships.

To find the component of **a** in the direction of an arbitrary vector **b**, we take **b** as the direction of one of a new set of coordinate axes, e.g., take $\mathbf{I} = (1/b)\mathbf{b}$. The terminology is consistent with that already used if we define:

the component of **a** in the direction of **b** = a cos (angle between the directions of **a** and **b**).

Note that the unqualified word 'component' as habitually used in the phrase 'component of **a** in the direction of **b**' tacitly implies a frame of mutually perpendicular base vectors **I**, **J**, **K**, i.e. an orthogonal frame. We adopt the convention that all reference frames are orthogonal unless the contrary is explicitly stated; oblique frames and oblique components will not concern us in this book.

2.42 Evaluation of the component in an arbitrary direction

Let the direction of **b** have direction cosines l_2, m_2, n_2. It follows that

the component of **i** in direction $\mathbf{b} = l_2$

the component of **j** in direction $\mathbf{b} = m_2$

the component of **k** in direction $\mathbf{b} = n_2$

and the component of $a_1\mathbf{i} + a_2\mathbf{j} + a_3\mathbf{k}$ in direction $\mathbf{b}$ (by equation 1–11) is equal to

$$a_1 l_2 + a_2 m_2 + a_3 n_2 \qquad \text{2–21}(a)$$

The component in the direction of $\mathbf{b}$ (direction cosines l_2, m_2, n_2) of the unit vector $\mathbf{l} \equiv \mathbf{a}/a = l_1\mathbf{i} + m_1\mathbf{j} + n_1\mathbf{k}$ is $l_1 l_2 + m_1 m_2 + n_1 n_2$.

> i.e. cos (angle between two vectors $\mathbf{a}$ and $\mathbf{b}$)
> $$= l_1 l_2 + m_1 m_2 + n_1 n_2 \qquad \text{2–21(b)}$$
> where l_1, m_1, n_1 and l_2, m_2, n_2 are the direction cosines of $\mathbf{a}$, $\mathbf{b}$.

Equations 2–21(a) and (b) are of great use in routine evaluation of components. Of the two, 2–21(b) is probably the easier to memorize, and, used in conjunction with the definition of a component, serves the same purpose as 2–21(a).

Exercise 4

1. Use equation 2–21(b) to show that for any direction l, m, n, $l^2 + m^2 + n^2 = 1$.
2. Show that $\mathbf{e}_1 = (1/\sqrt{3}, 1/\sqrt{3}, 1/\sqrt{3})$ and $\mathbf{e}_2 = (0, 1/\sqrt{2}, 1/\sqrt{2})$ are both unit vectors. Use equation 2–21(b) to find the angle between them.
3. Find unit vectors parallel, respectively, to $\mathbf{a} = (1, 3, 2)$ and $\mathbf{b} = (0, -2, 2)$, and hence find the angle between $\mathbf{a}$ and $\mathbf{b}$.

The following example does not demand the use of vectors. The result displayed below will be needed in § 2.55.

4. In FIG. 2.12, p_1 and p_2 are two planes intersecting along line RS. A, B and C are any three points in p_1; AA', BB', CC' are the perpendiculars from A, B and C on to plane p_2.

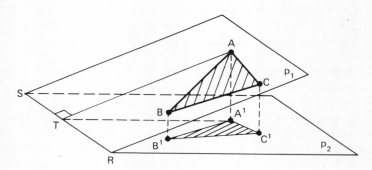

Fig. 2.12

(i) Prove that the plane through A perpendicular to RS contains AA'. (**Hint:** consider the line through T perpendicular to p_2, and use the fact that any two parallel lines are coplanar.)

(ii) If V is the point at which the plane through B perpendicular to RS cuts RS, prove that the area of trapezium A'B'VT is equal to (area ABVT) $\cos \alpha$ where $\alpha = \angle A'TA$.

(iii) Deduce that

> $$\text{area A'B'C'} = (\text{area ABC}) \cos \alpha \qquad \text{2–22}$$

Differentiation of vectors; accelerations; scalar and vector products

2.5 Use of scalar and vector product notations

2.51 Angles, components and scalar product

From section 2.42 we know that, given vectors **a**, **b** of magnitudes a, b respectively and directions $\mathbf{I}_1$, $\mathbf{I}_2$ respectively, where $\mathbf{I}_1 = l_1\mathbf{i} + m_1\mathbf{j} + n_1\mathbf{k} = (l_1, m_1, n_1)$ and $\mathbf{I}_2 = l_2\mathbf{i} + m_2\mathbf{j} + n_2\mathbf{k} = (l_2, m_2, n_2)$, the angle θ between $\mathbf{I}_1$ and $\mathbf{I}_2$ is given by $\cos\theta = l_1l_2 + m_1m_2 + n_1n_2$.

Of very similar structure is the formula for the component of **a** in the direction of **I**. Given $\mathbf{a} = a_1\mathbf{i} + a_2\mathbf{j} + a_3\mathbf{k} = (a_1, a_2, a_3)$ and $\mathbf{I} = l\mathbf{i} + m\mathbf{j} + n\mathbf{k} = (l, m, n)$, the desired component is
$$a\cos\theta = a_1l + a_2m + a_3n$$

In general, given that
$$\mathbf{a} = a_1\mathbf{i} + a_2\mathbf{j} + a_3\mathbf{k}$$
$$\mathbf{b} = b_1\mathbf{i} + b_2\mathbf{j} + b_3\mathbf{k}$$
then
$$ab\cos\theta = a_1b_1 + a_2b_2 + a_3b_3 \qquad \text{2-23(a)}$$

[for $ab\cos\theta = ab(l_1l_2 + m_1m_2 + n_1n_2) = (l_1a)(l_2b) + (m_1a)(m_2b) + (n_1a)(n_2b)$
$$= a_1b_1 + a_2b_2 + a_3b_3]$$

Quantities of the form we have been discussing in this paragraph are of sufficiently common occurrence and of sufficient importance to justify the introduction of a special notation. $ab\cos\theta$ is determined by the magnitudes and directions of **a** and **b**, so only the symbols **a** and **b** need to be mentioned, together with some arbitrarily chosen sign to denote the particular combination or function of those vector entities. The conventional sign is a dot – dots have not previously been attributed any meaning in vector algebra – thus

$$\mathbf{a} \cdot \mathbf{b} \equiv ab\cos\theta = a_1b_1 + a_2b_2 + a_3b_3 \qquad \text{2-23(b)}$$

This quantity is a number, not a vector, and it is given the name of 'the scalar product' of **a** and **b**. The decision to call it a product does not confer on it the properties associated with products of numbers (scalars). Its definition confers specific properties, which will be discussed in § 2.6, which allow corresponding laws of algebraic manipulation to be **derived**.

2.52 Examples of the use of scalar products

Example 1 Evaluation of the components of a vector
Given a vector $\mathbf{a} = a_1\mathbf{i} + a_2\mathbf{j} + a_3\mathbf{k}$, we observe that $\mathbf{a} \cdot \mathbf{i} = a_1$; so, to find the components of any vector **a** we merely evaluate

$$a_1 = \mathbf{a} \cdot \mathbf{i}; \quad a_2 = \mathbf{a} \cdot \mathbf{j}; \quad a_3 = \mathbf{a} \cdot \mathbf{k} \qquad \text{2-24}$$

Example 2
Given that the vertices of a triangle ABC are the points A (2, 1, 3), B (−1, 1, 1), C (1, 0, 3), find the magnitude of angle C.

56 An introduction to mechanics and modelling

We make no attempt in FIG. 2.13a to plot the actual coordinates of A, B and C but use the figure only to clarify completely general relationships. The coordinates of A, B and C give us the values of the position vectors, r_A, r_B, r_C ($r_A = 2i+j+3k$; $r_B = -i+j+k$; $r_C = i+3k$).

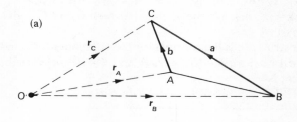

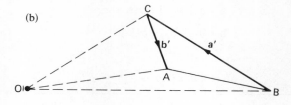

Fig. 2.13

From r_A, r_B, r_C we deduce

$$\mathbf{b} = r_C - r_A \;(= -1, -1, 0)$$
and
$$\mathbf{a} = r_C - r_B \;(= 2, -1, 2)$$

To rotate the direction of **b** into that of **a** requires rotation through angle C, hence

$$\mathbf{b} \cdot \mathbf{a} = ba \cos C$$

whence
$$\cos C = (\mathbf{b} \cdot \mathbf{a})/ba$$

Now using 2-23(b), $\mathbf{b} \cdot \mathbf{a} = -2 + 1 = -1$; $b = \sqrt{2}$; and $a = \sqrt{9} = 3$. So

$$\cos C = -1/3\sqrt{2} = -0.2357$$

i.e. C is an obtuse or reflex angle, $180 \pm 76°22'$.
The angle of a triangle is necessarily less than 180°, so we deduce $C = 103°38'$.

We note that, had we allotted vector symbols to $\overrightarrow{BC}$ and $\overrightarrow{CA}$ (instead of to $\overrightarrow{BC}$ and $\overrightarrow{AC}$, for example $\overrightarrow{BC} = \mathbf{a}'$, $\overrightarrow{CA} = \mathbf{b}'$ as in FIG. 2.13b, the angle required to rotate the direction of $\mathbf{b}'$ into that of $\mathbf{a}'$ would be $C + 180°$. In this case

$$\mathbf{b}' \cdot \mathbf{a}' = b'a' \cos(C + 180°)$$
$$= -b'a' \cos C$$

i.e.
$$\cos C = -(\mathbf{b}' \cdot \mathbf{a}')/b'a'$$

This would again yield $\cos C = -0.2357$ and $C = 103°38'$.

Differentiation of vectors; accelerations; scalar and vector products 57

A definition is needed for the following example.

Power: the power expended when a force **F** is applied to a body moving with velocity **v** at the point where the force is applied, is $\mathbf{F} \cdot \mathbf{v}$ (see § 10.3).

Example 3

A dinghy with an outboard motor is travelling SSE at 3 km h^{-1} through initially still water, with its engine developing its maximum power of 5 kW (where 1 kW = 1000 m s^{-1}N). The wind comes from due west. The rudder is so turned that the force exerted by the water on the boat because of the propeller action is due south. This force is only $\frac{1}{3}$ of that which would be exerted if the water was itself immovable. Find (a) the force exerted by the water, (b) the force exerted by the wind on the boat.

(**Note:** the reader is expected only to follow the arithmetical steps; he/she is not yet adequately prepared to analyse the concealed assumptions.)

If only the boat, not the water, were set in motion, the magnitude of the force **P** exerted on the boat could be calculated by equating the scalar product of force and velocity to the power developed. Thus, since

$$\text{magnitude of velocity} = 3 \text{ km h}^{-1} = \frac{3000}{60 \times 60} \text{ m s}^{-1} = \tfrac{5}{6} \text{ m s}^{-1};$$

$$\text{magnitude of force } \mathbf{P} = P;$$

and the angle between force and velocity = $22\tfrac{1}{2}°$; it follows that

$$\tfrac{5}{6}P \cos 22\tfrac{1}{2}° = 5000; \quad P = 6494 \text{ (N)}$$

The actual force exerted on the boat by the water is therefore 2165 (N) (to the nearest integer).

The wind comes from the west so exerts a force, Q say, towards the east. From FIG. 2.14 we conclude that the wind force

$$Q = 2165 \tan 22\tfrac{1}{2}° = 897 \text{ (N)}$$

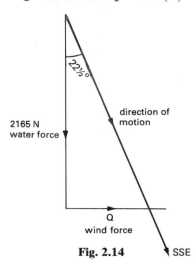

Fig. 2.14

58 An introduction to mechanics and modelling

2.53 Vectors at right angles to a given plane; vector product

Given orthogonal vectors **i**, **j**, the third mutually orthogonal unit vector **k** is automatically determined, for there is only one direction perpendicular to the plane of **i** and **j** in the required sense. Since **k** is determined by **i** and **j**, it can appropriately be written as a function of **i** and **j**.

There are other situations in which a physically meaningful quantity is associated with the direction at right angles to the plane of two given vectors. Thus, if, as in FIG. 2.15, P is a point of a rigid body free to swivel about O; if $\overrightarrow{OP} = \mathbf{r}$; and if force **F** is applied at P, then the body will begin to rotate about an axis through O perpendicular to the plane of **r** and **F**. Let us denote this direction by the unit vector **e** (not shown in FIG. 2.15 because at right angles to the paper). The rate of change of angular speed about this axis is determined by the magnitude of $rF \sin \theta$, where θ is the angle between **r** and **F**, as shown in FIG. 2.15. We may explain this expression, $rF \sin \theta$, to ourselves in the following way: the component $F \cos \theta$ of the force **F** is neutralized by the rigid connection OP. The remaining component $F \sin \theta$ is in the direction of the unrestricted motion of P relative to O; it is moreover at right angles to **r**, so the most rudimentary form of the 'principle of the lever' is enough to suggest that $r(F \sin \theta)$ is likely to measure the effectiveness of the turning action.

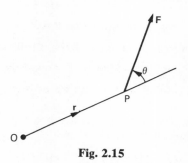

Fig. 2.15

Because it is a useful quantity in discussing the motion of rigid bodies (and this is further confirmed in Chapter 6), a special notation is introduced to denote the vector of magnitude $rF \sin \theta$ associated with an axis of rotation parallel to **e**. We write

$$\mathbf{r} \wedge \mathbf{F} = (rF \sin \theta)\mathbf{e} \qquad \qquad 2\text{--}25(a)$$

This quantity is known as the **vector product** of the two vectors **r** and **F**. Its algebraic properties flow from the definition, equation 2–25(a), and only as these properties are investigated will the student know the extent to which the word 'product' evokes helpful or false expectations. In general, given vectors **a**, **b**, their vector product is

$$\text{(definition)} \quad \mathbf{a} \wedge \mathbf{b} = ab \sin \theta \, \mathbf{e} \qquad \qquad 2\text{--}25(b)$$

where θ is the angle the direction of **b** makes with **a**; and **e** is that unit normal to the

Differentiation of vectors; accelerations; scalar and vector products 59

plane of **a** and **b** whose sense corresponds to the longitudinal motion of a screw (into or out of a block of wood, say) the groove in whose head is rotated from the direction of **a** into the direction of **b** through the angle θ.

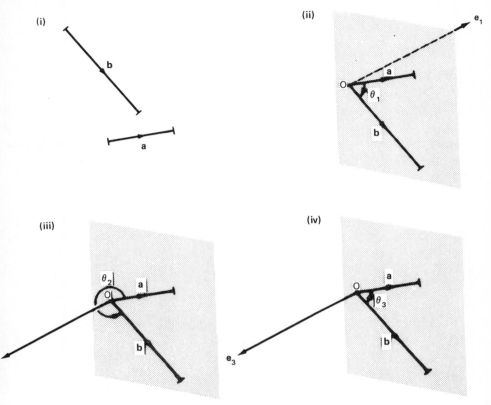

Fig. 2.16

The reader will observe, cf. FIG. 2.16(ii) and (iii), that no restriction need be made about whether θ is to be measured clockwise or anticlockwise (even if these terms were meaningful in three-dimensional space). This is the case because reversal of the sense of measurement of the angle ensures that, in the notation shown on these figures,

$$\mathbf{e}_2 = -\mathbf{e}_1$$

while

$$\theta_2 = 360° - \theta_1$$

i.e.

$$\sin \theta_2 = -\sin \theta_1$$

whence

$$(ab \sin \theta_2)\mathbf{e}_2 = (ab \sin \theta_1)\mathbf{e}_1$$

On the other hand, we see from FIG. 2.16(iv) that

$$\mathbf{b} \wedge \mathbf{a} = (ba \sin \theta_3)\mathbf{e}_3$$

where

$$\theta_3 = \theta_1 \quad \text{i.e.} \sin \theta_3 = \sin \theta_1$$

and

$$\mathbf{e}_3 = \mathbf{e}_2 = -\mathbf{e}_1$$

Hence

$$\mathbf{b} \wedge \mathbf{a} = -\mathbf{a} \wedge \mathbf{b} \qquad 2\text{-}26$$

Note that, both in vector products and in scalar products, **a**, **b** commonly belong to different vector sets (e.g. **r**, **F**) which have different units.

2.54 Examples of vector products

Example 1 Relations between **i**, **j**, **k**

The definition of a vector product ensures that the third unit vector of any right-handed set of orthogonal unit vectors may be written explicitly in terms of the other two, thus

$$\mathbf{i} = \mathbf{j} \wedge \mathbf{k} \qquad \mathbf{j} = \mathbf{k} \wedge \mathbf{i} \qquad \mathbf{k} = \mathbf{i} \wedge \mathbf{j} \qquad 2\text{-}27(a)$$

and we note that

$$\mathbf{k} \wedge \mathbf{j} = -\mathbf{i} \qquad \mathbf{i} \wedge \mathbf{k} = -\mathbf{j} \qquad \mathbf{j} \wedge \mathbf{i} = -\mathbf{k} \qquad 2\text{-}27(b)$$

Example 2 Vector product $\mathbf{a} \wedge \mathbf{a}$

Since the angle between any vector and itself is zero,

$$\mathbf{a} \wedge \mathbf{a} = 0 \qquad 2\text{-}27(c)$$

In particular,

$$\mathbf{i} \wedge \mathbf{i} = \mathbf{j} \wedge \mathbf{j} = \mathbf{k} \wedge \mathbf{k} = 0$$

Example 3

The area of the triangle ABC in FIG. 2.17 is

$$\tfrac{1}{2}ab \sin \theta = \tfrac{1}{2}|\mathbf{a} \wedge \mathbf{b}| \qquad 2\text{-}28$$

The orientation of the plane of the triangle is conveniently specified by the direction of its normal, i.e. the direction of $\mathbf{a} \wedge \mathbf{b}$.

Definition: the quantity $\tfrac{1}{2}(\mathbf{a} \wedge \mathbf{b})$ specifying both the area and its orientation is given the name 'vector area'

$$\mathbf{A} = \tfrac{1}{2}\mathbf{a} \wedge \mathbf{b} \qquad 2\text{-}29$$

Differentiation of vectors; accelerations; scalar and vector products

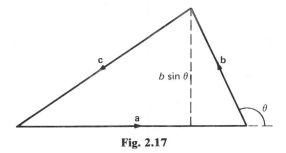

Fig. 2.17

Since it is equally true that $\mathbf{A} = \frac{1}{2}\mathbf{b} \wedge \mathbf{c}$, and $\mathbf{A} = \frac{1}{2}\mathbf{c} \wedge \mathbf{a}$, it follows that if $\mathbf{a}, \mathbf{b}, \mathbf{c}$ can be represented as the sides of a triangle, i.e.

| if $\mathbf{a} + \mathbf{b} + \mathbf{c} = 0$, then $\mathbf{a} \wedge \mathbf{b} = \mathbf{b} \wedge \mathbf{c} = \mathbf{c} \wedge \mathbf{a}$ | 2–30 |

2.55 Vector product expressed in components

The **k**-component of the vector area $\frac{1}{2}\mathbf{a} \wedge \mathbf{b}$ is of magnitude $\frac{1}{2}|\mathbf{a} \wedge \mathbf{b}| \cos \alpha$, where α is the acute angle between **k** and $\mathbf{a} \wedge \mathbf{b}$. But the angle between these two normals (for **k** and $\mathbf{a} \wedge \mathbf{b}$ are respectively normal to the x-y plane and the plane of ABC) is the same as the angle between those planes. Hence the magnitude of the selected component is the area of the projection (A'B'C') of △ABC on the x-y plane (FIG. 2.18a).

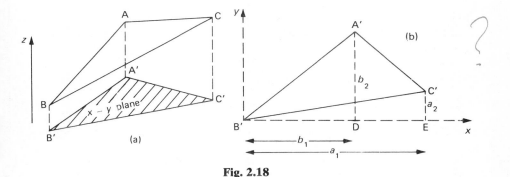

Fig. 2.18

If $\mathbf{a} = a_1\mathbf{i} + a_2\mathbf{j} + a_3\mathbf{k}$, then side B'C' of the projected triangle $= a_1\mathbf{i} + a_2\mathbf{j}$. Similarly side C'A' $= b_1\mathbf{i} + b_2\mathbf{j}$. We now consider the area of △A'B'C', which lies wholly in the x-y plane. Area A'B'C' = A'B'D + A'DEC' − C'B'E in FIG. 2.18b (where B'E is the line through B' parallel to the x-axis)

$$\text{Area A'B'C'} = \frac{1}{2}[b_1 b_2 + (a_1 - b_1)(a_2 + b_2) - a_1 a_2]$$

$$= \frac{1}{2}(a_1 b_2 - a_2 b_1)$$

Hence the **k**-component of $\mathbf{a} \wedge \mathbf{b} = a_1 b_2 - a_2 b_1$. Writing down the corresponding

62 An introduction to mechanics and modelling

expressions for the **i**-, **j**-components, we find

$$\mathbf{a} \wedge \mathbf{b} = (a_2 b_3 - a_3 b_2)\mathbf{i} + (a_3 b_1 - a_1 b_3)\mathbf{j} + (a_1 b_2 - a_2 b_1)\mathbf{k} \qquad \text{2-31(a)}$$

which may also be written

$$\mathbf{a} \wedge \mathbf{b} = \begin{vmatrix} \mathbf{i} & \mathbf{j} & \mathbf{k} \\ a_1 & a_2 & a_3 \\ b_1 & b_2 & b_3 \end{vmatrix} \qquad \text{2-31(b)}$$

2.56 Components of the moment of a force

The vector product $\mathbf{r} \wedge \mathbf{F}$ introduced in § 2.53, equation 2–25, is called the moment **M** of the force **F** about the origin O. Writing $\mathbf{r} = x\mathbf{i} + y\mathbf{j} + z\mathbf{k}$; $\mathbf{F} = X\mathbf{i} + Y\mathbf{j} + Z\mathbf{k}$, we see from equation 2–31(a)

$$\begin{aligned}\mathbf{M} &\equiv \mathbf{r} \wedge \mathbf{F} \\ &= (yZ - zY)\mathbf{i} + (zX - xZ)\mathbf{j} + (xY - yX)\mathbf{k}\end{aligned} \qquad \text{2-32}$$

Whether or not the vector law of addition of quantities like $\mathbf{r} \wedge \mathbf{F}$ has any relevance to mechanics should not be assumed, but will be discussed later in the context of equivalent systems of forces (§§ 6.73, 8.2).

2.6 Algebraic manipulation of scalar and vector products

2.61 Algebraic laws for the scalar product

Since the components of $\mathbf{b} + \mathbf{c}$ are $b_1 + c_1$, $b_2 + c_2$, $b_3 + c_3$ (cf. equation 1–11)

$$\begin{aligned}\mathbf{a} \cdot (\mathbf{b} + \mathbf{c}) &= a_1(b_1 + c_1) + a_2(b_2 + c_2) + a_3(b_3 + c_3) \\ &= (a_1 b_1 + a_2 b_2 + a_3 b_3) + (a_1 c_1 + a_2 c_2 + a_3 c_3) \\ &= \mathbf{a} \cdot \mathbf{b} + \mathbf{a} \cdot \mathbf{c} \quad \text{(Law of distribution)}\end{aligned} \qquad \text{2-33}$$

Two laws which follow immediately from the definition of the scalar product are

$$\mathbf{a} \cdot \mathbf{b} = \mathbf{b} \cdot \mathbf{a} \quad \text{(Law of commutation)} \qquad \text{2-34}$$

and

$$\mathbf{a} \cdot (k\mathbf{b}) = (k\mathbf{a}) \cdot \mathbf{b} = k(\mathbf{a} \cdot \mathbf{b}) \quad \text{(Law of association for scalar multiples)} \qquad \text{2-35}$$

Special notice should be taken of the following:

Negative property A

$(\mathbf{a} \cdot \mathbf{b})\mathbf{c}$ is a vector parallel to **c**

$\mathbf{a}(\mathbf{b} \cdot \mathbf{c})$ is a vector parallel to **a**.

Differentiation of vectors; accelerations; scalar and vector products 63

Hence if **c** is not parallel to **a**,

$$(\mathbf{a} \cdot \mathbf{b})\mathbf{c} \neq \mathbf{a}(\mathbf{b} \cdot \mathbf{c})$$

i.e. **there is no general law of association valid for this kind of 'product' of three vectors.**

Negative property B
Given $\mathbf{a} \cdot \mathbf{b} = \mathbf{a} \cdot \mathbf{c}$, we may **not** deduce $\mathbf{b} = \mathbf{c}$, i.e. **division is not an allowable operation** in the way that it is for numbers. For $\mathbf{a} \cdot \mathbf{b} = \mathbf{a} \cdot \mathbf{c}$ implies only

$$ab \cos \theta_1 = ac \cos \theta_2$$

i.e. in general

$$b \cos \theta_1 = c \cos \theta_2,$$

a relation which determines neither the magnitude nor the direction of **b** in terms of **c**.

2.62 Algebraic laws for the vector product
From equation 1–11 and 2–31(a)

$$\mathbf{a} \wedge (\mathbf{b} + \mathbf{c}) = [a_2(b_3 + c_3) - a_3(b_2 + c_2)]\mathbf{i} + [\quad]\mathbf{j} + [\quad]\mathbf{k}$$
$$= (a_2 b_3 - a_3 b_2)\mathbf{i} + (\quad)\mathbf{j} + (\quad)\mathbf{k}$$
$$+ (a_2 c_3 - a_3 c_2)\mathbf{i} + (\quad)\mathbf{j} + (\quad)\mathbf{k}$$
$$= \mathbf{a} \wedge \mathbf{b} + \mathbf{a} \wedge \mathbf{c} \quad \text{(Law of distribution)} \qquad 2\text{–}36$$

Two further laws follow immediately from the definition of vector product, viz.

$$\mathbf{a} \wedge \mathbf{b} = -\mathbf{b} \wedge \mathbf{a} \quad \text{(Anticommutative law)} \qquad 2\text{–}37$$

and

$$(k\mathbf{a}) \wedge \mathbf{b} = \mathbf{a} \wedge (k\mathbf{b}) = k(\mathbf{a} \wedge \mathbf{b}) \quad \text{(Law of association for scalar multiples in a vector product)} \qquad 2\text{–}38$$

Negative properties

(*A*). As for scalar products, so for vector products, division is not an allowable operation; $\mathbf{a} \wedge \mathbf{b} = \mathbf{a} \wedge \mathbf{c}$ does **not** imply $\mathbf{b} = \mathbf{c}$. This need occasion no surprise in view of the arbitrariness of the choice of the name 'product' for the function of **a** and **b** which we choose to write as $\mathbf{a} \wedge \mathbf{b}$.

(*B*). There is also no general associative law for vector triple products,

i.e. in general $\qquad (\mathbf{a} \wedge \mathbf{b}) \wedge \mathbf{c} \neq \mathbf{a} \wedge (\mathbf{b} \wedge \mathbf{c})$

This is easily demonstrated by particular examples, e.g.

$$(\mathbf{i} \wedge \mathbf{i}) \wedge \mathbf{j} = \mathbf{0} \wedge \mathbf{j} = \mathbf{0}$$

but $\qquad \mathbf{i} \wedge (\mathbf{i} \wedge \mathbf{j}) = \mathbf{i} \wedge \mathbf{k} = -\mathbf{j} \neq \mathbf{0}$

64 An introduction to mechanics and modelling

Exercise 5
(The results stated in questions 8, 9, 14, 15 are of importance.)

1. Given that **i**, **j**, are unit vectors in two perpendicular directions, draw diagrams representing the following vectors,
$$p = 3i - j$$
$$q = -i + 2j$$
Calculate the magnitudes of **p** and **q**. (Note: $p \cdot p = |p|^2$.)

2. Given that **i**, **j**, **k** are three mutually orthogonal unit vectors, and that
$$a = 3i + j - k$$
$$b = i - 2j + k$$
state (i) the components in the **i**, **j**, **k** directions, (ii) the magnitudes, of the vectors **a**, **b**, (**a**+**b**), (**a**−**b**).

3. Given that $c = 2i - j$; $d = 4i + 2j$; $f = j$, verify that $(c + 2f) \cdot (d - c) = 7$.

4. Given that $p = 3i - j$; $q = -i + 2j$, calculate the angle between **p** and **q**. (Check your answer by reference to the diagram of question 1.)

5. Find the angle between the vectors $a = 3i + j - k$; $b = 2j + k$.

6. Find the component of force $P = i - j$ in the direction of the line joining $(-2, 1)$ to $(3, 3)$.

7. With the values of **a**, **b**, **c**, ... **q** specified in questions 1–3 above, evaluate: (i) $p \wedge q$; (ii) $a \wedge b$; (iii) $(a \wedge d) \cdot f$; (iv) $a \cdot (d \wedge f)$; (v) $d \cdot (f \wedge a)$.

8. Given that **r**, **s**, **t** are non-coplanar vectors, write down vectorial expressions for:
    (i) the vector area of a parallelogram whose sides are **r**, **s**;
    (ii) the unit vector perpendicular to the above parallelogram;
    (iii) the component of **t** perpendicular to the parallelogram's surface. Verify that $(r \wedge s) \cdot t$ represents the volume of a parallelepiped whose sides are **r**, **s**, **t**. Now explain why the values obtained in (iii), (iv) and (v) of question 7 are equal.

9. Prove, by considering the x-component, and 'similarly' the y, z-components of each side, that for any vectors **a**, **b**, **c**
$$a \wedge (b \wedge c) = (a \cdot c)b - (a \cdot b)c.$$
and
$$(a \wedge b) \wedge c = (a \cdot c)b - (b \cdot c)a$$

> Note that in both formulae the two vectors which appear inside brackets on the left appear outside the brackets on the right-hand-side; and that the vector **b** which occupies the middle position on the left, is the vector part of the first term on the right.

10. The coordinates of three points A, B, C in a right-hand rectangular cartesian frame Oxyz are, respectively, $(2, 1, 0)$, $(1, 2, -1)$, $(3, 4, -2)$. Express the vectors $a = \overline{BC}$, $b = \overline{CA}$, $c = \overline{AB}$ in terms of the unit vectors **i**, **j**, **k** lying along Ox, Oy, Oz respectively. Deduce the value of the (interior) angle ABC. Also evaluate the perpendicular distance of A from BC.

11. Form the vector product of both sides of the equation
$$a + b = c$$
with **a**. Deduce the relation $a/\sin A = b/\sin B = c/\sin C$ between the angles and sides of any triangle ABC. (Note that the angle between the positive directions of **a** and **b** is the **exterior** angle of the triangle illustrating the relationship $a + b = c$.)

12. Two sides $\overrightarrow{AB}$, $\overrightarrow{BC}$ of a triangle are represented by the vectors $3i + 6j - 2k$ and $4i - j + 3k$. Determine the angles of the triangle, and find a unit vector perpendicular to the plane of the triangle.

Differentiation of vectors; accelerations; scalar and vector products 65

13 Three non-coplanar vectors **a**, **b**, **c** each have unit magnitude, and the angles between **b** and **c**, **c** and **a**, **a** and **b** are respectively α, β, γ. If the vector **u** is defined by $\mathbf{u} = \mathbf{b} - (\mathbf{a} \cdot \mathbf{b})\mathbf{a}$, prove that **u** is perpendicular to **a** and has magnitude $\sin \gamma$.

Given also that $\mathbf{v} = \mathbf{c} - (\mathbf{a} \cdot \mathbf{c})\mathbf{a}$, and that the angle between **u** and **v** is A, prove that

$$\cos \alpha = \cos \beta \cos \gamma + \sin \beta \sin \gamma \cos A$$

14 To describe the rotation of a rigid body in two dimensions we may write the position vector of any point of the body relative to a fixed point in the body as

$$\mathbf{r} = X\mathbf{I} + Y\mathbf{J}$$

where X and Y are constant, but **I** and **J** change their directions as time goes on. By differentiating **r** and using the formulae 2–10, 2–11 for $\dot{\mathbf{I}}$ and $\dot{\mathbf{J}}$, deduce that

$$\mathbf{v}(\equiv \dot{\mathbf{r}}) = \boldsymbol{\omega} \wedge \mathbf{r} \quad \text{where} \quad \boldsymbol{\omega} = \dot{\theta}\mathbf{K}$$

(**Note:** although the above proof is particular to two dimensions, the result is true for three dimensions: the instantaneous angular velocity of a rigid body moving in any way in three dimensions may be represented by a vector $\boldsymbol{\Omega}$, and the velocity of any point of the body relative to an arbitrarily chosen origin embedded in the body is correctly given by $\mathbf{v} = \boldsymbol{\Omega} \wedge \mathbf{r}$.)

15 By use of components or otherwise, verify that if **a**, **b** are vectors which vary with t,

$$\frac{d}{dt}(\mathbf{a} \cdot \mathbf{b}) = \mathbf{a} \cdot \frac{d\mathbf{b}}{dt} + \frac{d\mathbf{a}}{dt} \cdot \mathbf{b} \qquad 2\text{–}39$$

and that

$$\frac{d}{dt}(\mathbf{a} \wedge \mathbf{b}) = \mathbf{a} \wedge \frac{d\mathbf{b}}{dt} + \frac{d\mathbf{a}}{dt} \wedge \mathbf{b} \qquad 2\text{–}40$$

16 The position of a small mass m moving in a plane is specified by polar coordinates (r, θ). Given that the vector sum of all forces acting on the particle has radial component R and transverse component S, use equation 2–18 to express its equation of motion ($m\ddot{\mathbf{r}} = \mathbf{F}$) in terms of **I**, **J**. Deduce separate radial and transverse equations of motion for the particle. (**Hint:** this question may be approached in a number of ways. The simplest is to recognize that if two vectors **F**, $m\ddot{\mathbf{r}}$ are equal, then their components are equal. This is generalized in (ii) of § 2.82. Alternatively, by writing $\mathbf{F} = R\mathbf{I} + S\mathbf{J}$, the equation $m\ddot{\mathbf{r}} = \mathbf{F}$ takes the form $a\mathbf{I} + b\mathbf{J} = R\mathbf{I} + S\mathbf{J}$, whence

$$(a\mathbf{I} + b\mathbf{J}) \cdot \mathbf{I} = (R\mathbf{I} + S\mathbf{J}) \cdot \mathbf{I}, \quad \text{i.e. } a = R$$

Hence also $b = S$.)

17 The position of a small mass m moving in a plane is specified by intrinsic coordinates s, ψ. Given that the resultant force on the mass has tangential and (inward) normal components T, N respectively, use equation 2–16 to deduce separate tangential and normal equations of motion for the mass. (See also § 4.51.)

2.7 Applications of vector algebra to geometry in three dimensions

The remaining paragraphs of this chapter extend the modelling capabilities of vector geometry, and are a preliminary to Chapter 3. Less advanced students who are interested only in mechanics, may, if desired, proceed immediately to Chapter 4.

66 An introduction to mechanics and modelling

2.71 Conventions

a, b, etc. denote constant, presumably known, vectors, whose cartesian components are (a_1, a_2, a_3), (b_1, b_2, b_3), etc. respectively.

r denotes the position vector of a point P whose cartesian coordinates are (x, y, z). The point P may be at any point in space consistent with the restrictions which we shall choose to specify.

λ, μ, etc. denote real (scalar) variables; λ may take all values in the range $-\infty < \lambda < \infty$, and similarly for μ, etc.

2.72 Equations of a straight line

Given **a** as in FIG. 2.19a all the points $P_1, P_2, \ldots$ in FIG. 2.19b satisfy the equation

$$\mathbf{r} = \lambda \mathbf{a} \qquad \qquad 2\text{–}41$$

for some value of λ. Every point $\mathbf{r} = \lambda \mathbf{a}$ lies on the broken line shown, and conversely every point of that line corresponds to $\mathbf{r} = \lambda \mathbf{a}$ for some value of λ; hence $\mathbf{r} = \lambda \mathbf{a}$ is the equation of the straight line through the origin O parallel to the vector **a**.

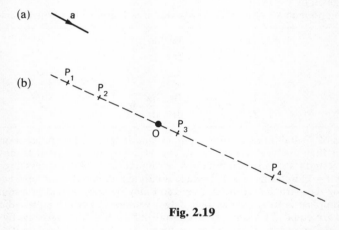

Fig. 2.19

The equation of a straight line through a given point B (write $\overrightarrow{OB} = \mathbf{b}$) and parallel to **a** may be constructed by reference to FIG. 2.20. For each of the points P′, P″, P‴, etc. $\mathbf{r} = \mathbf{b} + \lambda \mathbf{a}$ (for some λ). Conversely every point satisfying this

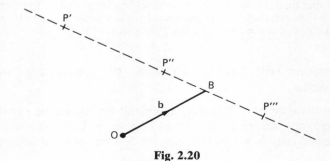

Fig. 2.20

Differentiation of vectors; accelerations; scalar and vector products 67

relation lies on the required line, so

$$\mathbf{r} = \mathbf{b} + \lambda \mathbf{a} \qquad 2\text{-}42$$

is the equation of a straight line parallel to $\mathbf{a}$ and passing through point $\mathbf{r} = \mathbf{b}$.

Note 1. $\mathbf{r} = \mathbf{b}$ may be regarded as 'the equation of a point'!
2. The directions of $\mathbf{a}, \mathbf{b}$ are not restricted to the plane of the paper: any line in space may be represented by an equation of form 2–42.

2.73 Cartesian equations derived from vector equations

2.731 Two-dimensional cartesian equation for a line through the origin (equation 2–44)

When the line lies in the x-y plane, $\mathbf{a}$ also lies in that plane, so $a_3 = 0$. Consequently, from equation 2–41,

$$\mathbf{r} = \lambda(a_1\mathbf{i} + a_2\mathbf{j})$$

i.e. $\qquad x = \lambda a_1; \quad y = \lambda a_2 \qquad 2\text{-}43$

whence $\qquad y = (a_2/a_1)x, \quad \text{or, writing } a_2/a_1 = m$

$$= mx \qquad 2\text{-}44$$

Equation 2–44 is the explicit, equation 2–43 the parametric cartesian equation of the required straight line.

2.732 Equation of the straight line joining points $B(\mathbf{r}=\mathbf{b})$ and $C(\mathbf{r}=\mathbf{c})$, and its translation into three-dimensional cartesian form

We note that $\overrightarrow{BC} = \mathbf{c} - \mathbf{b}$, so the required equation is (cf. equation 2–42)

$$\mathbf{r} = \mathbf{b} + \lambda(\mathbf{c} - \mathbf{b}) \qquad 2\text{-}45(a)$$

Example
Find cartesian equations for the straight line joining the points $B(2, -1, 3)$ and $C(6, 0, 1)$.

$$\mathbf{c} - \mathbf{b} = 4\mathbf{i} + \mathbf{j} - 2\mathbf{k}$$

so the required line has equation

$$\mathbf{r} = (2\mathbf{i} - \mathbf{j} + 3\mathbf{k}) + \lambda(4\mathbf{i} + \mathbf{j} - 2\mathbf{k}) \qquad 2\text{-}45(b)$$

i.e. it has parametric cartesian equations

$$x = 2 + 4\lambda; \quad y = -1 + \lambda; \quad z = 3 - 2\lambda; \qquad 2\text{-}46$$

from which elimination of the indeterminate parameter λ gives the pair of equations

$$\frac{x-2}{4} = \frac{y+1}{1} = \frac{z-3}{-2} \qquad 2\text{-}47$$

Note that the denominators 4, 1, −2 are determined by the vector **c−b**: they characterize the direction of the line; they are proportional to the direction cosines.

Exercise 6 (Coordinate geometry using vector methods)

1. Draw a directed line segment to represent **a**. Through some origin not on this segment, draw the line **r** = λ**a**. Indicate the range of values of λ corresponding to 3 portions of the line as follows:
   - (i) the segment starting at the origin and ending at the point **r** = **a**;
   - (ii) the segment from the latter point to the end of the line (infinity);
   - (iii) the remaining semi-infinite portion.
2. A unit cube is situated with one vertex at the origin, 3 edges coinciding with the axes of x, y, z, and the body of the cube in the region in which x, y, z are all positive (the 'positive octant'). Find the equation:
   - (i) of the diagonal of the cube which passes through the origin;
   - (ii) of each of the 3 concurrent diagonals of faces with a corner at the origin;
   - (iii) What are the direction cosines (ref. § 1.6) of each of the four diagonals mentioned above?
   - (iv) Using tables, or pocket calculator, determine the angle between the long diagonal of the cube and each of the coordinate axes (ref. also § 2.76).
3. Repeat question 2 for a rectangular box whose length, breadth and height are 4, 3, 2, respectively. Also calculate the angle between each pair of concurrent face diagonals (3 different angles).
4. Find (i) the vector, (ii) the cartesian, equation of the straight line joining the points (0, 6, 2) and (1, −1, −7).
5. Find (i) the vector, (ii) the cartesian, equations of the three face-diagonals of the unit cube of question 2 which pass through the point (1, 1, 1).
6. Find the position vector of point P which divides BC externally in the ratio 3:2 where B = (4, 2, 3) and C = (6, 3, 4). Write down the equation (i) of the line joining the origin to P, and (ii) of the line joining B and C.
7. The position vectors of points P, Q are given by **r** = 2**i**+3**j**−**k**; **r** = 4**i**−3**j**+2**k**. Determine $\overline{PQ}$ in terms of **i, j, k**, and find the length of PQ. Write down the equation of the line PQ.
8. Show that the line through A(3, −4, −2) parallel to the vector 9**i**+6**j**+2**k** has equations $(x-3)/9 = (y+4)/6 = (z+2)/2$. What are the coordinates of the two points on the line which are at distance 22 units from A?
9. Show that the line joining the points A(2, −3, −1) and B(8, −1, 2) has equations $(x-2)/6 = (y+3)/2 = (z+1)/3$, and rewrite this pair of equations in a form which shows that it passes through point B, i.e. $(x-8)/6 = ? = ?$
10. Given that **a** is a vector of variable direction in space, but of constant magnitude a, prove that the vector $d\mathbf{a}/dt$ is perpendicular to **a**. [**Hint:** consider $(d/dt)(\mathbf{a} \cdot \mathbf{a})$.]
11. C is a smooth curve in three dimensions; O is a fixed origin on C and s ($= \pm$arc OP) is the coordinate of any point P on C. Three vectors **t, n, b** (corresponding to the initials of the words tangent, normal and binormal) are defined as follows:
    - (i) unit vector in the direction of the tangent at P is denoted **t**;
    - (ii) unit vector in the direction of $d\mathbf{t}/ds$ (evaluated at P) is denoted **n**;
    - (iii) **b** = **t** ∧ **n**.

    Verify the following:
    - (a) the three vectors **t, n, b** are mutually perpendicular, and, like **t** and **n**, **b** is of unit length; [**Hint:** just as $\mathbf{I} = \dot{\theta}\mathbf{J}$, so $d\mathbf{I}/ds = (d\theta/ds)(\mathbf{J})$, i.e. $d\mathbf{I}/ds$ is perpendicular to **I**; similarly $d\mathbf{t}/ds$ is perpendicular to **t**.]
    - (b) $d\mathbf{b}/ds$ is parallel to −**n**; (**Hint:** first prove $d\mathbf{b}/ds = \mathbf{t} \wedge d\mathbf{n}/ds$, and hence that it is in the plane of **t** and **n**, and perpendicular to **t**.)
    - (c) $d\mathbf{n}/ds$ lies in the plane of **b** and **t**.

Writing the scalar multiples in (ii) and (b) as $1/\rho$ and $1/\sigma$ (where ρ is called the radius of curvature and σ the radius of torsion) we have:
(d) $d\mathbf{t}/ds = \mathbf{n}/\rho$; $d\mathbf{b}/ds = -\mathbf{n}/\sigma$.
With these definitions, verify:
(e) $d\mathbf{n}/ds = \mathbf{b}/\sigma - \mathbf{t}/\rho$.
The results (d) and (e) are known as the Frenet-Serret formulae.

2.74 Equations of planes

In FIG. 2.20, any point on the broken line could be represented by $\mathbf{r} = \mathbf{b} + \lambda \mathbf{a}$. Now consider the set of points which can be represented by

$$\mathbf{r} = \mu \mathbf{b} + \lambda \mathbf{a}$$

where $-\infty < \mu < \infty$. Instead of first proceeding to point B, and then travelling an arbitrary distance parallel to $\mathbf{a}$, we now start by going an arbitrary distance towards, beyond or away from B, followed by a displacement of arbitrary magnitude parallel to $\mathbf{a}$. Any point in the plane may be reached by this prescription.

$$\mathbf{r} = \lambda \mathbf{a} + \mu \mathbf{b} \qquad 2\text{-}48$$

is the equation of a plane through the origin, oriented so that it can contain representations of both the vectors $\mathbf{a}$, $\mathbf{b}$. Similarly

$$\mathbf{r} = \mathbf{c} + \lambda \mathbf{a} + \mu \mathbf{b} \qquad 2\text{-}49$$

is the plane through the point $\mathbf{r} = \mathbf{c}$, containing $\mathbf{a}$, $\mathbf{b}$.

2.75 Equation of a plane through a given point perpendicular to a given line

Let the given point be C ($\mathbf{r} = \mathbf{c}$) and the given line be $\mathbf{r} = \mathbf{b} + \lambda \mathbf{a}$ (which has

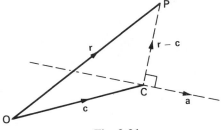

Fig. 2.21

direction $\mathbf{a}$). For any point $\mathbf{r}$ on the required plane, $\mathbf{r} - \mathbf{c}$ must be perpendicular to $\mathbf{a}$ (FIG. 2.21). The required equation is thus

$$(\mathbf{r} - \mathbf{c}) \cdot \mathbf{a} = 0 \qquad 2\text{-}50$$

i.e. in cartesian form

$$a_1 x + a_2 y + a_3 z = \mathbf{c} \cdot \mathbf{a}$$
$$= \beta, \quad \text{say} \qquad 2\text{-}51$$

70 An introduction to mechanics and modelling

where β is a constant, and the coefficients a_1, a_2, a_3 in equation 2–51 are components of a vector normal (i.e. perpendicular) to the plane.

Exercise
Write down the three component equations ($x = ?$, $y = ?$ and $z = ?$) of equation 2–49, and verify that elimination of λ, μ leads to a linear relation between x, y, z (cf. form of equation 2–51).

2.76 Angles between lines and planes

The angle between two planes is the angle between their normals.
The angle between two lines is the angle between the vectors specifying their directions.
The angle between a line and a plane is 90° – the angle between the line and the normal to the plane.

Having identified the two relevant directions, by vectors **a**, **b** say, and evaluated the cartesian components of **a**, **b**, the required angle can be calculated by use of

$$\cos\theta = \frac{\mathbf{a}\cdot\mathbf{b}}{|\mathbf{a}||\mathbf{b}|} = \frac{a_1 b_1 + a_2 b_2 + a_3 b_3}{ab}$$

where $a = \sqrt{(a_1^2 + a_2^2 + a_3^2)}$ and $b = \sqrt{(b_1^2 + b_2^2 + b_3^2)}$. If $\mathbf{a}\cdot\mathbf{b}$ is positive, θ is an acute angle; if $\mathbf{a}\cdot\mathbf{b}$ is negative, θ is an obtuse angle (cf. FIG. 2.22). It is not customary to refer to the reflex angle $360° - \theta$ between the same two lines; indeed, for many purposes the sign of $\mathbf{a}\cdot\mathbf{b}$ may be ignored, and only the acute angle used.

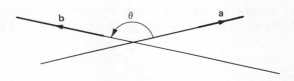

Fig. 2.22

Example
Find the angle between the plane $3x - y + z = 0$ and the line $(x-1)/2 = (y-2)/4 = (z+3)/-1$.

The normal to the plane is in the direction $\mathbf{a} = (3, -1, 1)$ and the direction of the line is given by $\mathbf{c} = (2, 4, -1)$ (cf. comment about equation 2–47). The angle between the line and the normal is thus given by

$$\cos\theta = \frac{\mathbf{a}\cdot\mathbf{c}}{ac} = \frac{6-4-1}{\sqrt{11\times 21}} = \frac{1}{\sqrt{231}} = 0.0658$$

The required angle is therefore equal to $90° - 86.2° = 3.8°$.

Exercise 7

1 Points A, B, C have coordinates (3, 1, 2), (2, 2, 3) and (1, 0, 2) respectively. Find (i) the equation of line AB, (ii) the equation of plane ABC.

Differentiation of vectors; accelerations; scalar and vector products

2 Find the equation of the straight line through the points with cartesian coordinates $(1, -2, 1)$ and $(0, -2, 3)$; and the equation of the plane which contains the origin and the points $(2, 4, 1)$ and $(4, 0, 2)$. Find also the point in which the line meets the plane. (**Hint:** $\mathbf{r}$ must satisfy both the foregoing relations.)

3 Show that the plane through the points $A(3, -5, -1)$ and $B(-1, 5, 7)$ and parallel to the vector $3\mathbf{i}-\mathbf{j}+7\mathbf{k}$ is $3x+2y-z = 0$. (**Hint:** Use the vector equation for a plane to deduce parametric equations for x, y, z; then evaluate $3x+2y-z$ from these parametric equations.)

4 (For those conversant with basic ideas about forces.) Forces $\mathbf{F}_1$ and $\mathbf{F}_2$ with components $(6, 2, -3)$ and $(-1, 3, 4)$ act at point $(1, -1, 2)$. Find the equation of the line of action of the resultant of $\mathbf{F}_1$ and $\mathbf{F}_2$. If this is expressed in the form $\mathbf{r}=\mathbf{r}_0+\lambda\mathbf{F}$, what are the units of λ? What are the direction cosines of this line?

5 Force $\mathbf{F}_1=2\mathbf{i}+3\mathbf{j}+\mathbf{k}$ acts at point $A(1, 1, 1)$ and force $\mathbf{F}_2=\mathbf{i}-4\mathbf{k}$ acts at the origin. Do the forces intersect? What conclusion do you draw about their action on any rigid body? (You may return to this question after reading Chapter 8.)

6 Identify any three non-collinear points on the plane $\mathbf{r}=\lambda(\mathbf{i}+\mathbf{j})+\mu(2\mathbf{i}-3\mathbf{k})$. (**Hint:** You could choose $\lambda = 53$, $\mu = -149$! for one of the points, but the simpler the values of λ, μ chosen, the less work you will have.)

2.77 Vector area and its use

$\frac{1}{2}\mathbf{a} \wedge \mathbf{b}$ is of magnitude $\frac{1}{2}ab \sin \theta$ and direction perpendicular to plane of $\mathbf{a}$ and $\mathbf{b}$. It therefore gives information both about the magnitude and the orientation of the triangle (FIG. 2.23) defined by vectors $\mathbf{a}$, $\mathbf{b}$ sprouting from the same vertex.

$$\text{We write } \mathbf{A} = \tfrac{1}{2}\mathbf{a} \wedge \mathbf{b} = \text{vector area of OAB} \qquad 2\text{-}52$$

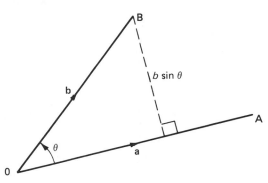

Fig. 2.23

The definition may be extended to any other surface, whether in two or three dimensions, provided that its bounding curve lies in a plane. The vector area $\mathbf{A}$ of any such surface has (by definition)

> magnitude = plane area enclosed by the boundary;
> direction = that of the normal to the plane area.
> $\qquad$ 2-52(a)

Thus the two surfaces S_1 and S_2 shown in FIG. 2.24 have the same vector area $\mathbf{A}$. The surface S_3, which lies on the opposite side of the boundary curve has vector area $-\mathbf{A}$.

72 An introduction to mechanics and modelling

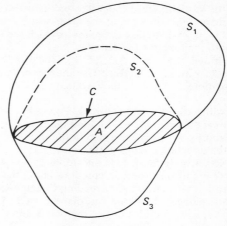

Fig. 2.24

A closed surface (FIG. 2.25) may be considered as the limit of a sequence of surfaces whose boundaries shrink to a point, so a closed surface S has zero vector area. We note that the area of the closed surface formed by S_2 and S_3 in FIG. 2.24 is zero. For this particular case, then, the vector law of addition has significance, $\mathbf{A} - \mathbf{A} = \mathbf{0}$.

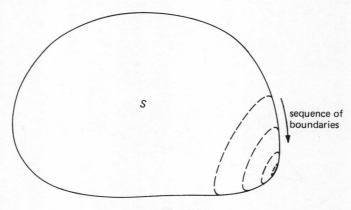

Fig. 2.25

In question 2 of Exercise 8 we shall find further indication that sets of quantities with magnitudes and directions as defined in 2–52(a) have vectorial characteristics, but a complete investigation of the vectorial character of area will not be undertaken in this volume.

Example: *Area of a triangle*
Find the area of the triangle whose vertices are at $(0, 0, 2)$, $(1, 1, 0)$ and $(3, -1, 2)$.

As an aid to thought, we first write down an expression for the vector area of a triangle whose vertices are prescribed by general position vectors **c**, **d**, **e**

Differentiation of vectors; accelerations; scalar and vector products

say. The two sides sprouting from the vertex **c** correspond to vectors $\mathbf{d}-\mathbf{c}$, $\mathbf{e}-\mathbf{c}$; the vector area is therefore

$$\tfrac{1}{2}(\mathbf{d}-\mathbf{c}) \wedge (\mathbf{e}-\mathbf{c})$$

In our case

$$\mathbf{d}-\mathbf{c} = (1, 1, 0) - (0, 0, 2) = (1, 1, -2)$$

and

$$\mathbf{e}-\mathbf{c} = (3, -1, 0)$$

whose vector product

$$(\mathbf{d}-\mathbf{c}) \wedge (\mathbf{e}-\mathbf{c}) = (0-2)\mathbf{i} + (-6-0)\mathbf{j} + (-1-3)\mathbf{k}$$
$$= -2(\mathbf{i} + 3\mathbf{j} + 2\mathbf{k})$$

The vector area is thus

$$= (-)(\mathbf{i} + 3\mathbf{j} + 2\mathbf{k})$$

(the minus sign has no relevance to the questions asked).

The magnitude of the area of the triangle is therefore

$$\sqrt{(1^2 + 3^2 + 2^2)} = \sqrt{14}.$$

The volume of a tetrahedron, triangular prism or parallelepiped can be conveniently calculated by use of the vector area concept.

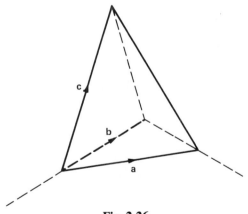

Fig. 2.26

The volume of the tetrahedron with concurrent sides, **a**, **b**, **c**, as shown in FIG. 2.26, is equal to

$\tfrac{1}{3}$ base area × perpendicular height

$= \tfrac{1}{3}|\tfrac{1}{2}\mathbf{a} \wedge \mathbf{b}| \times$ component of **c** at right angles to the base, i.e. parallel to its vector area

$= \tfrac{1}{6}(\mathbf{a} \wedge \mathbf{b}) \cdot \mathbf{c}$ \hfill 2-53

74 An introduction to mechanics and modelling

In the same way, the volume of a parallelepiped with sides **a**, **b**, **c** is equal to

$$(\mathbf{a} \wedge \mathbf{b}) \cdot \mathbf{c} \qquad \qquad 2\text{--}54$$

This formula for the volume provides a neat way of deriving the general identities

$$(\mathbf{a} \wedge \mathbf{b}) \cdot \mathbf{c} = (\mathbf{b} \wedge \mathbf{c}) \cdot \mathbf{a} = (\mathbf{c} \wedge \mathbf{a}) \cdot \mathbf{b}$$
$$= -(\mathbf{a} \wedge \mathbf{c}) \cdot \mathbf{b} = -(\mathbf{c} \wedge \mathbf{b}) \cdot \mathbf{a} = -(\mathbf{b} \wedge \mathbf{a}) \cdot \mathbf{c} \qquad 2\text{--}55$$

each of which may also be written:

$$\begin{vmatrix} a_1 & a_2 & a_3 \\ b_1 & b_2 & b_3 \\ c_1 & c_2 & c_3 \end{vmatrix} \qquad \qquad 2\text{--}56$$

> *Note* that if $(\mathbf{a} \wedge \mathbf{b}) \cdot \mathbf{c} = 0$, **c** is perpendicular to $\mathbf{a} \wedge \mathbf{b}$, i.e. **c** lies in the plane of **a** and **b**; and the volume of the parallelepiped is zero. Conversely for any three coplanar vectors **a**, **b**, **c** the determinant 2–56 is zero.

2.78 Perpendicular distances from points on to lines or planes

In FIG. 2.27 we show the points A, B, C. The area of $\triangle ABC$ may be written either as $\frac{1}{2}p\,BC$ or as $\frac{1}{2}|(\mathbf{c}-\mathbf{b}) \wedge (\mathbf{a}-\mathbf{b})|$. The required perpendicular distance, p, is therefore calculated

$$p = \text{magnitude of vector area}/\tfrac{1}{2}BC$$

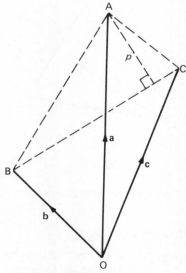

Fig. 2.27

Example
Perpendicular distance from A(1, 1, −1) on to the join of B(3, 0, 1) and C(3, 1, 0)

$$\mathbf{a} - \mathbf{b} = -2, 1, -2$$
$$\mathbf{c} - \mathbf{b} = 0, 1, -1 \quad (\text{so } BC = \sqrt{2})$$

Vector area of triangle $= \frac{1}{2}[(-1+2)\mathbf{i} + (0-2)\mathbf{j} + (-2-0)\mathbf{k}]$ of which the magnitude $= \frac{1}{2}\sqrt{(1^2 + 2^2 + 2^2)} = 3/2$. Since $\frac{1}{2}BC = \frac{1}{2}\sqrt{2}$, we get

$$p = \frac{3}{\sqrt{2}} = \frac{3\sqrt{2}}{2}$$

2.79 Shortest distance between two skew lines

Two non-intersecting, non-parallel lines AC, BD as in FIG. 2.28, i.e. two non-coplanar lines, are called skew. The shortest line which can be drawn to join them, PQ say, must be perpendicular to both. For, if PQ were not perpendicular to AC, then it would be longer than the perpendicular QX from Q on to AC; similarly if it were not perpendicular to BD, it could not be the shortest distance. To determine the position of the line PQ we may proceed as follows.

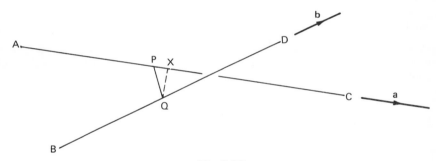

Fig. 2.28

Let $\mathbf{a}$, $\mathbf{b}$ be vectors parallel respectively to $\overrightarrow{AC}$ and $\overrightarrow{BD}$, and let $\overrightarrow{AB} = \mathbf{c}$. These three vectors $\mathbf{a}$, $\mathbf{b}$, $\mathbf{c}$ may be presumed known. $\overrightarrow{AP} = \lambda \mathbf{a}$; $\overrightarrow{BQ} = \mu \mathbf{b}$; where λ, μ have to be determined.

$$\overrightarrow{PQ} = \overrightarrow{PA} + \overrightarrow{AB} + \overrightarrow{BQ} = -\lambda \mathbf{a} + \mathbf{c} + \mu \mathbf{b}$$

To ensure PQ is perpendicular to AC and BD respectively

$$\left. \begin{array}{r} \mathbf{a} \cdot (-\lambda \mathbf{a} + \mu \mathbf{b} + \mathbf{c}) = 0 \\ \mathbf{b} \cdot (-\lambda \mathbf{a} + \mu \mathbf{b} + \mathbf{c}) = 0 \end{array} \right\} \quad \text{2-57}$$

For known $\mathbf{a}$, $\mathbf{b}$, $\mathbf{c}$, the scalar products $\mathbf{a} \cdot \mathbf{a}$, $\mathbf{a} \cdot \mathbf{b}$, $\mathbf{a} \cdot \mathbf{c}$, etc. are determinable ordinary numbers, and equations 2–57 are therefore two ordinary simultaneous equations for λ and μ. They have a unique solution provided

$$\frac{\mathbf{a} \cdot \mathbf{a}}{\mathbf{b} \cdot \mathbf{a}} \neq \frac{\mathbf{a} \cdot \mathbf{b}}{\mathbf{b} \cdot \mathbf{b}} \quad \text{i.e. provided } a^2 b^2 \neq (ab \cos \theta)^2$$

where θ is the angle between **a**, **b**. Since AC, BD are not coplanar, $\cos\theta \neq \pm 1$; hence equations 2–57 determine unique values of λ, μ, i.e. the points P, Q are uniquely determined.

Exercise 8
1. Prove that $(\mathbf{a} \wedge \mathbf{b}) \cdot \mathbf{c} = \mathbf{a} \cdot (\mathbf{b} \wedge \mathbf{c})$.
2. Prove that the projections of any plane area on the coordinate planes are the components of the corresponding vector area in the directions of the x-, y-, z-axes.
3. Find the perpendicular distance from the point $A(3, 0, 1)$ to the line through the points $B(1, 0, -1)$ and $C(1, 2, 3)$. Find also the angles of the triangle ABC.
4. Find the perpendicular distance from the point $P(2, 2, 1)$ on to the plane $x - y + z = 0$. (**Hint:** the plane passes through the origin; the required distance is the component of OP in the direction of the normal to the plane.)
5. Find the perpendicular distance of the point $(-1, 7, 3)$ from the plane which passes through the three points $(1, 0, 2)$, $(2, 1, -1)$ and $(0, 3, 1)$.
6. The vertices of a triangle are at points with position vectors **a**, **b**, **c** respectively. Express its vector area in a form which is symmetrical in **a**, **b** and **c**.
7. A tetrahedron has one vertex at the origin and the other three vertices have position vectors **a**, **b**, **c**. Find the vector area of each face of the tetrahedron, choosing that order of the factors of each vector product which, according to your figure, attributes the direction of the outward normal to each face. Verify that the sum of the four vector areas is zero.
8. Discuss the generalization of question 7 to any closed surface. How does the result relate to the theorem of question 2 above?

Comment: the student should always try to recognize clearly the property which he wishes to reflect in his mathematical relations. In that way he will not need to memorize innumerable formulae; he should be able to create his own methods – no matter if they are a bit long at times.

2.8 Linear dependence and independence of vectors

2.81 Multiples of a single vector a
(i) $k\mathbf{a}$ in general represents a vector parallel to **a**
(ii)   If $k\mathbf{a} = \mathbf{0}$, either $\mathbf{a} = \mathbf{0}$

         or $k = 0$

(iii)   If $k_1\mathbf{a} = k_2\mathbf{a}$,   i.e. $(k_1 - k_2)\mathbf{a} = \mathbf{0}$,   either $\mathbf{a} = \mathbf{0}$ or $k_1 = k_2$.

2.82 Linear combinations of two vectors a and b
(i) $k_1\mathbf{a} + k_2\mathbf{b}$ in general represents a vector in the plane of **a** and **b**.
(ii)   If
$$k_1\mathbf{a} + k_2\mathbf{b} = \mathbf{0}$$

and either of **a**, **b** is zero, we return to the situation of § 2.81;

otherwise, either $k_1 = 0$ *whence also* $k_2 = 0$; or $\mathbf{a} = -\left(\dfrac{k_2}{k_1}\right)\mathbf{b}$, i.e. **a** is parallel to **b**

Vectors **a**, **b** satisfying $k_1\mathbf{a} + k_2\mathbf{b} = \mathbf{0}$ with non-zero k_1, k_2 are called linearly dependent. Two linearly dependent vectors are parallel; one is just a multiple of the other.

Differentiation of vectors; accelerations; scalar and vector products 77

(iii) If
$$k_1\mathbf{a} + k_2\mathbf{b} = \lambda_1\mathbf{a} + \lambda_2\mathbf{b}$$
i.e. $(k_1 - \lambda_1)\mathbf{a} + (k_2 - \lambda_2)\mathbf{b} = \mathbf{0}$,

either $k_1 = \lambda_1$ and $k_2 = \lambda_2$ or $\mathbf{a}$, $\mathbf{b}$ are linearly dependent. Given that $\mathbf{a}$, $\mathbf{b}$ are non-parallel (and non-zero) vectors, i.e. that they are linearly independent, we may deduce from the equation
$$k_1\mathbf{a} + k_2\mathbf{b} = \lambda_1\mathbf{a} + \lambda_2\mathbf{b}$$
that $k_1 = \lambda_1$ and $k_2 = \lambda_2$, i.e. under the stated circumstances we may 'equate coefficients'. This is the same as saying that equal vectors in a plane have equal (oblique) components.

2.83 Linear combinations of three vectors a, b and c
(i) $k_1\mathbf{a} + k_2\mathbf{b} + k_3\mathbf{c}$ in general represents a vector in three-dimensional space.
(ii) If $k_1\mathbf{a} + k_2\mathbf{b} + k_3\mathbf{c} = \mathbf{0}$ and either of $\mathbf{a}$, $\mathbf{b}$, $\mathbf{c}$, k_1, k_2, k_3 is zero, we return to the situation of § 2.82, otherwise
$$\mathbf{a} = -\frac{k_2\mathbf{b} + k_3\mathbf{c}}{k_1}$$

i.e. $\mathbf{a}$ can be represented in the plane of $\mathbf{b}$, $\mathbf{c}$, i.e. $\mathbf{a}$, $\mathbf{b}$, $\mathbf{c}$ are coplanar. (**Note:** in vector work, coplanar signifies 'can be represented in a single plane' not 'must be' so represented.)

Vectors $\mathbf{a}$, $\mathbf{b}$, $\mathbf{c}$ which satisfy $k_1\mathbf{a} + k_2\mathbf{b} + k_3\mathbf{c} = \mathbf{0}$ with non-zero k_1, k_2, k_3 are called linearly dependent. Three linearly dependent vectors are coplanar. (If any **one** of k_1, k_2, k_3 is zero, the same conclusion may be drawn because of the consequent collinearity of two of the three vectors.)
(iii) If $k_1\mathbf{a} + k_2\mathbf{b} + k_3\mathbf{c} = \lambda_1\mathbf{a} + \lambda_2\mathbf{b} + \lambda_3\mathbf{c}$, i.e.
$$(k_1 - \lambda_1)\mathbf{a} + (k_2 - \lambda_2)\mathbf{b} + (k_3 - \lambda_3)\mathbf{c} = \mathbf{0}$$

then either the vectors $\mathbf{a}$, $\mathbf{b}$, $\mathbf{c}$ are coplanar or $k_1 = \lambda_1$, $k_2 = \lambda_2$, and $k_3 = \lambda_3$,	2-58

i.e. if non-zero vectors $\mathbf{a}$, $\mathbf{b}$, $\mathbf{c}$ are not coplanar, we may equate coefficients in any relation of the form
$$k_1\mathbf{a} + k_2\mathbf{b} + k_3\mathbf{c} = \lambda_1\mathbf{a} + \lambda_2\mathbf{b} + \lambda_3\mathbf{c}$$

2.84 Linear combinations of four vectors a, b, c and d
(i) $k_1\mathbf{a} + k_2\mathbf{b} + k_3\mathbf{c} + k_4\mathbf{d}$ (where $\mathbf{a}$, $\mathbf{b}$, $\mathbf{c}$, $\mathbf{d}$ are all vectors in three dimensions) represents a vector in three-dimensional space. Note this is not the analogue of point (i) of §§ 2.81–2.83.
(ii) If $k_1\mathbf{a} + k_2\mathbf{b} + k_3\mathbf{c} + k_4\mathbf{d} = \mathbf{0}$,
i.e. $\qquad -k_4\mathbf{d} = (k_1\mathbf{a} + k_2\mathbf{b} + k_3\mathbf{c})$

then k_1, k_2, k_3 are the (oblique) components of the vector $-k_4\mathbf{d}$ using the base vectors $\mathbf{a}$, $\mathbf{b}$, $\mathbf{c}$.

78 An introduction to mechanics and modelling

(iii) If $k_1\mathbf{a}+k_2\mathbf{b}+k_3\mathbf{c}+k_4\mathbf{d}=\lambda_1\mathbf{a}+\lambda_2\mathbf{b}+\lambda_3\mathbf{c}+\lambda_4\mathbf{d}$ then we have no justification for equating coefficients. Such justification could be provided only by postulating that **a, b, c, d** were not all vectors in the same three-dimensional space – a situation with which we are not concerned in this book.

Exercise 9

1 (i) $\mathbf{A}=2\mathbf{i}-\mathbf{j}-\mathbf{k}$; $\mathbf{B}=\mathbf{i}-4\mathbf{k}$; $\mathbf{C}=4\mathbf{i}-3\mathbf{j}-\mathbf{k}$. Determine whether **A, B, C** are linearly dependent or independent.
   (ii) $\mathbf{D}=\mathbf{i}-3\mathbf{j}-2\mathbf{k}$; $\mathbf{E}=2\mathbf{i}-4\mathbf{j}-\mathbf{k}$; $\mathbf{F}=3\mathbf{i}-2\mathbf{j}-\mathbf{k}$. Determine whether **D, E, F** are linearly dependent or independent.
   (iii) $\mathbf{P}=\mathbf{i}-3\mathbf{j}-2\mathbf{k}$; $\mathbf{Q}=\mathbf{i}-4\mathbf{k}$; $\mathbf{R}=\mathbf{i}-6\mathbf{j}$. Determine whether **P, Q, R** are linearly dependent or independent.

3

Modelling in the real world: a particular problem and some general comments

Unlike school examination questions, which the examinee is expected to solve correctly at the first attempt, the modelling of real-life problems usually involves a succession of rethinks. Within each round of thought there is a sequence of steps; the sequences have a similar pattern in all investigations. Several real-life problems will be discussed in this book, and, in order that the reader may become accustomed to their common pattern, the main stages in each round of thought will be labelled $A, B, \ldots E$. In problems involving several rethinks, the labels will be $A_1, B_1, \ldots E_1$ for the first round; $A_2, B_2, \ldots E_2$ for the second round; and so on.

3.1 A problem in plumbing and geometry

The floor of a factory office is a 6 m × 3 m rectangle, and the room is 3.6 m high. Two pipes, which have to be connected, run obliquely, one on the face of each of the two long walls. Looking from one end of the room, the pipe on the right-hand wall starts at floor level and rises to ceiling level at the other end, following a diagonal of the wall. From the same viewpoint, the pipe on the left-hand wall starts at a height of 2.7 m and descends to a height of 0.9 m at the far end of the room. What recommendations can be given to the plumber about the best line along which to join the pipes (any point of the first to any point of the second) by a route within the room?

3.11 First round of thought

Step A_1
The first step is to draw a diagram and identify a possible line of approach to the problem. The room is shown in FIG. 3.1; PQ and FH are the two pipes which have to be connected. The problem is primarily geometrical: geometry and vector algebra are the mathematical tools most likely to be useful. Before plunging into calculations it is a good idea to consider the practical alternatives which are open to the plumber.

The pipe PQ may be connected to FH either by a single straight pipe, or by a pipe with one or more elbows. Pipes must either keep to the walls and ceiling,

80 An introduction to mechanics and modelling

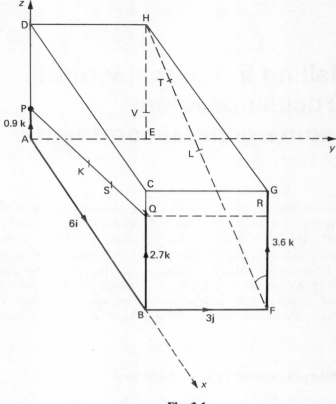

Fig. 3.1

or, if they cross the room, they must be at a sufficiently high level to leave adequate headroom. First consider a single straight pipe.

Step B_1
A single pipe running along the face of a wall can provide the necessary connection only if the path is PH or QF. Mathematically speaking, QF can be rotated into the position of PH by a half-turn about a lengthwise axis through the middle point of the room; consequently the pipe length and angles involved are the same for each of the two routes. The choice between QF and PH would be determined by the position in the room of the door, windows, etc.

Step C_1
For route QF, the plumber will require a length of pipe QF, a junction to allow QF to leave PQ at angle PQF with pipe PQ, and a second junction to allow QF to join FH at angle QFH. Little difficulty will arise over the supply of any chosen length of pipe, but the angles at the junctions may be non-standard, and must be calculated before making recommendations to the plumber.

Modelling in the real world 81

Taking x-, y-, z-directions as indicated in FIG. 3.1,

$$\vec{PQ} = 6\mathbf{i} + 1.8\mathbf{k}; \quad |\vec{PQ}| = \sqrt{39.24}$$
$$\vec{FQ} = -3\mathbf{j} + 2.7\mathbf{k}; \quad |\vec{FQ}| = \sqrt{16.29}$$
$$\vec{PQ} \cdot \vec{FQ} = 1.8 \times 2.7 = 4.86$$

and the angle between PQ produced and FQ produced is equal to ∠PQF, so

$$\cos(\angle PQF) = 4.86/(\sqrt{39.24} \times \sqrt{16.29}) = 0.192(2)$$
$$\angle PQF = 78.9°$$

3-1

Similarly

$$\vec{FQ} = -3\mathbf{j} + 2.7\mathbf{k}$$
$$\vec{FH} = -6\mathbf{i} + 3.6\mathbf{k}; \quad |\mathbf{FH}| = \sqrt{48.96}$$
$$\vec{FQ} \cdot \vec{FH} = 9.72$$
$$\cos(\angle QFH) = 9.72/(\sqrt{16.29} \times \sqrt{48.96}) = 0.344(2)$$
$$\angle QFH = 69.9°$$

3-2

Step D_1
The fitting of a single straight pipe against the wall would require, besides the straight section of pipe, two junctions, with angles of approximately 79° and 70° respectively.

Step E_1
Junctions with angles 79° and 70° are not standard fittings. They would have to be specially made, at considerable expense. There is need for a rethink of the problem. At this stage it would be a good idea to have a word with a plumber or supplier to find out what standard fittings are available. For the purposes of this discussion we shall assume that only 90°, 60° and 45° junctions are available for the type of pipe under consideration.

3.12 Second round of thought. Use of two pipes and an elbow

Steps A_2 and B_2
By the use of a single elbow (right-angle), the connection could follow the route QRF. This still leaves a junction of angle RFH which we expect to be non-standard.

Step C_2
Calculation shows

$$\cos(\angle RFH) = 0.9/\sqrt{3.06} = 0.5144$$
$$\angle RFH = 59.05°$$

3-3

82 An introduction to mechanics and modelling

Step D_2

This is so near to 60° that it is unlikely that use of a 60° junction would cause unacceptable, or indeed perceptible, distortion.

Step E_2

The route QRF, or its counterpart PVH on the opposite wall, are feasible alternatives. Before finally settling for them, it is worthwhile considering single-pipe overhead routes.

3.13 Third round of thought. Overhead routes

Steps A_3 and B_3

The shortest distance between PQ and FH is the join of their mid-points, KL. This is the only line making a right angle with both PQ and FH (§ 2.79). The mid-points of PQ and FH are both 1.8 m above floor level, so a connection between pipes PQ and FH along a line at right angles to both would be everywhere 1.8 m above the floor. This gives insufficient headroom for a work-room (1.8 m is less than 6 ft), and would not be recommended.

The plane through the mid-point of PQ, and also perpendicular to PQ, cuts FH at its mid-point. A parallel plane through any other point S of PQ will cut HF between L and F if the point S lies between K and Q. In other words, the parallel plane through any point of PQ higher than 1.8 m will cut FH at a point lower than 1.8 m, and vice versa. Any connecting pipe making an angle of 90° with PQ (with whatever angle is necessary at the other end) will thus be lower than 1.8 m at one end or the other, and therefore undesirable. The same argument excludes connections with the right angle at the point of connection with FH.

To check whether or not it is worth investigating a join ST, making angles of 60° or 45° at its ends, with PQ and FH, we make a short preliminary calculation of angles PQH and QHF. If these angles prove to be greater than 60°, any join ST lying inside the room would make angles greater than 60° with the existing pipes, and the proposed investigation would not be pursued.

By scalar product calculations it is found that $\angle$PQH = 35.8° and $\angle$FHQ = 34°. A pipe ST with 60° junctions at each end (or one with 45° junctions at each end) looks *a priori* quite feasible.

Step C_3

We shall consider the join of the two arbitrary points S and T on PQ and FH respectively, and write down the conditions that the junction angles at the two ends shall both be 60°.

Since
$$\overrightarrow{PQ} = 6\mathbf{j} + 1.8\mathbf{k},$$
the position vector of S is of the form
$$\mathbf{r}_1 = 0.9\mathbf{k} + \lambda(6\mathbf{i} + 1.8\mathbf{k}) = 6\lambda\mathbf{i} + (0.9 + 1.8\lambda)\mathbf{k}$$

The position vector of T is of the form
$$\mathbf{r}_2 = 6\mathbf{i} + 3\mathbf{j} + \mu(-6\mathbf{i} + 3.6\mathbf{k})$$
$$= 6(1-\mu)\mathbf{i} + 3\mathbf{j} + 3.6\mu\mathbf{k}$$

Modelling in the real world 83

(and we note that both $\lambda > \tfrac{1}{2}$ and $\mu > \tfrac{1}{2}$ in order that S, T shall both be higher than 1.8 m).

$$\vec{ST} = \mathbf{r}_2 - \mathbf{r}_1 = 6(1 - \mu - \lambda)\mathbf{i} + 3\mathbf{j} + (3.6\mu - 0.9 - 1.8\lambda)\mathbf{k}$$
$$= a\mathbf{i} + \mathbf{j} + b\mathbf{k}$$

where $a = 6(1 - \mu - \lambda)$; $b = 0.9(4\mu - 1 - 2\lambda)$.

It will be convenient to do the calculations in terms of a, b in the first place, and only subsequently to evaluate λ, μ from the simultaneous equations

$$1 - \mu - \lambda = a/6; \quad 4\mu - 1 - 2\lambda = b/0.9 \qquad 3\text{--}4$$

The requirement $\lambda, \mu > \tfrac{1}{2}$ implies $a/6 < 0$, i.e. a negative.
We require $\angle PST = 60°$ i.e. $\cos \angle PST = \tfrac{1}{2}$.

$$\vec{ST} = a\mathbf{i} + 3\mathbf{j} + b\mathbf{k}; \quad |\vec{ST}| = \sqrt{(a^2 + b^2 + 9)}$$
$$\vec{QP} = -6\mathbf{i} - 1.8\mathbf{k}; \quad |\vec{QP}| = \sqrt{39.24} = 6.26$$

Equating the two available expressions for the scalar product,

$$-6a - 1.8b = \tfrac{1}{2} \times 6.26 \times \sqrt{(a^2 + b^2 + 9)} \qquad 3\text{--}5$$

We likewise require $\angle STF = 60°$

$$\vec{ST} = a\mathbf{i} + 3\mathbf{j} + b\mathbf{k}$$
$$\vec{FH} = -6\mathbf{i} + 3.6\mathbf{k}; \quad |\vec{FH}| = \sqrt{48.96} = 6.997$$

(Scalar product equation) $-6a + 3.6b = \tfrac{1}{2} \times 6.997 \times \sqrt{(a^2 + b^2 + 9)} \qquad 3\text{--}6$

Eliminating the square root between equations 3–5 and 3–6 gives

$$6.26(-6a + 3.6b) = 6.997(-6a - 1.8b)$$
i.e. $$a = -7.94b \qquad 3\text{--}7$$

since a has to be negative, it follows that b must be positive. By squaring either equation 3–5 or 3–6, and using 3–7

$$b^2 = 0.0598; \quad b = +0.2446$$
$$a = -1.942$$

Finally, solution of equations 3–4 gives

$$\mu = 0.653; \quad \lambda = 0.671$$
$$\text{Height of } S = 0.9 + 1.8\lambda = 2.10(7) \text{ m}$$
$$\text{Height of } T = 3.6\mu = 2.35(1) \text{ m} \qquad 3\text{--}8$$
$$\text{Length of pipe } ST = 3.58(2) \text{ m}$$

The parallel calculations, with 45° angles at both ends, give

$$\text{Height of } S = 2.37 \text{ m}$$
$$\text{Height of } T = 2.82 \text{ m} \qquad 3\text{--}9$$
$$\text{Length of pipe } ST = 4.73 \text{ m}$$

84 An introduction to mechanics and modelling

Step D_3

Either of the alternatives expressed by equations 3–8 and 3–9 might be acceptable. The former involves a shorter length of pipe and less severe bends, but provides less headroom than the second. In either case it should be borne in mind that the pipe may need to be supported from the ceiling at one or more points.

Step E_3

On the face of it, we now have three feasible alternatives between which to choose, namely the solutions indicated in equations 3–8 and 3–9 above, and that of § 3.12. There is need for on-site consultation, to establish the positions of doors, windows, etc. and immovable fittings; to find out whether there is any functional reason why sharp (e.g. 45°) bends or long pipes should be avoided. Without such practical information, the mathematician cannot give an opinion about which is his preferred solution.

One important reservation must be made about the accuracy of the calculations, viz., no allowance has been made for the diameter of the pipes. If diameters are appreciable, and accuracy of fit is needed, the positions of S and T for the chosen pipe-route should be recalculated making the appropriate corrections for pipe-diameter. No more can usefully be done without on-the-spot information.

3.2 General comments

The reader of § 3.1 will have noticed that the obviously mathematical work of Step C was sandwiched between Steps A and B (in which the real-world situation was weighed up and a mode of mathematical attack decided on), and Steps D and E in which provisional conclusions were drawn and the current round of thought subjected to criticism.

Steps A, B, D and E required both an appreciation of the spatial nature of rooms, the practicalities of plumbing, etc. and a background knowledge of the tools of geometry to funnel the line of thought into a useful channel. The two requirements went hand in hand; one without the other would have provided an insufficient platform from which to embark on the process of modelling.

Modelling in any field of study consists of steps parallel to those labelled A, B, D and E in the examples of this book. Step C is usually a hard centre of mathematical technique; although its contents are sometimes called 'the mathematical model', it is the step which least involves modelling skills!

Problems of apparently similar type are not necessarily all appropriately modelled by the same mathematics. For example, progressive changes like the changes of amplitude of an oscillation, or the growth of populations may be well-represented by certain simple relationships when time-intervals of, say, a few days are in question. The same mathematical formulae might be very misleading if the relevant time-scale were either years or minutes. Modelling is not a matter of learning the 'right' equations: it is an adventurous activity whose course cannot be fully foreseen at the outset.

Modelling in the real world

In classical fields of mathematical inquiry, such as geometry, mechanics and physics, the mathematician who has studied the material can be reasonably confident of his ability to recognize most contributory factors, and can choose to disregard those he deems to be of lesser importance. In fields as complex as those of biology and economics, not even the specialist can rely on being so happily placed when he tackles a novel real-life problem.

The chances of successful modelling in any field are poor unless the investigator has a more-than-superficial appreciation of the subject matter under discussion as well as a good background knowledge of mathematical techniques. It would be frivolous to attempt within the pages of one book to give a grounding in more than one field of study. In this book the chosen field is mechanics, the study of geometry which we have already undertaken being a necessary preliminary.

Just as in our light-weight investigation of § 3.1 there were several distinguishable rounds of thought, so too the natural course of major scientific discovery proceeds through a succession of trial formulations before a definitive formulation is arrived at (if it ever is!). The following chapter, Chapter 4, may be regarded as a first round of thought about mechanics. It is an introduction to the more intellectually satisfying formulation of mechanics which starts at Chapter 5, and thereafter maintains a continuous thread throughout the book.

Suggestions for a short modelling assignment
Either find a solution of the plumbing problem of this chapter when the connecting-pipe material comes only in 1 m lengths which cannot be cut, junctions being available with angles of 0°, 10°, 20°, 30°, ...
Or examine the structure of an umbrella, and decide whether the ribs, etc. could be provisionally modelled by straight lines and/or circular arcs. Identify the measurements which will determine the shape of the umbrella when raised, and write down relations which will predict characteristics of shape from the quantities you have identified as determining factors. Check your predictions by inspection of other umbrellas, and revise or extend your calculations as appropriate.
Or discuss solutions to the following (more artificial, and largely pre-modelled) problem.

A circular table-top of diameter 1 metre has a metallic rim. The table is to be supported solely by struts attached at one end to the vertical sides of the metallic rim, and at the other end passing through a fixed point Q at ground level (and thereafter secured under the ground). The desired height of the table above the ground is 0.75 m, and point Q is vertically below a point of the **table's rim**. If the struts can only be conveniently attached to the rim at angles of 30°, 45°, 60° or 90° to the circumference, suggest alternative arrangements for a suitable number of struts, determining the points and angles of attachment at the rim in each case.

If it were further necessary to connect the centre of the table to one or more of the struts by a connection perpendicular to the strut, calculate the length of the connecting piece appropriate to one of the off-centre struts which you suggested above.

4
Rudimentary ideas of force and mass; conventional modelling in elementary statics and dynamics; Newton's definitions and laws of motion

Introduction

This chapter gives answers to the questions 'What are the main ideas and tools of mechanics?' and 'How do real objects fit into the framework of those ideas?' The first idea to be discussed is that of force. As has already been said, the general idea of force may be conveyed by saying that

> a force is a push or a pull or a rub.

This general idea can be converted into a mathematical concept relevant to the world around us by means of the modelling steps outlined in §§ 4.1 and 4.5 below. A precise quantitative definition of force emerges in Chapter 5.

Application of the theory of mechanics to problems of real life requires a continuation of the modelling process to include all the main characteristics of physical objects which affect their physical motion, e.g. weight, roughness, flexibility etc. This chapter starts the necessary processes of interpretation.

4.1 Chosen mathematical attributes of force

(1) A push, pull or rub has a direction. Since the idea of a push is insecurely located in space, the direction of the force is defined as the direction of motion of a small object under the action of the given force only, starting from rest.
(2) A push (or pull or rub) may be strong or gentle, i.e. it has magnitude, though the best prescription for the determination of that magnitude is not self-evident.
(3) Having ascribed magnitude and direction to the concept of force, it is natural to ask 'is force a vector?' i.e. do forces form sets within which the vector law of addition is meaningful? Consideration of this question brings

Rudimentary ideas of force and mass 87

us back to physical reality with a bump. The idea of a disembodied 'set of forces' dissolves as we scrutinize it.

> Forces act between physical objects. One or more material bodies must be envisaged/identified before the idea of mathematical relationships between forces becomes meaningful.

Some thought about forces on objects is therefore necessary before question (3) can be answered (in § 4.25).

4.2 Physical attributes of force: forces in relation to physical objects

4.21 Interaction between two bodies

Whenever a force acts, two bodies (i.e. material objects) are involved – typically the pusher and the pushed. In FIG. 4.1 the boy is pushing the barrow. In FIG. 4.2, the plank PQ is pushing the barrow; but, equally, the barrow is (partly) supporting, i.e. pushing back on the plank. Returning to FIG. 4.1, we can recognize the possibility of describing the retarding action which the barrow has on the boy by saying that the barrow is pushing back on him.

Fig. 4.1

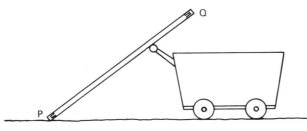

Fig. 4.2

Clearly a decision has to be made: **either** the definition of force will mirror the asymmetrical aspect of interaction – the active boy, the passive barrow – **or** we shall choose a form of definition of force which is free from reference to

88 An introduction to mechanics and modelling

active/passive characteristics. The accepted definition of force conforms to the latter choice.

> The definition of force carries no implication about which of the two bodies concerned is the active and which the passive partner. The descriptive term 'reaction' is used for a force which is called into being by an instigating force, but the mathematical status of a reaction as a force is no different from that of any other force.

To distinguish between active/passive or winner/loser partners in a quantitative way, a different mathematical concept – that of 'work', see Chapter 10 – is introduced.

4.22 Use of diagrams to model sets of forces

> Diagrams indicating the forces on a body are an indispensable first modelling step in mechanics. For reasons which will emerge in § 4.25, we group in a set **all the forces on one body** (not all the forces between pairs of bodies).

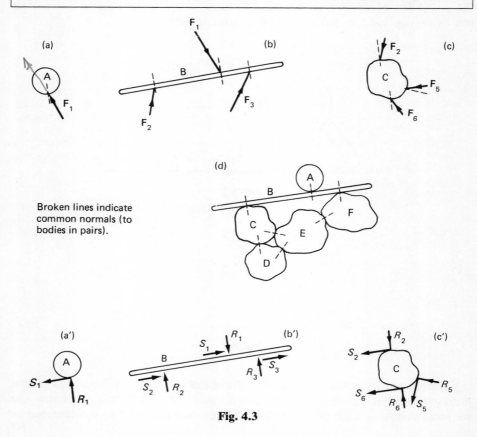

Broken lines indicate common normals (to bodies in pairs).

Fig. 4.3

Rudimentary ideas of force and mass 89

Given several bodies A, B, C, ... etc. (FIG. 4.3d) in contact, we select and represent by arrows:

(a) the set of contact forces acting on body A (FIG. 4.3a),
(b) the set of contact forces acting on body B (FIG. 4.3b),
(c) the set of contact forces acting on body C (FIG. 4.3c),

and so on.

The true magnitudes and exact directions of the forces will frequently be unknown at the preliminary diagram stage. No significance attaches to the precise length of the directed line segments in FIG. 4.3a–c, nor to whether any one of them is located inside (FIG. 4.3a) or outside (FIG. 4.3a') the boundary. As for direction, contact forces between solid bodies are ordinarily a combination of push (directed **into** the body acted on, parallel to the dotted normals) and rub (contributing a component parallel to the surface at the point of application); so the forces are represented by line segments of indeterminate length directed inwards at ill-defined angles, as in FIG. 4.3c, or alternatively (not supplementarily!) by components of unknown magnitude, as in FIG. 4.3c'.

If two surfaces in contact would separate in the absence of a nail, screw or adhesive, we adduce that the nail or adhesive effectively supplies a pull-force between the two surfaces.

Anticipating § 4.25, the force which any body A exerts on a neighbouring body B should be drawn in the direction antiparallel to that of the force exerted by body B on body A.

Exercise 1
(Where a problem is ill-defined, consciously idealize it in the way **you** think appropriate.)
1 The arms of three sky-divers form a horizontal equilateral triangle, the men's heads being at the vertices. Each man now pushes against his neighbours' hands. Indicate on a diagram the contact forces acting on one man.
2 A postal packet slides down a chute. Indicate on one diagram the push and rub components of force exerted by the packet on the chute, and on a second diagram the push and rub components exerted by the chute on the packet.
3 Indicate by arrows the general nature of the contact forces acting on the keystone of an arch (in the absence of mortar).
4 The roof BAC in FIG. 4.4 is strengthened by crossbeam DE. Indicate in separate diagrams (a) the contact forces exerted on beam DE, (b) the contact forces on roof section BA. Comment on the importance of fixing screws/nails for member DE. (This question is suitable for verbal discussion.)

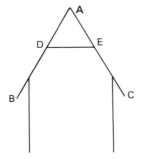

Fig. 4.4

90 An introduction to mechanics and modelling

5 A fast-spinning wheel, held by its axle in bearings, is lowered gently on to loose gravel. Show in diagrams:
   (a) the force exerted by the wheel on a piece of the gravel,
   (b) the force exerted by the gravel, and that exerted by the hand which supports the axle-bearings, on the wheel/axle system,
   (c) the force exerted on the supporting hand by the axle.
6 By analogy with question 5, discuss the relation between the forces exerted on the wheels of a motor car by the road surface, the action of the engine, and the motion of the car.
7 A shelf is supported by a bracket fixed to a wall. On separate diagrams show (a) contact forces on the shelf, (b) contact forces on the bracket, including those necessarily provided by screws or nails. (You must imagine for yourself a simplified version of a shelf which will enable you to discuss these features.)
8 A ladder stands with its foot on horizontal ground and its upper end against a wall; a man stands on a rung three-quarters of the way up the ladder. Indicate in separate diagrams (a) the force on the ground, (b) the force on the wall, (c) the contact forces on the ladder, (d) the contact force exerted on the man.

4.23 Deformation of non-rigid bodies, and its provisional use as a measure of force

If, under the action of forces, the distance between two points of a body can change, the body is called **non-rigid**, or deformable. If, when the forces are removed, the body returns to its original size and shape, its material is said to be **elastic**.

One can make the hypothesis that the magnitude of the deformation (measured by length, perhaps, or by volume) is a measure of the magnitude of the force which causes the deformation.

The extension of a spring is a case in point. One may make the hypothesis that the extension of a spring (one end of which is fixed) is a direct measure of the magnitude of the force applied at the other end to produce that extension.

Since, among other objections, actual springs yield and break if overloaded, the stated hypothesis cannot furnish us with an ultimately satisfactory definition of the nature or magnitude of a force. Nevertheless

> we adopt springs with appropriate scales, i.e. spring balances, as a temporary means of associating magnitudes with forces.

4.24 Experiments (actual or notional) for the student

1 Provide yourself with three or more spring balances with scales in consistent units, and with similar ranges. Check their consistency by weighing the same object on each (several objects).
2 Hook two spring balances A and B together (FIG 4.5), pull apart to extend the springs by any convenient amount, and record the readings on the two balances.
3 Repeat with the 'active' and 'passive' ends interchanged; repeat with different extensions and with different pairs of springs.
4 Hook three spring balances A, B and C to a ring (FIG. 4.6) or other *light* rigid body. (By specifying 'light' we sidestep complications due to gravity.) Pull to extend all three springs, and, while the ring is steady, record the

Rudimentary ideas of force and mass 91

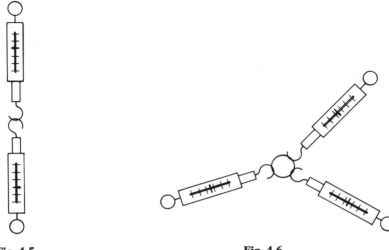

Fig. 4.5 Fig. 4.6

readings on the three balances, and the directions of the axes of the three springs.
5 Plot directed line segments end to end, the first in the direction of spring A and of length proportional to the reading on scale A; the second similarly related to the position and reading of B; the third for C.
6 Repeat steps 4 and 5 for different extensions of all three springs, and different bodies D. Repeat using four springs, if available.
7 Consider whether your observations – bearing in mind experimental inaccuracies – are consistent with the basic hypotheses of mechanics stated immediately below.

4.25 Two basic hypotheses of mechanics

> I. The force exerted by body A on body B is equal in magnitude, but opposite in direction to the force exerted by body B on body A (cf. experiments 2 and 3 above). This law is often expressed, 'To every action there is an equal and opposite reaction', and is Newton's 'third law'.
> II. A necessary condition for the equilibrium of any rigid body (i.e. condition that it stays still) is that the vector sum $(\mathbf{F}_1 + \mathbf{F}_2 + \mathbf{F}_3 + \cdots + \mathbf{F}_n)$ of the forces acting on the body is zero.

Hypothesis II has the further consequence that the vector sum of any number of forces $\mathbf{F}_1, \mathbf{F}_2, \ldots, \mathbf{F}_s$ acting on a single body has a physical significance: for the set of forces with vector sum $\mathbf{F}$ must be supplemented by force $-\mathbf{F}$ if the body (ring D say in the experiments 4–6 above) on which they all act is to remain in equilibrium. The condition $\mathbf{F} = \mathbf{0}$ in fact ensures that one central point of the rigid body remains at rest (or, more generally, has zero acceleration); it provides no information about possible rotation.

92 An introduction to mechanics and modelling

4.26 Discussion of Newton's third law

Any reader who finds Newton's third law, hypothesis I above, difficult to accept should consider two hypothetical experiments.

1. Try to push against nothing, FIG. 4.7. Your hands wave ineffectually in the air; you cannot exert a force of, say, 5 kg-f against nothing; an opposing force (reaction) is a condition of being able to exert the desired force.

Fig. 4.7

2. Imagine barbed wire or powdered glass between your hands and those of a friend you are trying to push, FIG. 4.8. Whether he moves backwards or you, as the result of the match, your hands and his will be similarly lacerated.

Fig. 4.8

Newton's third law postulates **the equality of action and reaction forces not only when the contact point is at rest, but in any motion whatsoever of the point of contact**. This hypothesis is basic to mechanics.

4.27 Mathematical consequences of the vector law of addition for forces and the extension to bodies in motion

Forces acting on a single body are meaningfully combined according to the vector (i.e. triangle) rule of addition. The significance of the vector sum for

Rudimentary ideas of force and mass   93

discussing **equilibrium** is stated in hypothesis II. The significance of the vector sum in discussions of the **motion** of a body is part of the content of Newton's second law as given in § 4.5 below. [In Chapter 6, we shall see that a certain significance attaches to the vector sum of forces acting on any identifiable group of bodies, but whenever the separate motions of different constituent bodies are under discussion, forces which act on different bodies belong to different sets of vectors, between which vector addition is **not** to be regarded as a meaningful operation.]

Even when the vector sum is meaningful, it does not reflect **all** the characteristics of the original set of forces. Thus, neither the stresses inside a rigid body nor the relative motions of the parts of a rotating or deformable body are discussable in terms of vector sums of forces.

> A set of forces is equivalent to its vector sum (only) in the context of Newton's second law and applications and extensions thereof.

4.3. Modelling the downward force of gravitation: weight and 'weightlessness'

When a substantial body is suspended from the hook of a spring balance, it hangs vertically downwards, and a force is registered on the balance. The downward force is attributed (Newton) to the gravitational pull of the earth on the body. The gravitational force depends only on the body chosen and its position relative to the earth and other celestial bodies; but the balance reading is affected by the state of motion (notably the common acceleration) of the body and balance relative to the frame of reference provided by the earth and stars. It is the gravitational force on the body, not the force exerted on it by the spring balance, which is signified in this book by the word 'weight'.

The term 'weightless' as commonly used of orbiting spacemen, does not signify zero weight (on our definition of weight), but zero force registered on a spring balance travelling with the body. Any body in free fall is 'weightless' in this sense.

Near the surface of a large celestial object, e.g. the earth or the moon, it is often convenient to regard the weight as a constant characteristic of the test body; but its numerical value depends on which celestial object is nearby. A body of weight W on earth, is of weight $\frac{1}{6}W$ on the moon; and on the journey between earth and moon its weight takes a range of values which, if the small contribution made by the attraction of the sun is ignored includes the value zero. At the point of zero weight the pull of the moon on the test body is equal and opposite to that of the earth.

In any diagram showing the forces on a body in a terrestrial setting, the weight of the body must be included: it should be shown as a force, W say, acting vertically downwards through a roughly central point of the body. A force measured on a spring balance marked kg has units called kgf (=kg force). Gravitation, weight and centre of gravity are further discussed in Chapters 5, 6 and 8.

94   An introduction to mechanics and modelling

4.4 Modelling of strings and tension; Hooke's law

A girl (FIG. 4.9) holds a light string PQ at end P, and wishes to exert force on a body B attached to the string at Q. It is a matter of common experience that she must first pull the string until it is nearly straight, and that she can then exert on the body, by means of the string, a force directed from Q towards P, more or less along the length of the string. She can neither exert any appreciable push force, nor transverse (sideways) component of force through the string.

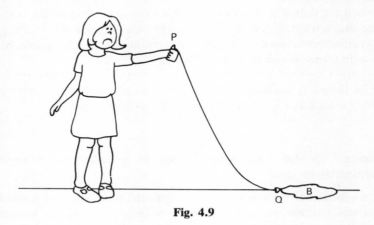

Fig. 4.9

These observations are captured and made mathematically available by choosing to consider a model of the properties of the girl's string, wherein:

(1) the force exerted by the string on body B is exactly zero unless the portion PQ is straight;

(2) when PQ is straight, the string can exert only a direct pull force, i.e. the force component on body B in direction $\overrightarrow{QP} \geq 0$, and in any direction perpendicular to QP is exactly zero. The pull force is called a 'tension', of magnitude T, say.

[It is further convenient to identify characteristics of the string which give rise to (1) and (2). Precise meanings are attributed to words like 'flexible' and 'light' when used to prescribe features of a model. The reader who is already familiar with mechanics may proceed to consider this, as follows:

> To say that a body is 'light' means that the approximation mass = 0 is acceptable. Since, anticipating § 4.5, $\mathbf{F} = m\mathbf{a}$, it follows that the vector sum of all the forces acting on a 'light' body is necessarily zero at all times

(except perhaps for the infinitesimal period of some instant of infinite acceleration corresponding to an instantaneous change of position). In the case of a light object in contact with other objects at only two points, $m = 0$ necessitates that the forces acting at those two points are equal and opposite (otherwise the instantaneous change of position would occur). Hence, if a light string is acted on only by forces at its ends, those two forces are equal and opposite.

Rudimentary ideas of force and mass   95

> The word 'string' or 'cord' implies that the object in question has length but negligible cross-section, so that its geometrical character is represented by a curved or straight line; further, that at neither end can it exert a push force; and finally, that it is 'perfectly flexible', i.e. that at neither end can it exert a force at right angles to the tangent to the curve along which it lies.

Thus a 'light string' in contact with other objects only at its two ends is **either** slack, i.e. exerting no force at either end, in which case the string may take up any curvilinear position, **or** it is straight, in which case it can exert equal and opposite pull forces on the objects to which it is attached at the two ends (this follows from the definition of 'light' together with Newton's third law).]

(3) Returning to the model of the girl holding the string, the end P of the string exerts on the girl a pull T in the direction $\overrightarrow{PQ}$ i.e. a force equal and opposite to the force exerted by the string on B.

Not every real string will be appropriately modelled as 'light' or even as a 'string', e.g. the string which pushes aside the eye of the needle through which we are attempting to thread it. When 'light string' is used as a modelling specification, properties (1), (2) and (3) above are specific attributes of the model.

A light **inextensible** string, moreover, **exerts precisely that tension T at its two ends, which will ensure that the objects attached at those ends never get further apart than the unstretched length l_0 of the string.**

A light **elastic** string, on the other hand, when stretched so that its ends are distance l apart, satisfies **Hooke's law**, namely

$$T = kx \qquad (x \geqslant 0)$$

where x $(= l - l_0)$ is the extension of the string, and k is an elastic constant. When $x < 0$, $T = 0$.

The words 'light elastic **spring**' imply that the object in question may be considered to be linear, flexible, and capable of exerting a thrust when compressed as well as a tension when extended, so that Hooke's law $T = kx$ applies for negative as well as positive values of x. For real springs, Hooke's law is an approximation appropriate only in a comparitively small range of x. For springs, the constant k is known as the 'stiffness' or 'force constant'. The unstretched length l_0 of spring or string is also known as its natural length.

On joining two similar strings together end to end, the same tension produces double the extension, i.e. the value of k is halved. The quantity $\lambda = kl$ is characteristic of the linear material of the string or spring, and is called *Hooke's modulus* of elasticity. In terms of λ, Hooke's law takes the form $T = \lambda x/l$.

The usefulness of formulae in terms of k and λ is restricted by the fact that in practical situations the values of k and λ are seldom known, and it is not always convenient to have to resort to experimental determination. For homogeneous materials there is a convenient alternative source of information.

When the material of a string or wire is homogeneous, a doubling of the cross-section necessitates a doubling of the tension in order to maintain the same extension per unit length. The nature of the bulk material determines the tension

96 An introduction to mechanics and modelling

per unit area (T/A) required for a given x/l. Writing $E = \lambda/A$,

$$T/A = E(x/l)$$

The quantity E, called **Young's modulus**, may be looked up in tables of physical constants for many materials. Typical values of E in Nm^{-2} are: steel: 20×10^{10}; rubber: 0.05×10^{10}; wood: 1×10^{10}.

To find out numerical values for theoretical parameters, such as k or λ above, is a very salutary exercise. It is not only that the assignment of values makes mechanics usable; equally important is the check which it provides on the acceptability of the model. If numerical values of ostensibly empirical parameters are not given in standard reference books, it may well be that the proposed empirical law has little validity for the circumstances being considered. If numerical values are available, the source of information usually also indicates how much reliance can be placed on them.

Exercise 2

1. In FIG. 4.10, solid lines denote 'light strings'; A, B, etc. are fixed points; broken lines denote applied forces and dotted lines indicate horizontal or vertical directions. In each case a small object of weight W is attached at point P. Complete the representation of the forces acting on the body at P (each case separately) by inserting the relevant gravitational and tension forces. From the condition of equilibrium $\mathbf{F} = 0$ (expressed by the two most convenient component equations), deduce all the unknown tensions and the angles marked α.
2. For each of the cases of question 1, draw diagrams to show the force exerted by the strings on the fixed points A, B, etc.

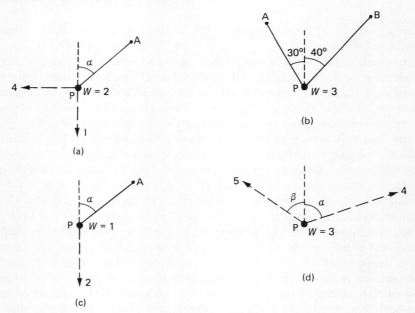

Fig. 4.10

Rudimentary ideas of force and mass 97

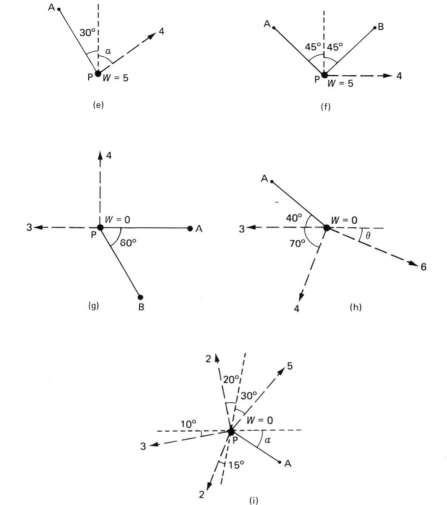

Fig. 4.10 (*cont.*)

3 The girl of FIG. 4.9 uses her real string to prevent a balloon flying away, as in FIG. 4.11. Is it still appropriate to model her string as light and flexible? (You may ask 'what do you want to investigate?' Good question. What **can** you usefully calculate, given the simple principles which are known to you?)
4 In FIG. 4.12, P is a small object of weight 0.5 kg f resting on the smooth plane OX and restrained by the light string AP. In view of the 'smoothness' – which we interpret as zero rub component (parallel to the plane) – represent the contact force on P due to the plane, by a single component, R say, and complete the diagram of forces on P. Deduce the magnitudes of R and of the tension in the string.
5 A small object of weight 0.5 kg f is at rest on a rough plane inclined at 30° to the horizontal. Find the push and rub (i.e. normal and tangential) components of the force exerted by the plane on the object.

98 An introduction to mechanics and modelling

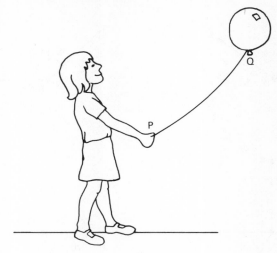

Fig. 4.11

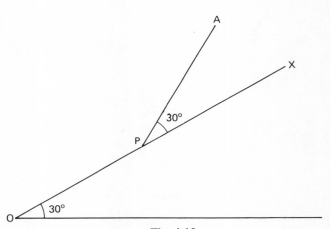

Fig. 4.12

6 A light string passes over a free-running pulley, and each of the ends is pulled – with forces F_1 and F_2 respectively, say. What does experience suggest as the condition for equilibrium of the string? Express this as a modelling postulate regarding the tensions at the two ends of a string in equilibrium when the intervening section is in contact only with free-running pulleys. Given that your modelling postulate is acceptable when the string and pulleys are at rest, will it necessarily remain acceptable when they are in motion? (It is convenient to distinguish between small and large, light and heavy pulleys. The reader is not yet equipped to give mathematical reasons for any hunches he may have.)

7 Find all the unknown tensions in the equilibrium situations shown in FIG. 4.13a–d, using (i) the 'light string' model of all strings; (ii) the modelling postulate 'the tensions at the two ends of a string which passes round a free-running pulley at rest, are equal'; and (iii) the further assumption that 'the force exerted on the pulley by the string is the same as the force which would be exerted if the string were attached to the pulley only at the two ends of the arc of contact, the straight sections of string being tangents to the circumference at these points'.

Rudimentary ideas of force and mass 99

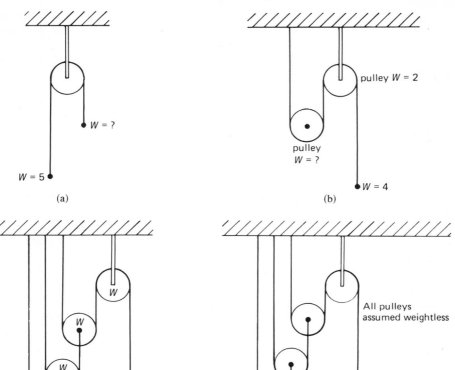

Fig. 4.13

8 A light string hangs over a free-running pulley. A fly alights on one end of the string, and the system remains at rest. Comment.
9 A real table of weight W stands on real horizontal ground. What does the equation of equilibrium enable you to say about the forces exerted by the ground on the four legs of the table?
10 An idealized table, weight W, with four symmetrically placed legs, stands on ground which is modelled as a perfectly smooth horizontal plane. What force is exerted by the ground on any one leg?
11 State briefly (i) the advantages and (ii) disadvantages of modelling as in question 10 the situation referred to in question 9.
12 AB and BC are two strings of lengths 3 m and 4 m respectively, and points A and C are on the same horizontal level. A 1 kg mass (weight 1 kg f) hangs from B.
   (i) Given that angle ABC is a right angle, calculate the tension in each of the strings AB and BC.
   (ii) A horizontal force of magnitude 0.2 kg f is now applied to the hanging mass. Given that the direction of the force is parallel to AC, find the new tensions in strings AB and BC.

100 An introduction to mechanics and modelling

(iii) The horizontal force is subsequently increased to 20 kg f. Find the tensions which would be required in the strings to sustain the mass B in its original position. State the significance of your result. (**Note** postulate (2) of § 4.4.)

4.5 Newton's second law: the nature of mass

If suitable experiments were devised, the student could convince himself, through observation, of the acceptability of the following conjecture, appropriate to any body of constant material composition: the acceleration **a** of a central point of the body is given by:

$$\mathbf{a} = \text{constant} \times \mathbf{F} \qquad 4\text{--}1$$

where **F** is the vector sum of all the forces acting on the body; and the value of the constant of proportionality is, apart from its obvious dependence on the units chosen for force and acceleration, an invariable characteristic of the body. When the body is symmetrical, the 'central point' is its centre of symmetry or geometrical centre; for a uniform plane triangle, it is the centroid; for any body, it is the point usually thought of (even though it may be rather imprecisely) as the **middle** of the body; it is known as the **centre of mass**, and a mathematical formula for its position will become available as we examine the consequences of equation 4–1 in the context of the particle model of matter (Chapter 6).

The constant in equation 4–1 is small for a massive body like an elephant, and large for, say, a peanut. The **reciprocal** of the constant is large for an elephant and small for a peanut, and seems to be closely related to the familiar, if imperfectly defined, concept of mass which we proceed to talk about below. We may write $1/\text{constant} = M$ (whether or not this quantity is to be identified with 'mass'); equation 4–1 then takes the form

$$M\mathbf{a} = \mathbf{F} \qquad 4\text{--}2$$

which is the simplest form of Newton's second law to use and remember.

The concept of mass is best introduced in the context of its most familiar uses. Every housewife who talks of $1\frac{1}{2}$ kg of flour or $\frac{1}{4}$ kg of tea appreciates the general nature of the quantity, namely mass, which is measured by the scalars $1\frac{1}{2}$ (kg) or $\frac{1}{4}$ (kg). If she receives a gift of $\frac{1}{4}$ kg tea from the Himalayas, she will expect it to weigh $\frac{1}{4}$ kg on her kitchen scales at home, in Europe, US or elsewhere, notwithstanding the fact that her scales might need radical re-adjustment or recalibration if they were transported to the Himalayas. It is the measuring device which has to be corrected for changes of location: the mass of any object of unchanging constitution remains constant whatever its location in the universe. The task of checking the calibration of balances and scales so that this consistency is maintained is a matter for the Inspectorate of Weights and Measures.

In this chapter we regard mass as **the invariable quantity which is determined by the reading, e.g. in kg, on properly calibrated domestic or commercial scales (of any type)**. This will not preclude the formulation of a more intellectually satisfying definition later (Chapter 5).

We shall provisionally identify the constant M in equation 4–2 with the mass determined by scales. This lays a constraint on the units chosen for mass,

Rudimentary ideas of force and mass

acceleration and force. Matching systems of units of distance, time, acceleration, mass, force, etc. are called absolute units. The SI system is based on the metre, second and kg and matching units of acceleration, force, etc. The matching unit of force, kg m s^{-2} is called the newton (N).

4.51 Component forms of Newton's second law

Equation 4–2 is a vector equation, so it is equivalent to three component equations. One, two or all of the three may be needed for the solution of any particular problem.

In rectangular cartesian coordinates the component equations take the form:

$$M\ddot{x} = X; \quad M\ddot{y} = Y; \quad M\ddot{z} = Z. \qquad 4\text{–}3$$

If polar coordinates r, θ are used in the x-y plane, the three component equations may be written (cf. § 2.33):

$$\left. \begin{array}{ll} M(\ddot{r} - r\dot{\theta}^2) = R & \text{(radial component)} \\[4pt] \dfrac{M}{r}\dfrac{d}{dt}(r^2\dot{\theta}) = S & \text{(transverse component)} \\[4pt] M\ddot{z} = Z & \end{array} \right\} \qquad 4\text{–}4$$

If s, ψ coordinates are used in the x-y plane, and the z-component equation is not of interest, the relevant equations are:

$$\begin{array}{ll} M\ddot{s} = T & \text{(tangential component)} \\[4pt] M\dot{s}^2/\rho = N & \text{(inward normal)} \end{array} \qquad 4\text{–}5$$

where $\rho = ds/d\psi$, cf. § 2.32.

4.52 Interpretation of the acceleration a

A very small body may be sufficiently located by the coordinates of a single point (at which the whole mass of the body is notionally concentrated). The acceleration of this point is then identified with the quantity **a** in equation 4–2; and – according to the coordinate system chosen – any of equations 4–3, 4–4 or 4–5 may be used. A small body modelled as a point mass is called a particle.

For a large body, there are two situations where a simple identification of **a** may be made,

(1) When the body moves without rotating: at any time, all points of the body have the same velocity, and hence the same acceleration; it is this common acceleration at time t which is denoted by **a**.

(2) When the body has a centre of symmetry, or point of intersection of three planes of symmetry: then **a** is the acceleration of this point, the centre of mass.

4.53 Equations of motion and auxiliary geometrical conditions

The single symbols X, Y, Z, R, S, T, N in equations 4–3, 4–4, and 4–5 denote quantities each of which will, in problem work, usually be written as the sum of

102 An introduction to mechanics and modelling

components of known or postulated forces. When the forces are so written, in terms appropriate to the problem in hand, the equations expressing Newton's second law (e.g. 4–3, 4–4 or 4–5 above) are known as the **equations of motion** for the problems.

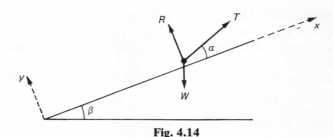

Fig. 4.14

For example, FIG. 4.14 shows a particle of mass m moving up a smooth inclined plane under the action of a known applied force T, in addition to the anticipated normal reaction R, and its own weight W. Taking x, y-axes parallel and perpendicular to the inclined plane, the equations of motion in these two directions are:

$$\left. \begin{array}{l} m\ddot{x} = T \cos \alpha - W \sin \beta \\ m\ddot{y} = R + T \sin \alpha - W \cos \beta \end{array} \right\} \qquad 4\text{–}6$$

When unknown forces appear in the equations of motion, a corresponding geometrical condition is typically implied, and provided this is recognized and written down as part of the formulation of the problem, the geometrical relation commonly supplies the information necessary to determine the unknown force. Thus, in equations 4–6, the unknown R arises because the plane constrains the particle to move only along its surface – a geometrical condition written down as $y = 0$. Because $y = 0$ at all times t applicable to the motion being considered, $\ddot{y} = 0$, and this zero value may be substituted into equation 4–6 to give

$$R = W \cos \beta - T \sin \alpha$$

4.54 Conversion of units
Since Newton's second law can be applied only if the units of force, mass and acceleration are consistent, it is important to be able to convert units quickly and accurately. In the case of force and mass, it is only necessary to look up the appropriate conversion factor, thus:

$$1 \text{ lb f} = 4.448 \text{ N}; \qquad 1 \text{ ton f} = 9.964 \text{ kN}$$
$$1 \text{ lb} = 0.4536 \text{ kg}; \qquad 1 \text{ ton} = 1016 \text{ kg}$$

The conversion of accelerations involves the conversion factors of both length and time. The method of conversion is illustrated as follows:

Example
Express an acceleration of 10 miles per hour per minute in SI units.

Rudimentary ideas of force and mass   103

We need to know that 1 mile = 1609 m and that 1 hour = 60 min = 3600 sec. We proceed:

$$10 \text{ mi} = 16\,090 \text{ m}$$
$$1 \text{ h} = 3600 \text{ s}$$

so

$$10 \text{ mi h}^{-1} = 16\,090/3600 \text{ m s}^{-1}$$
$$1 \text{ min} = 60 \text{ s}$$

whence

$$10 \text{ mi h}^{-1} \text{ min}^{-1} = 16\,090/(3600 \times 60) \text{ m s}^{-2}$$

The albebraic usage explained in § 0.51 allows these steps to be more compactly written

$$a = v/t = 10 \text{ mi h}^{-1}/1 \text{ min}$$
$$= \frac{16\,090 \text{ m}}{3600 \text{ s}} \div 60 \text{ s}$$
$$= 16\,090/(3600 \times 60) \text{ m s}^{-2}.$$

The mathematically gifted reader should not underestimate the importance of being able to carry out such conversions with good grace as well as ease, even when his/her mind is mainly concentrated on larger problems.

4.6 Examples of the use of Newton's second law

4.61 Motion in a circle
A small mass m is observed to be moving in a circle of radius b with constant angular speed ω. Deduce the force on the mass.

The equation of a circle is simpler in s, ψ or in polar coordinates than in cartesians: $s = b\psi$ or $r = b$ respectively, rather than $x^2 + y^2 = b^2$; so we use equations 4–5 or 4–4 rather than 4–3.

Using equations 4–5, with $s = b\psi$ and $\dot{\psi} = \omega =$ constant, we deduce that the force on the given mass has tangential and normal components

$$\left. \begin{array}{l} T = m\ddot{s} = mb\ddot{\psi} = 0 \\ N = m(b\dot{\psi})^2/b = mb\omega^2 \end{array} \right\} \qquad 4\text{--}7$$

i.e. the required force is $mb\omega^2$ directed towards the centre.

Variations on the same problem
Equations 4–7 could supply a quick answer to some of the following questions, but the student will be well-advised to go back to basic principles each time he tackles a problem.

104 An introduction to mechanics and modelling

Variant 1. In FIG. 4.15 a particle of mass m moves on a smooth horizontal table. A light inextensible string is attached at one end to a fixed point O of the table, and at the other to the particle, thereby constraining its motion to lie inside a circle.

Show that the equations of motion predict that, so long as the string is taut, the angular velocity of the particle about O is constant. Also determine the tension in the string, and the magnitude of the normal reaction between the table and the particle.

R_1 and T_1 denote unknown forces, as in FIG. 4.15. The corresponding geometrical constraints are

$$z = \text{constant}$$
$$\text{and } r = b \ (=\text{constant}),$$
(whence $\ddot{z} = 0$ and $\ddot{r} = 0$).

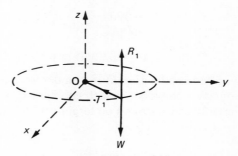

Fig. 4.15

Expressing the equations of motion in terms of polar coordinates (cf. equations 4–4 of § 4.51)

$$\left. \begin{array}{r} m(\ddot{r} - r\dot{\theta}^2) = -T_1 \\ \dfrac{m}{r} \dfrac{d}{dt}(r^2\dot{\theta}) = 0 \\ m\ddot{z} = R_1 - W \end{array} \right\} \quad 4\text{--}8$$

whence, by use of the geometrical conditions,

$$mb\dot{\theta}^2 = T_1$$
$$\frac{d}{dt}(b^2\dot{\theta}) = 0 \quad\quad 4\text{--}8(a)$$

and

$$0 = R_1 - W$$

From the second of these equations it follows that $b^2\dot\theta$ is constant, and hence that $\dot\theta$ is constant, $=\omega$ say. From the other two equations we deduce that throughout the motion, $T_1 = mb\omega^2$; and $R_1 = W$.

Variant 2. In FIG. 4.16, mass m is suspended from the fixed point P by a light inextensible string of length l, and describes a horizontal circle under the action of no forces other than its weight and the tension in the string. Prove that the tension in the string is proportional to the square of the angular velocity (ω say) with which the particle describes the horizontal circle.

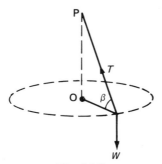

Fig. 4.16

By writing $r = b$ and using r, θ coordinates in the horizontal plane, we find that the equations of motion reduce to:

$$\dot\theta = \text{constant} = \omega \text{ say}$$

$$T \cos \beta = mb\omega^2 \text{ where } \cos \beta = b/l$$

$$\text{and } 0 = T \sin \beta - W$$

The second of these three relations gives us the required result, for from it

$$Tb/l = mb\omega^2$$

i.e. $$T = ml\omega^2$$

A mass and string moving as we have described constitute a 'conical pendulum'.

Variant 3. The time taken by a satellite in a circular orbit to circle the earth is known to depend on the height of the satellite above the earth's surface, but not on the mass of the satellite. Deduce from this fact that the weight of any body in orbit is proportional to its mass.

The only force acting on the satellite (we must assume) is its weight, and this acts in the line joining the satellite to the earth's centre. Does the plane of the satellite's orbit necessarily pass through the earth's centre? The reader is invited to write down the equation of motion in the direction perpendicular to the plane of motion, i.e. the z-direction in the nomenclature used in variant 1 above. He will find that the z-equation of motion can be satisfied provided the plane of the orbit passes through the centre of the earth.

106 An introduction to mechanics and modelling

With this information, the other two equations of motion (and it is recommended that the reader writes them down specifically) tell us that $\dot{\theta} = \text{constant} = \omega$ say, and that $W = mb\omega^2$ where b depends only on the height above the earth's surface and, of course, the earth's radius.

Now, the time taken for an orbit in which ω is constant is

$$\text{Time for orbit, } T = 2\pi/\omega \quad (\text{angle} \div \text{angular speed})$$

It follows that ω like T depends on the height of the satellite above the earth, but not on the satellite's mass. Hence

$$W = mb\omega^2 = km$$

where k depends only on the height of the orbit above the earth's surface, showing that the weight of any satellite body is proportional to its mass.

4.62 Mass and weight

An empirical method, used by British walking enthusiasts, of determining the depth of a pot-hole, disused well, empty gully, or other vertical drop is – after ascertaining that no-one is below – to measure the time (t) in seconds for a stone to fall, from rest, to the bottom. This value of the time is multiplied by 4, the result is squared and the answer is interpreted as the depth of the hole in feet. The formula amounts to

$$s = (4t)^2 = 16t^2$$

From it we infer that the acceleration of a stone of arbitrary size near the earth's surface is

$$\ddot{s} = \frac{d^2}{dt^2}(16t^2) = 32 \text{ (ft sec}^{-2})$$

If the same acceleration is expressed in SI units, its value is 9.8 (m s^{-2}). The symbol g is used to denote the value of the acceleration due to gravity near the earth's surface in general discussions; g must be interpreted as 9.8 when the other quantities involved are evaluated in SI units, and as 32 if old British units are in use etc. The important thing about g is that at a given location it is the same for all bodies. This is the algebraic interpretation of Galileo's empirical observation that all bodies appear to fall to earth with the same acceleration. The equation of motion of a falling body, viz.

$$m\ddot{s} = W \qquad 4\text{-}9$$

used in conjunction with the empirical law $\ddot{s} = g$, tells us that near the earth's surface,

$$W = mg \qquad 4\text{-}10$$

This relation can be used to inform us about the magnitude of the force unit which must be adopted in order that Newton's second law shall be applicable in the form $m\mathbf{a} = \mathbf{F}$ from which equation 4–9 derives. The SI unit is called the newton (N), and equation 4–10 tells us that in these units $W = 9.8\,m$, i.e. a mass of 1 kg has a weight of approximately 10 N. Expressing this in another way, 1 newton is a force equal to the weight of a 1/10 kg mass ($\simeq 4$ oz), which, as admirers of Newton have been happy to point out, is a typical weight for an apple!

By writing $W = mg$ in dynamical problems in locations 'near the earth's surface' we (usually) reduce the number of unknowns in the working.

4.63 Motion of a projectile

4.631 Relevance of air resistance

A body, thrown up into the air, with a velocity in any direction, and thereafter allowed to move without human intervention, is called a **projectile**. A gravitational force $W = mg$ acts on the body; there is usually no other non-contact force of appreciable magnitude. The only material substance with which the body is in contact is the air itself. The possibility that air resistance forces may be important must never be overlooked; in our modelling of statical situations in §§ 4.2 and 4.3 we were guilty of neglect of contact forces between objects and the air; fortunately in statical situations these forces usually cancel out. In modelling objects in motion through the air, exact cancellation does not normally occur, and the greater the speeds, the more important are air resistance forces.

A suitable strategy is as follows. First investigate the motion on the assumption that air resistance forces are negligible; using the velocity values predicted by this simple model, estimate the magnitude of the air resistance forces which have been neglected (this will be discussed in more detail in Chapter 7); reappraise the assumption that air resistance may be neglected.

Air resistance effects are discussed in § 7.4. Here we discuss projectile motion when air resistance may be neglected.

4.632 Equations of motion for a projectile

If the earth's surface is modelled as a sphere (FIG. 4.17), $r = r_0$ say, a polar coordinate system (as in variant 3 of § 4.61) may well be the most appropriate.

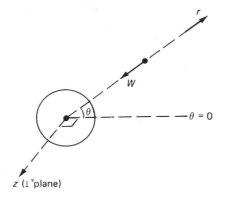

Fig. 4.17

If the problem in hand relates to the distances of the projectile from an inclined plane (FIG. 4.18), cartesian axes parallel and perpendicular to the slope will

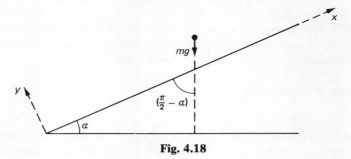

Fig. 4.18

probably be most convenient, the curvature of the earth being neglected! Horizontal and vertical cartesian axes will be best for problems concerning the height of the projectile above a horizontal plane (FIG. 4.19).

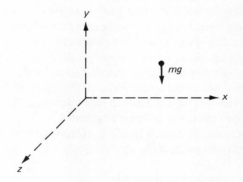

Fig. 4.19

In terms of polar coordinates, Fig. 4.17, the equations of motion may be written

$$m(\ddot{r} - r\dot{\theta}^2) = -W$$
$$\frac{m}{r}\frac{d}{dt}(r^2\dot{\theta}) = 0 \qquad 4-11$$
$$m\ddot{z} = 0$$

(on the scale where the earth's curvature is relevant, the particle is not 'near the earth's surface', so the approximation $W = mg$ is not acceptable).

When the tilted cartesian axes of FIG. 4.18 are used, the equations of motion take the form

$$\left.\begin{array}{l} m\ddot{x} = -mg \sin \alpha \\ m\ddot{y} = -mg \cos \alpha \\ m\ddot{z} = 0 \end{array}\right\} \qquad 4-12$$

Rudimentary ideas of force and mass

And in the simple case of FIG. 4.19 we have

$$\left.\begin{array}{l} m\ddot{x} = 0 \\ m\ddot{y} = -mg \\ m\ddot{z} = 0 \end{array}\right\} \qquad 4\text{-}13$$

Each of the equations in 4–12 and 4–13 is of the form

$$m\ddot{s} = m \times \text{constant (in some cases zero)}$$

where s may represent any of the variables x, y, z in either of FIGS. 4.18, or 4.19, i.e. each is of the form

$$\frac{d^2 s}{dt^2} = \text{constant} = a \text{ say} \qquad 4\text{-}14$$

4.633 Integration of the equation $\ddot{s} = a$

Given that $\ddot{s} = \text{constant}$ ($=a$ say) for all $T_1 < t_0$, $t < T_2$ and that when $t = t_0$, $s = s_0$ and $v = v_0$, one integration gives

$$(\dot{s} \equiv) \, v = at + C \qquad (T_1 < t < T_2) \qquad 4\text{-}15$$

in particular, provided t_0 lies in the stated range $T_1 < t_0 < T_2$,

$$v_0 = at_0 + C \qquad 4\text{-}16$$

so by subtraction

$$v - v_0 = a(t - t_0); \; v = v_0 + a(t - t_0) \qquad 4\text{-}17$$

Since equation 4–17 is true for all t in the range, it can be further integrated, giving

$$s - s_0 = v_0(t - t_0) + \tfrac{1}{2} a(t - t_0)^2 \qquad 4\text{-}18$$

> The formulae 4–17 and 4–18 and their particular cases 4–20 and 4–21 below, are true only in the closed interval $T_1 \leqslant t \leqslant T_2$ for the interior of which a is constant. The formulae are false if the acceleration is variable in the range (t_0, t).   4-19

In principle the author would like to discourage memorization of formulae which are false in the general case, viz. non-constant acceleration. In practice, it is convenient to be able to quote results equivalent to equations 4–17 and 4–18. The expressions are neatened by shifting the origins of s, t to s_0, t_0 respectively, so that, in the new coordinates, we have: when $t = 0$, $s = 0$ and $v = v_0$ which we shall now call u.
Equations 4–17, 4–18 then become

$$v = u + at \qquad 4\text{-}20$$

$$s = ut + \tfrac{1}{2} at^2 \qquad 4\text{-}21$$

110 An introduction to mechanics and modelling

We may further eliminate t from 4–20 and 4–21:
From 4–20

$$at = v - u$$

From 4–21

$$2as = 2u(at) + (at)^2$$

So

$$2as = 2u(v-u) + (v-u)^2$$

i.e.

$$v^2 = u^2 + 2as. \qquad 4\text{–}22$$

The corresponding equation derived from equations 4–17 and 4–18 is

$$v^2 = v_0^2 + 2a(s - s_0) \qquad 4\text{–}22(a)$$

Problems involving the use of constant acceleration formulae are found in §§ 4.65 and 4.66

4.64 Sign conventions and their consequences

Let O be a fixed and P be a variable point, both lying on the curve C_1C_2 in FIG. 4.20. The coordinate s is defined as

$$s = +(\text{arc length OP}) \text{ when P lies between O and } C_2$$
$$= -(\text{arc length OP}) \text{ when P lies between } C_1 \text{ and O}.$$

The coordinate s is distinguished from a mere distance by the attribution of a sign.

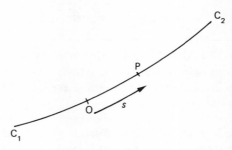

Fig. 4.20

The positive senses of the time derivatives $\dot{s}$, $\ddot{s}$ of any spatial coordinate s are the same as the positive sense for the coordinate itself: for $\dot{s} > 0$ implies s increasing, i.e. s, $\dot{s}$ have the same sense. It is immaterial (apart from convenience) which sense of the original coordinate is defined as positive, or what origin O on C_1C_2 is chosen.

4.65 Parallel discussions of vertical motion, using different choices of coordinate

Problem

A mass is thrown vertically upwards, reaches a height h, and is caught at height $\tfrac12 h$ on its descent. Find its velocity of projection, its velocity just before it is caught, and the duration of the entire flight.

We shall show the working for two alternative choices of coordinates. Instead of reading the solutions, the reader is invited to carry out the calculations for himself preferably **not** with the coordinate system of version 1. He may refer to the printed solutions if/when in difficulty.

Solution

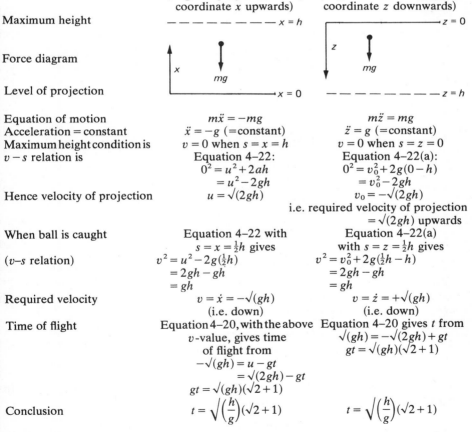

Item	Version 1 (origin at projection level, coordinate x upwards)	Version 2 (origin at maximum height, coordinate z downwards)
Maximum height	$x = h$	$z = 0$
Force diagram		
Level of projection	$x = 0$	$z = h$
Equation of motion	$m\ddot{x} = -mg$	$m\ddot{z} = mg$
Acceleration = constant	$\ddot{x} = -g$ (=constant)	$\ddot{z} = g$ (=constant)
Maximum height condition is	$v = 0$ when $s = x = h$	$v = 0$ when $s = z = 0$
$v - s$ relation is	Equation 4–22: $0^2 = u^2 + 2ah$ $= u^2 - 2gh$	Equation 4–22(a): $0^2 = v_0^2 + 2g(0 - h)$ $= v_0^2 - 2gh$
Hence velocity of projection	$u = \sqrt{(2gh)}$	$v_0 = -\sqrt{(2gh)}$ i.e. required velocity of projection $= \sqrt{(2gh)}$ upwards
When ball is caught (v–s relation)	Equation 4–22 with $s = x = \tfrac12 h$ gives $v^2 = u^2 - 2g(\tfrac12 h)$ $= 2gh - gh$ $= gh$	Equation 4–22(a) with $s = z = \tfrac12 h$ gives $v^2 = v_0^2 + 2g(\tfrac12 h - h)$ $= 2gh - gh$ $= gh$
Required velocity	$v = \dot{x} = -\sqrt{(gh)}$ (i.e. down)	$v = \dot{z} = +\sqrt{(gh)}$ (i.e. down)
Time of flight	Equation 4–20, with the above v-value, gives time of flight from $-\sqrt{(gh)} = u - gt$ $= \sqrt{(2gh)} - gt$ $gt = \sqrt{(gh)}(\sqrt{2} + 1)$	Equation 4–20 gives t from $\sqrt{(gh)} = -\sqrt{(2gh)} + gt$ $gt = \sqrt{(gh)}(\sqrt{2} + 1)$
Conclusion	$t = \sqrt{\left(\dfrac{h}{g}\right)}(\sqrt{2} + 1)$	$t = \sqrt{\left(\dfrac{h}{g}\right)}(\sqrt{2} + 1)$

The reader should take note of the transition from the v–s relation $v^2 = gh$ (both versions) to $v = -\sqrt{(gh)}$ in version 1, but $v = +\sqrt{(gh)}$ in version 2. When taking square roots, the investigator has to decide which one corresponds to the situation he is concerned with.

112 An introduction to mechanics and modelling

4.66 Example involving projectile and inclined plane

Example
A hillside may be modelled as a plane making angle α with the horizontal. A path runs directly uphill and ends at a gate of height h in a plane at right angles to the slope of the hill and at right angles to the path. A small stone (whose air resistance may be neglected) is thrown from the bottom of the hill with velocity V at such an angle that it just grazes the top of the gate. Given that the gate is distance $100h$ up the path from the point of projection (which is at ground level), find a relation between V, α, h and the angle of projection.

If horizontal and vertical axes are chosen, the condition that the stone grazes the gate is clumsy to express. If axes parallel and perpendicular to the plane are chosen, as in FIG. 4.21, the condition reduces to $x = 100h$, $y = h$ for some common value of t.

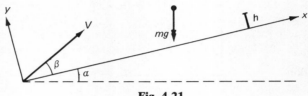

Fig. 4.21

With the axes shown, the equations of motion are (cf. equations 4–12):

$$m\ddot{x} = -mg \sin \alpha$$
$$m\ddot{y} = -mg \cos \alpha$$

i.e.
$$\ddot{x} = -g \sin \alpha \ (=\text{constant}) \qquad 4\text{–}23$$
$$\ddot{y} = -g \cos \alpha \ (=\text{constant}) \qquad 4\text{–}24$$

With initial velocity of projection making angle β with the x-direction, the conditions at $t = 0$ are $u_1(=\dot{x}) = V \cos \beta$; $u_2(=\dot{y}) = V \sin \beta$. Use equation 4–20 (for equation 4–23 with $a = -g \sin \alpha$ and $u = u_1$, and for equation 4–24 with $a = -g \cos \alpha$ and $u = u_2$), thus

$$\left. \begin{array}{l} 100h = V \cos \beta\, t - \tfrac{1}{2} g \sin \alpha\, t^2 \\ h = V \sin \beta\, t - \tfrac{1}{2} g \cos \alpha\, t^2 \end{array} \right\} \text{ for the same value of } t \qquad 4\text{–}25$$

One way of eliminating the common value of t is as follows. Eliminate h (to find t in terms of V, α, β), using $100h - 100h = 0$

$$V(\cos \beta - 100 \sin \beta)t = \tfrac{1}{2} g t^2 (\sin \alpha - 100 \cos \alpha)$$

Since $h \neq 0$, we reject the solution $t = 0$ and deduce

$$t = \frac{2V}{g} \frac{(\cos \beta - 100 \sin \beta)}{(\sin \alpha - 100 \cos \alpha)}$$

Rudimentary ideas of force and mass   113

Substitution of this value into the second of equations 4–25 gives the required relation.

4.7 Identifying forces

4.71 How to enumerate the forces on a body

It helps to identify three categories of force.

(1) 'Invisible forces', like gravity, which are exerted on a body by objects which may be at a distance and perhaps unseen. The force of gravity (weight) is the most common such force, but electric or magnetic attractions/repulsions are also of this type.
(2) Known forces, deliberately applied. In exercises, these will be explicitly mentioned.
(3) Other contact forces, not already accounted for in (2). To recognize these, one must look all round the body in question, and note everything with which it is in contact: the air; any support, like a table; any string joined to the body; any other object or fluid which it touches. How to represent or determine such contact forces will be a recurring theme in Chapters 5–8.

4.72 A note on pulleys in motion

In Exercise 2, questions 6 and 7 the student was encouraged to adopt, for pulleys at rest, the modelling hypothesis: if the pulley is free-running, i.e. frictionless at the bearings, the tensions T_1, T_2 on the two sides of the pulley are equal.

> In examples involving pulleys whose angular acceleration is not negligible, subsequent theory will vindicate the same hypothesis ($T_1 = T_2$) only if the pulley is 'light' or 'small' as well as being frictionless at the bearings.

Those words, 'light' and 'small', identify the characteristics which ensure that the force required to produce the angular acceleration of the pulley will be small compared with other forces in the problem.

Example

A light inextensible string passes over a small, free-running pulley whose axis is fixed, and carries masses of 4 kg and 5 kg at its two ends. Find the tension in the string, also find the velocity of the 4 kg mass, one second after release from rest.

Since the string is inextensible, the same variable x may be used to denote the position coordinate of the 4 kg mass measured upwards, and the position coordinate of the 5 kg mass measured downwards, as in FIG. 4.22a. The solution follows:
Equations of motion (see FIG. 4.22b)

$$4\ddot{x} = T - 4g$$
$$5\ddot{x} = 5g - T$$

114   An introduction to mechanics and modelling

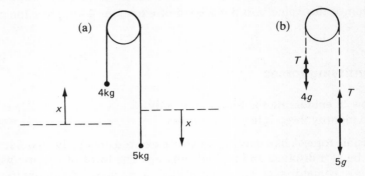

Fig. 4.22

Adding,
$$9\ddot{x} = g, \text{ i.e. } \ddot{x} = g/9 (=\text{constant}) \qquad 4\text{-}26$$
whence
$$T = 4g/9 + 4g = 40g/9,$$
and, using the constant acceleration formula, $v = u + at$, for $t = 1$ sec,
$$v(=\dot{x}) = g/9 (\simeq 1.1 \text{ ms}^{-1})$$

The predictions made in §§ 4.6 and 4.7 were all derived by use of Newton's second law in conjunction with additional modelling assumptions. Predictions like these are capable of being tested against experimental observation and by means of such tests in the past, massive corroboration of the fabric of Newtonian mechanics has been built up.

Exercise 3
1. In the situation of the above (§ 4.7) example, find the distance which the 5 kg mass has moved after 2 seconds (from release).
2. Work the problem of § 4.65 taking the origin at the highest point and measuring coordinate s upwards.
3. A mass m is thrown vertically upwards with velocity 4 m s^{-1}, and moves freely under gravity until, after reaching its highest point, it descends to the ground, whose level is 1 m below the level of projection. Taking $g \simeq 10 \text{ m s}^{-2}$, find the time of flight and the speed with which the mass reaches the ground.
4. In question 3, identify the **mathematical** reason why the formula for the position and velocity of the stone are inapplicable after the stone makes contact with the ground.
5. A mass moves under the action of no forces. Write down its cartesian equations of motion, and draw a conclusion about its velocity at any time. (This conclusion is Newton's first law – a radical discovery in his day, which we more coolly regard as a natural consequence of his indispensible second law.)
6. A bead of mass m slides along a smooth horizontal wire bent into the form $s = a \exp(\psi)$ starting from rest at the point $s = a$, under the action of a constant tangential force of magnitude k. Apply a convenient component form of Newton's second law from § 4.51; deduce the bead's velocity at time t, and the magnitude of the normal reaction between the wire and the bead.

Rudimentary ideas of force and mass

7 A perfectly smooth long narrow tube rotates in a horizontal plane about one end (O) which is fixed in the plane. A small fragment of the tube, situated at distance a from O, separates itself from the inside wall. The tube itself is constrained to continue to rotate with constant angular velocity ω, the fragment remaining inside it. Prove that the fragment's radial distance from O after time t is

$$r = a \cosh \omega t \; [\equiv \tfrac{1}{2} a (\exp(\omega t) + \exp(-\omega t))],$$

and find the magnitude of the force of interaction between the particle and the tube at the same instant. (If you cannot **deduce** the given form for r, just **verify** that it satisfies the equation of motion. Use polar equations of motion with $\dot{\theta} = \omega$.)

8 A packing case weighing 20 kg is placed on a plane which is inclined at angle 30° to the horizontal. In motion, the friction component of the plane-package reaction is one-tenth of the normal component. Find the speed of the package after $\tfrac{1}{2}$ second, and the speed when the package has travelled 1 metre ($g \approx 9.8$). State why the 'particle' equation of motion $m\mathbf{a} = \mathbf{F}$ is (a) valid for, (b) sufficient for, the solution of this problem.

9 One end of a light inextensible string of length l is attached to a fixed point O. At the other end P a mass m is attached. The mass starts with OP making an angle α with the downward vertical. Write down equations of motion for the subsequent motion of the mass in a circle (acted on by two forces T and $W = mg$). Deduce that the particle's velocity increases until the string is vertical and thereafter decreases. Show also that the tension is at a maximum when the particle is at its lowest point. Given that the maximum tension is $3mg/2$, deduce the particle's velocity at the lowest point of its path.

10 A mass m moves along the curve $s = a \exp(\psi)$ with constant speed c, starting from the point $s = a$. Find the force acting on the mass as a function of s.

11 A particle describes the curve $xy = a^2$ under the action of a force parallel to the y-axis. Show that the magnitude of the force is proportional to y^3.

12 OP is a uniform plank of mass M and length l, attached at one end, O, to a fixed freely-turning hinge whose axis is horizontal (and at right angles to the length of the plank). The plank is released from rest in the position in which it makes angle β with the downward vertical. Express the components of reaction exerted by the hinge on the plank (in directions parallel and perpendicular to the plank) at any instant in the subsequent motion, in terms of the angular velocity $\dot{\theta}$ and the angular acceleration $\ddot{\theta}$ of the plank at that instant. (**Note**: the plank is 'uniform', i.e. symmetrical about its mid-point, so the acceleration $\mathbf{a}$ relates to this mid-point. β will not appear explicitly in your answer.)

13 Two particles, each of mass m, are connected by a light inextensible string of length $2a$. Initially the string is taut, and the system is in a gravity-free environment. One mass is projected at right angles to the string, with velocity V. The string remains taut, so the whole system consisting of the string and the two particles moves as a single rigid body.
 (i) Deduce the subsequent velocity of the mid-point of the string.
 (ii) Express the acceleration of one of the particles in terms of its acceleration relative to the mid-point of the string and the acceleration of that mid-point.
 (iii) Deduce the angular acceleration $\ddot{\theta}$ of the string, its angular velocity $\dot{\theta}$ and its tension T at any instant after its release.
 (iv) What is the direct distance of each particle from its starting point by the time the string has made a complete rotation ($\theta = 2\pi$)?

14 A string of length $2a$ with masses m at each end, as in question 13, starts from a horizontal (straight) position near the earth, gravity g. One of the masses is projected vertically upwards with velocity V. Write down expressions for the velocity and acceleration of this particle relative to the mid-point of the string, in terms of the angular velocity $\dot{\theta}$ and angular acceleration $\ddot{\theta}$ of the string when its inclination to the horizontal is θ. Deduce the true acceleration components of the particle, parallel and perpendicular to the instantaneous direction of the string (in terms of θ and its

116   An introduction to mechanics and modelling

derivatives as well as g). From the corresponding equations of motion deduce that the string remains taut, and has a tension of $mV^2/4a$ at all subsequent times until it touches the ground.

15 In FIG. 4.23, mass m slides down the smooth slope AB of a wedge ABC of mass M. The wedge rests on a smooth horizontal plane (i.e. there is no resistance to horizontal motion). Coordinate y specifies the position of m relative to the moving point B of the wedge. Write down equations of motion for the wedge and for mass m (one equation for M, two equations for m, three unknowns x, y, R), and deduce that $\mathrm{d}^2y/\mathrm{d}t^2 = -(M+m)g \sin \alpha /(M+m \sin^2\alpha)$. Find an expression for the distance the wedge has moved along the table after 0.1 second.

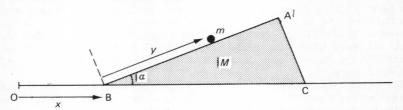

Fig. 4.23

16 A gun is pointed in a direction 20° W of N and is elevated to 70°. A shot is fired with speed 500 m s^{-1} and hits a missile which at the instant of firing was $1\frac{1}{2}$ km N of the gun, and travelling horizontally westward. Find the height and speed (assumed uniform) of the missile, neglecting air resistance. [**Hint**: first consider components in the horizontal plane.]

17 A number of particles are projected simultaneously from the same point with equal speeds in the same vertical plane. Prove that the direction of relative motion of any particle as viewed from any other is independent of time. Prove also that at any instant all the particles lie on a circle. (**Note**: an equation of the form $(x-a)^2 + (y-b)^2 = $ constant represents a circle.)

4.8   Newton's definitions, his laws of motion, and a useful consequence

Newton began his *Principia* (*Mathematical Principles of Natural Philosophy*) with definitions, the first four of which may be translated from the Latin, and slightly abridged as follows.

Definition 1.   Quantity of matter is measured by the product of density and volume, and is hereafter referred to as body or mass. It is proportional to weight.
Definition 2.   The quantity of motion is measured by the product of velocity and quantity of matter.
Definition 3.   The inertia or innate enforcement of matter is a power of resisting by which every body endeavours to persevere in its present state whether it be of rest or moving forward uniformly in a straight line.
Definition 4.   An impressed force is an action exerted on a body in order to change its state, either of rest or of moving uniformly forward in a straight line.

The characteristic property of matter or mass as far as Newtonian mechanics is concerned is partly captured by definition 3. Newton's intention in definition

Rudimentary ideas of force and mass

1 is not wholly clear, but since any practical measurement of density is based on weighing, it seems likely that his first definition is substantially equivalent to the kitchen scales definition which we introduced in the latter part of § 4.5.

The importance of definition 3 is that it points to an innate, i.e. constant, characteristic of matter. Together, definitions 1 and 3 served to justify the attribution to every material body of constant composition, of **a constant numerical magnitude** M, called the mass of that body. A route by which the 'innate characteristic' can be pursued and a value attributed to M, without recourse to weighing on scales, will be explored in Chapter 5.

Definition 2 defines the quantity which modern scientists call linear momentum $= M\mathbf{v}$. Definition 4 tells us that an impressed force $\mathbf{F}$ causes a change in $M\mathbf{v}$. Conversely, in view of definition 3, if the change in $M\mathbf{v}$ – which we may write as $[M\mathbf{v}]$ or as $\delta(M\mathbf{v})$ – is not equal to zero, then a force may be presumed.

The nature of the three **laws of motion** which Newton subsequently formulated, and which are given in full below, may be described.

The first law is a restatement of the insight contained in definition 3. The second law unites definitions 2 and 4 to give the quantitative relationship 'force equals mass times acceleration' which is at the centre of Newtonian mechanics; having stated the second law, the first law becomes redundant, as it just corresponds to the case of zero force in the second law. (Historically the recognition of the first law was a crucial step.) The third law states the equality of action and reaction, already discussed in § 4.26. The following is a fairly close translation of Newton's three laws of motion.

I Every body remains in its own original state of rest or of uniform motion in a straight line unless external forces oblige it to change that state.
II Change of motion, (which we interpret as $\delta(M\mathbf{v})$ or $[M\mathbf{v}]$) is proportional to the impressed force, and acts along the same straight line.
III Action is always equal and in the opposite direction to reaction.

Modern mathematical notation, the explicit recognition of the role of time, and the choice of units of force consistent with those of mass, length and time, enable us to state Newton's second law more concisely as

$$\frac{d}{dt}(M\mathbf{v}) = \mathbf{F} \qquad 4\text{-}27$$

or

$$[M\mathbf{v}] = \int \mathbf{F}\, dt \qquad 4\text{-}28$$

Since the mass M is constant for any body of constant composition, the second law may also be stated,

$$M\, d\mathbf{v}/dt = \mathbf{F} \qquad 4\text{-}29$$

i.e. mass times acceleration equals force.

So long as the student is dealing with separate identifiable bodies one at a time, the following statement of Newton's laws will serve his/her needs.

> Mass times acceleration equals force (Newton's second law); and
> Action and reaction are equal and opposite (Newton's third law). 4–30

Newton's second law may be re-expressed in different forms; the form given above is the simplest. From it many variations and extensions may be derived. There is no need for the beginner to abandon the basically familiar idea of acceleration in favour of the less familiar 'rate of change of momentum' which occurs in alternative formulations.

In applying Newton's second law, the first essential step is to identify the constant mass which is under discussion. This is particularly true in problems of rocketry and the like. Only when the constant body of matter has been identified can the total force acting on it be discussed with clarity, or the law be used.

In the statements 4–30 we are as yet using ill-defined terms. We have given no entirely satisfactory numerical definition of either mass or force. The standard scientific mode of advance in such circumstances has already been outlined in the Introduction, § 0.5. We stand the 'laws' on their head, and use them to create definitions. As two undefined quantities are involved in 4–30, a preliminary elimination of one is arranged for, as follows.

Consider a collision between two bodies of masses m_1, m_2 respectively, there being no other forces acting on them. From Newton's third law we may assert:

if mass m_1 exerts force $\mathbf{F}_2$ on the mass m_2, then
the mass m_2 exerts force $\mathbf{F}_1 = -\mathbf{F}_2$ on the mass m_1.

Hence, using Newton's second law in the form given in equation 4–28,

$$\text{The change in } m_1 \underline{v}_1 \equiv [m_1 \underline{v}_1] = \int \mathbf{F}_1 \, dt = -\int \mathbf{F}_2 \, dt$$

$$= -[m_2 \underline{v}_2] \qquad 4\text{–}31$$

whence

$$-[\mathbf{v}_1] = \mu [\mathbf{v}_2] \text{ where } \mu = m_2/m_1 = \text{constant} \qquad 4\text{–}32$$

Newton's laws thus imply the relation 4–32.

Whereas the appreciation of Newton's second and third laws in their original form was bound up with ideas of mass and force derived from *ad hoc* measuring devices (with their attendant errors and uncertainties), equation 4–32 involves only the idea of comparing two velocity changes.

All experimental corroboration of Newton's laws equally corroborates equation 4–32. As we shall proceed to show in Part II, the theory of (Newtonian) mechanics may with advantage be started off from the ground of the simple relation 4–32. The presentation of this logically coherent system – the successful overall model which enables us to 'explain' so much and to control a little of what happens in the world around us – is the theme material of the 'second round of thought' about mechanics which follows in Part II. Newton's second and third laws (enriched by vector treatments) are the seminal first round ideas from which the entire system springs.

Part II
Understanding the motion of real objects

5

Mass and force; momentum; forces of nature

Among the questions to which answers may be found in this chapter are: What is mass? How can the mass of the earth be discovered? How can unmeasurable contact forces be taken into account?

5.1 Mass and force

Although the discussions and examples of Chapter 4 should have given the reader some idea of the importance of mass, no satisfactory definition has yet been given of this basic concept.

The terms of a scientific definition must make it clear what measurements determine the quantity in question, for, without a recipe for measurement as touchstone, a quantitative symbol such as Q or m may represent a mirage rather than a scientific entity. Often, however, no single real experiment can exclude all extraneous factors, all sources of experimental error. For this reason, hypothetical experiments rather than real ones are often referred to in the formulation of definitions. A hypothetical experiment represents the essence which could be extracted from a series of real experiments: it is in a sense a model of a set of real investigations. We use such an experimental model for defining mass.

5.11 Definition of mass

Envisage two bodies in otherwise empty space. We postulate that the two bodies under consideration (FIG. 5.1) are free to move under their own mutual influence only. This may include collisions, gravitational, electric or magnetic attractions

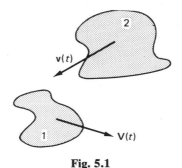

Fig. 5.1

or repulsions, etc. without restriction, except that each of the two bodies retains its own identity.

Let $\mathbf{V}(t)$ be the velocity of the first body at time t, and $\mathbf{v}(t)$ be that of the second body. The velocity of either body is unambiguously defined if **either** every point of the body moves with the same velocity (this common velocity is then the velocity of the body and the motion of the body is said to be a pure *translation*), **or** if the linear dimensions of the body are negligibly small, **or** if the body has a centre of symmetry, when the velocity of the body is the velocity of that centre. Let t_1 and t_2 be any two times at which the velocities of both bodies are unambiguously defined. Then Newton's laws can be reformulated (cf. § 4.8) to yield the following statements.

I The change in velocity of the first body, $\mathbf{V}(t_2) - \mathbf{V}(t_1)$ ($\equiv [\mathbf{V}]$ say), is antiparallel to the change in velocity, $\mathbf{v}(t_2) - \mathbf{v}(t_1)$ ($\equiv [\mathbf{v}]$), of the second body. We may therefore write

$$-[\mathbf{V}] = \mu[\mathbf{v}], \text{ where } \mu \text{ is a scalar multiple} \qquad 5\text{--}1$$

II The scalar multiple μ is constant for the two bodies, i.e. is independent of the times t_1 and t_2, and of the particular collision experiment which is conducted.

Whereas Newton's original laws involved imperfectly defined entities, and were therefore only indirectly verifiable, the two laws I and II are statements which lend themselves to direct testing; for example, by terrestrial experiment it may be seen whether there are any inexplicable discrepancies when only the horizontal components of equation 5–1 are considered, discrepancies for vertical components being attributed to vertical forces. In the case of symmetrical bodies, the laws may be tested, irrespective of whether the initial and final motions are pure translations, using the velocity of the centre as representative of the velocity of the body.

When the statements I and II are regarded as having been established by experiment, the following definition of mass becomes available.

Definition. If a standard lump of material (such as the standard lump of platinum–iridium alloy known as the kilogram, and kept for reference by the Bureau International des Poids et Mesures) is taken as the first body in an experiment of the kind described above, to which formula 5–1 applies, then the constant of proportionality μ is called the 'mass' of the second body, in the corresponding unit (kg).

It need hardly be said that the actual standard kilogram mass is not subjected to such indignities: the reference is to hypothetical experiment.

With this definition available, statement II above may be described as a law of constancy of mass. Statement I (equation 5–1) is a rudimentary form of the important law of 'conservation of linear momentum', which will be further discussed in § 5.22.

5.12 Particles

The *masses* of bodies of irregular shape can in principle be determined as in § 5.11, by arranging collisions wherein both initial and final motions are pure translations. The identification of the *acceleration* of a body of irregular shape is more difficult. Most motions are not pure translations, and there is no clear way (until a centre of mass has been identified) of attributing a single value of acceleration to a large non-symmetric body. No such difficulties arise for symmetric bodies or for bodies of negligibly small linear dimensions, hereafter referred to as particles.

For the solution of practical problems, the symmetric body often provides an appropriate and convenient model. For the purposes of theory-building, the simpler but less realistic abstraction, the point particle, proves useful: large bodies are regarded as connected sets of particles.

5.13 Acceleration and Newton's second law

If we restrict ourselves to the motion of identifiable points, whether these are the centres of symmetrical bodies or the positions of particles, sequences of values of $[\mathbf{v}]/(t_2 - t_1)$ can be constructed, and the limit value, $d\mathbf{v}/dt$ can be defined as in Chapter 2. Equation 5–1 may then conveniently be replaced by

$$-\frac{d\mathbf{V}}{dt} = \mu \frac{d\mathbf{v}}{dt} \qquad 5\text{--}2$$

a relation true at every instant of time, t.

5.14 Definition of force

As well as yielding equation 5–1, or its alternative 5–2, Newton's second and third laws imply the further statement of physical law:

> III If two bodies, of masses $\mu = m_1$ and m_2 respectively (defined using a standard mass as in § 5.11 above) are subjected to a collision or mutual interaction experiment with each other, then in it
>
> $$-m_1[\mathbf{v}_1] = m_2[\mathbf{v}_2] \qquad 5\text{--}3$$
>
> where $[\mathbf{v}_1]$, $[\mathbf{v}_2]$ denote the changes in velocity of the masses m_1, m_2 respectively.

When the velocities $\mathbf{v}_1$, $\mathbf{v}_2$ are the velocities of identifiable points (centres of symmetry or, subsequently, centres of mass) we can write 5–3 in the derived form,

$$-m_1\, d\mathbf{v}_1/dt = m_2\, d\mathbf{v}_2/dt$$

i.e. writing $d\mathbf{v}/dt = \mathbf{a}$,

$$-m_1 \mathbf{a}_1 = m_2 \mathbf{a}_2 \quad (= \mathbf{F} \text{ say}) \qquad 5\text{--}4$$

The quantity 'mass times acceleration' is thus of fundamental importance in any collision or interaction, and to it we attribute the name 'force'. What was originally stated as a law of nature (Newton's second law) is now used as a definition of the concept of force.

124   An introduction to mechanics and modelling

Because in § 5.11 we provided ourselves with a fundamental, quantitative definition of the mass of a particle/body, it is possible to proceed towards a similarly fundamental definition of force:

> Force is that which causes the velocity of a particle, or of the centre-point of a material body, to change. The magnitude and direction of the total force on a particle or body of mass m are, by definition, equal to the magnitude and direction of $m\mathbf{a}$. The unit of force-magnitude in the SI system is thus 1 kg m s^{-2}; this unit is called 1 newton (N); the direction of the force is described in the same way as is the direction of $\mathbf{a}$.

The definition of a force acting on any extended body of matter can be completed only after the definition of $\mathbf{a}$ for such a body has been elucidated (centre of mass property, § 6.2). Until then we shall confine our attention to particles, extended bodies in motions of pure translation, and symmetrical bodies, for which $\mathbf{a}$ is the acceleration of the geometrical centre. If several forces act simultaneously on a mass m, the magnitude and direction of any one of those forces is equal to its contribution to $m\mathbf{a}$, i.e. to (the value of $m\mathbf{a}$ with all the forces acting) − (the value of $m\mathbf{a}$ when all forces except the chosen one are acting); the sign − denoting vector subtraction. This definition is equivalent to postulating the applicability of the vector law of composition for the set of forces acting on the mass m whose motion is under consideration.

5.15 Use of equations of motion

The nature and precise theoretical definition of mass has now been explained: the reader may envisage every mass being determined by a collision experiment. In practice, of course, he/she will continue to measure mass on scales and balances − devices whose operation can be related to the fundamental definition only by use of mechanical laws which will be derived or enunciated later in this book.

In terrestrial applications of mechanics, the constant mass m of each body is often easy to determine beforehand, so, typically, m is regarded as a known quantity. The basic method of mechanics then consists of applying Newton's second law in the way already seen in the examples of §§ 4.5–4.6. The prescription may be summarized as follows (using the situation shown in FIG. 5.2 to exemplify it).

> Express the relation 'mass × acceleration = resultant force' (for each body separately identified) in a suitable component form, taking into account the salient features of the physical forces which contribute to the resultant (e.g. $m\ddot{z} = -kz + mg$). Then integrate the equations (e.g. $m\dot{z}^2/2 = -kz^2/2 + mgz + C$), and interpret the results (e.g. if at $z = 0$, $\dot{z} = u$, then $C = \tfrac{1}{2}mu^2$, and the greatest and least values of z in the subsequent motion, which occur when $\dot{z} = 0$, are the roots of the quadratic equation $kz^2 - 2mgz - mu^2 = 0$).

Mass and force; momentum; forces of nature   125

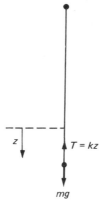

Fig. 5.2

5.151   Comments on this summarized prescription
(1) The identification and representation of the salient features of common natural forces is of prime importance; the discussions of § 5.3 should help the reader to achieve this crucial step.
(2) Sometimes, in a problem involving several different bodies, the mass of one or more is so small compared to the masses of the other bodies that the approximation $m = 0$ may be acceptable. In the parlance of mechanics, the body for which this approximation is acceptable in prescribed circumstances is described as 'light' in those circumstances. In different situations it may be decided that the mass of that same body should be taken into account: the body is then not regarded as light. The terminology which we use reflects a decision which has been made about what characteristics of reality to take into account, and what characteristics are going to be ignored, i.e. the use of words like 'light', 'small', 'rigid', etc. as technical terms indicates that initial modelling decisions have already been taken. The student is not being required to model real life, but only to proceed with an already partially modelled problem.

The word 'light' gives immediate information about the forces on the light body. For, from $m\mathbf{a} = \mathbf{F}$, we deduce that a light body's acceleration is infinite until the body has achieved a position in which the resultant $\mathbf{F}$ of all the forces acting on it is zero, i.e. a light body instantaneously takes up a position for which $\mathbf{F} = \mathbf{0}$, and, throughout its subsequent motion with finite velocities, the forces on it continue to have zero vector sum.
(3) By means of integration and other forms of mathematical manipulation, the equation $m\mathbf{a} = \mathbf{F}$ may be expressed in a variety of alternative forms; one variant, especially useful when discussing events of short duration, is dealt with in the next paragraph.

5.2   Momentum and impulse

Just as the occurrence of $m\mathbf{a}$ as a significant constituent in equation 5–4 prompted us to give it a name, **force**, so the occurrence on both sides of equation 5–3 of

126 An introduction to mechanics and modelling

quantities of the form

$$[m\mathbf{v}] \quad (\equiv m\mathbf{v}(t_2) - m\mathbf{v}(t_1))$$

suggests that $m\mathbf{v}$ is a mechanically significant entity to which a name may profitably be given. The name given is **momentum**.

$$m\mathbf{v} = \text{the momentum of mass } m \text{ moving with velocity } \mathbf{v}$$

The rate of change of momentum of any constant mass m,

$$\frac{d}{dt}(m\mathbf{v}) = m\frac{d}{dt}(\mathbf{v}) = m\mathbf{a}$$

$$= \mathbf{F} \qquad \text{(Newton's second law)}$$

Integration of this equation over the time interval $t_1 < t < t_2$ gives

$$m\mathbf{v}(t_2) - m\mathbf{v}(t_1) = \int_{t_1}^{t_2} \mathbf{F}\, dt = \mathbf{J} \quad \text{say} \qquad 5\text{-}5$$

i.e. the change of momentum of the mass $= \mathbf{J}$, and because of its significance in this context, the integral

$$\mathbf{J} = \int_{t_1}^{t_2} \mathbf{F}\, dt$$

is given a special name, viz. the **impulse** exerted by the force $\mathbf{F}$ (or suffered by the mass m) during the time interval.

The introduction of all this apparently trivial nomenclature is an initial hindrance to the student. It is justified only when the meaning of the terms, and the use of the relations in which they occur, have become thoroughly familiar. As we have already pointed out in the Introduction, using a quotation from Mach, understanding mechanics is partly a matter of familiarity; when we have reached the point where we are everywhere able to detect the same simple elements (e.g. force, mass, momentum, impulse) combining in the ordinary manner, the phenomena no longer perplex us, they are explained. More will be said about the notation $\int \mathbf{F}\, dt$ in § 9.25.

5.21 Example

A cricketer receives a ball of mass 160 gm travelling at $12\ \text{m s}^{-1}$ from an angle of $10°$ on his off-stump side, and returns it on the same side with a speed of $10\ \text{m s}^{-1}$ at $50°$ (Fig. 5.3). What impulse did the cricketer apply to the ball? Given that the duration of the contact between bat and ball was 0.02 second, what was the mean force during the contact interval? At what angle do you think a first-class batsman would have held his bat for this shot?

Comments
(i) The answer to the first question requires a direct evaluation following the definition of impulse, viz.

$$\mathbf{J} = m[\mathbf{v}] \equiv m(\mathbf{v}_2 - \mathbf{v}_1).$$

If you are a beginner, skip the rest of the comments and proceed straight to the solution below.

If you are more than a beginner, note that v_1 and v_2 are incompletely specified in the question, the inclination of the path of flight to the vertical being given neither before nor after the shot. To enable us to proceed with numerical calculations, we are forced to make assumptions about these elevations, and these must be stated explicitly. The calculations below apply when the vertical components of velocity before and after the shot may be ignored. The reader may, if he wishes, investigate different assumptions/situations.

(ii) The answer to the second question requires an interpretation of the phrase 'mean force'; in this context mean force $\bar{\mathbf{F}}$ is that **constant** force which satisfies

$$\int_{t=0}^{0.02} \bar{\mathbf{F}}\,dt = \mathbf{J} \quad \text{i.e.} \quad \bar{\mathbf{F}} \times 0.02 = \mathbf{J}$$

(iii) The answer to the third question requires judgement about what a first-class batsman would do. Any cricket-playing reader may be able to give a better answer than that proposed by the author, who is ignorant of the game. The proposed model serves the purpose of illustrating ideas about impulses. It is unlikely to cast much illumination on the game of cricket because of the writer's ignorance of the subject matter. Any reader who identifies important deficiencies of the model has taken the first step to becoming a good applied mathematician, especially if he/she can suggest how to construct an improved theory.

Solution

Take the line of the wicket as axis of x as in FIG. 5.3a. Then v_1 has components $12 \cos 10°$, $-12 \sin 10°$ and v_2 has components $-10 \cos 50°$, $10 \sin 50°$. Using SI units, $m = 0.16$, so

$$x\text{-component of } \mathbf{J}, \quad J_x = 0.16(-10 \cos 50° - 12 \cos 10°)$$
$$= -2.919$$
$$y\text{-component of } \mathbf{J}, \quad J_y = 0.16(10 \sin 50° + 12 \sin 10°)$$
$$= 1.559$$

whence, total impulse has magnitude $= \sqrt{(2.919^2 + 1.559^2)} = 3.31$ N s and is directed at an angle α with the line of the wicket, as shown in FIG. 5.3b, where

$$\tan \alpha = 1.559/2.919 = 0.535$$

i.e.

$$\alpha = 28.1°$$

From comment (ii), the mean force applied is of magnitude

$$= 3.311/0.02 = 166 \text{ N}$$

and it has the same direction as $\mathbf{J}$.

128 An introduction to mechanics and modelling

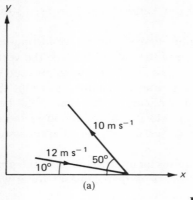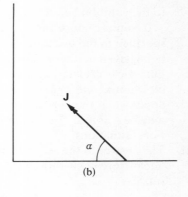

Fig. 5.3

I assume that a first-class cricketer would (a) want to do what he actually did, and (b) that he would hold his bat so that the impulse he wished to impart was perpendicular to the face of the bat. My answer, based on those assumptions, is that the face of his bat is likely to have been held at right angles to the direction of **J** found in part (i) above.

All these answers are subject to the acceptability of the modelling hypothesis that vertical components of velocity can be ignored. Alternative hypotheses might profitably be explored to find a **range** of results inside which the true answers will be expected to lie. See also comment (iii) above.

Assignment (for more advanced students)
1 Does the information given in the preceding example appear credible? (Consider, for instance, the fact that a standard cricket pitch is 22 yards long and that the bowler must deliver from a point within a prescribed distance of the wicket.) Is your assessment affected by the circumstance that the ball may have bounced before reaching the bat?
2 Bowlers use spin to create difficulties for the batsman. Would spin affect (a) the impulse required to produce the required over-all velocity change, (b) the impulse delivered by applying the bat in a stated way?

In both of the above questions, a verbal discussion between several students would be appropriate, backed up by any subsidiary calculations of which they are capable. (Many of the techniques which would assist a fuller discussion will receive mention in later chapters of this book.)

5.22 Principle of conservation of linear momentum
The direction of $m\mathbf{v}$ is associated with a particular line in real space, and $m\mathbf{v}$ is often called **linear** momentum to distinguish it from an analogous quantity called angular momentum, which is of significance for rotating bodies (§ 6.7).

Equation 5–3, referring to the interaction of two bodies, may be written,

$$[m_1\mathbf{v}_1 + m_2\mathbf{v}_2] = 0$$

Mass and force; momentum; forces of nature

i.e. the change in $m_1\mathbf{v}_1 + m_2\mathbf{v}_2$ is zero; or, the sum $m_1\mathbf{v}_1 + m_2\mathbf{v}_2$ is constant, or conserved, so long, that is, as the mutual interaction forces of these two masses are the only forces acting on them.

The equation

$$m_1\mathbf{v}_1 + m_2\mathbf{v}_2 = \text{constant} \qquad 5\text{--}6$$

expresses **the principle of conservation of linear momentum** for two bodies acted on by no external forces (i.e. forces due to bodies other than the two specified bodies). The principle may be extended to any number of bodies, thus:

> Given that bodies of masses $m_1, m_2, m_3, \ldots$ move under mutual forces and collisions but no external forces, then
>
> $$m_1\mathbf{v}_1 + m_2\mathbf{v}_2 + m_3\mathbf{v}_3 + \cdots = \mathbf{constant} \qquad 5\text{--}7$$
>
> i.e. the linear momentum of the system is conserved.

When the bodies involved are symmetrical, non-rotating or 'small', there is no ambiguity about the velocities concerned in equation 5–7. If the bodies are unsymmetrical, large and free to rotate, the velocities are those of their **centres of mass**, points whose identity is clarified in the next chapter.

Example
Two masses A and B lie on a smooth horizontal table. They are connected by a light inextensible string. The mass B is given an impulse **J** of magnitude $J = 0.07$ N s in a direction inclined at 60° to the line AB, and directed away from A. Find the velocities of the two masses immediately after the blow.

The problem is ill-defined until we have attributed values to the masses and specified whether or not the string is initially straight.

Let the two masses be M and m respectively. We consider the case in which the string is initially straight. (The alternative assumption would tell us that the mass m acquired a velocity J/m parallel to **J** and maintained it until such time as the string straightened.)

The velocity of A can be changed instantaneously only by an impulsive tension in the string, impulse J' say, in direction AB; from this, A will acquire velocity u parallel to AB (FIG. 5.4) and no component of velocity at right angles to AB. We have

$$Mu = J' \qquad 5\text{--}8$$

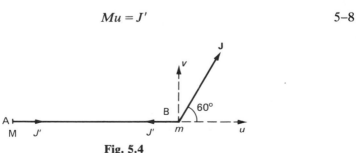

Fig. 5.4

130   An introduction to mechanics and modelling

The velocity of B relative to A can have no positive component in direction AB since the string is inextensible. The actual velocity component of B in direction AB is therefore less than or equal to u; the conventional interpretation of 'inextensible' embraces an assumption of inelasticity which precludes any spring back, or slackening, beyond that which is geometrically or dynamically inevitable. We therefore assume (a modelling hypothesis) that the actual velocity component of B in direction AB is equal to u. From equation 5–7,

$$mu = J \cos 60° - J' \qquad 5\text{–}9$$

and B's initial velocity component at right angles to AB is v where

$$mv = J \sin 60° \qquad 5\text{–}10$$

Equations 5–8, 5–9 and 5–10 contain three unknowns, viz. u, v and J'. Elimination of J' and solution for u, v allows us to conclude:

The initial velocity of A is $\frac{1}{2}J/(M+m)$ in direction AB.
The initial velocity of B is $(\frac{1}{2}J/(M+m), \sqrt{3}J/2m)$, all components having units of m s^{-1}.

Note that the sum of equations 5–8 and 5–9, viz. $(M+m)u = \frac{1}{2}J$, states that the change in the total component of momentum in the x-direction is equal to the x-component of impulse.

5.221   The additive property of mass

If, before collision, two masses m_1, m_2 have velocities $\mathbf{v}_1$, $\mathbf{v}_2$, and after collision they move together with velocity $\mathbf{v}$, then

$$m_1\mathbf{v}_1 + m_2\mathbf{v}_2 = m_1\mathbf{v} + m_2\mathbf{v} = (m_1 + m_2)\mathbf{v}$$

This may be interpreted as follows:

> The composite body formed when two masses m_1, m_2 move together as one, has mass $m_1 + m_2$.

The 'system of particles' model of a rigid body, in Chapter 6, will make use of this scalar property of mass.

Exercise 1

1. Before collision, two masses have velocities 10 m s^{-1} and 0. After collision they have velocities 2 m s^{-1} and 5 m s^{-1} respectively in the same direction. Find the ratio of the two masses.
2. Given $\mathbf{a}_1 = \ddot{x}_1\mathbf{i} + \ddot{z}_1\mathbf{k}$; $\mathbf{a}_2 = \ddot{x}_2\mathbf{i} + \ddot{z}_2\mathbf{k}$, express the collision equation $m_1\mathbf{a}_1 = -m_2\mathbf{a}_2$ as a pair of component equations. State the corresponding result for the overall change in (component) **velocities** of two bodies moving under the action of their mutual forces of gravitational attraction (only).
3. Two skaters approach each other at right angles with speeds 10 km h^{-1} and 12 km h^{-1} respectively. After colliding, they hold on to each other, and their momentum carries them in a direction making an angle of 55° with the original 10 km h^{-1} skater's course. Estimate the ratio of the skaters' masses, stating what assumption you need to make about the forces exerted by the ice.

Mass and force; momentum; forces of nature 131

4 A boy of mass 45 kg chases a truck of mass 175 kg which is moving at 3 m s^{-1}. When the boy is near enough, he takes a running jump on to the truck. If the truck is then observed to have a speed of 3.25 m s^{-1}, estimate the horizontal speed of the boy's running jump. State what assumptions you have made about the forces on the truck.

5 Brian (34 kg) and Charles (40 kg), both on roller skates, stand at rest, facing each other, with Brian holding a 1 kg package. Brian throws the package to Charles with a horizontal component of velocity equal to 10 m s^{-1}; after catching the package, Charles returns it with the same speed relative to the ground. Assuming that horizontal forces between the ground and the skates may be neglected, deduce (a) Charles' final speed, (b) the final speed of Brian, clutching the package.

6 Two aircraft collide head-on in mid-air, and fall to the ground as one body. Before the crash, the more massive aeroplane, A, had a horizontal speed 1300 km h^{-1}, and the lighter one, B, had horizontal speed 620 km h^{-1}. The crash occurred vertically above point O on the ground, and the wreckage reached the (horizontal) ground one minute later, at a distance 10 km from O. Given that the mass of aeroplane A and contents was 2×10^4 kg estimate the mass of aeroplane B and contents. Suggest a factor unmentioned in the above statement of the problem which could affect the accuracy of your estimate. Would inclusion of this factor increase or diminish your estimated value?

7 An excavator travelling at 8 km h^{-1} picks up 20 kg of earth per second. What horizontal force does the excavator apply to the earth which it collects. Could you, from this, estimate the magnitude of the driving force required from the engine to keep the excavator moving? (**Hint:** for the first part, decide what mass of earth you wish to consider, e.g. the mass picked up in any interval of time you care to specify.)

8 A hose pipe is pointed at a nearby vertical wall. The jet of water emerging from the pipe has a cross-section of 8 cm^2 and a horizontal speed of 12 m s^{-1}. Find the force on the wall: (i) if the water does not rebound, (ii) if the water rebounds at 3 m s^{-1}. (**Note:** the mass of 1 cm^3 of water is 1 g.)

9 A particle of mass 2 kg is acted on by a force $14\mathbf{i} - 18\mathbf{j}$ (its magnitude being expressed in newtons). Find the components of its acceleration in the **i**-, **j**-directions. Evaluate the magnitude of the total acceleration, and find the angle it makes with the **i**-direction. (Assume **i**, **j** are perpendicular unit vectors.)

10 The only two forces acting on a particle are equal in magnitude and opposite in direction. What can be said about the motion of the particle?

11 **To every impulse there is an equal and opposite impulse** – true or false?

12 A ball travelling with speed V strikes a fixed, smooth plane. Before impact the direction of travel makes an angle of 70° with the plane. The speed of the ball after impact is $\frac{1}{2}V$. Find the angle which the direction of travel after impact makes with the plane.

13 A ball impinges, as in question 12, on a smooth plane, but the final direction makes an angle of 40° with the plane. Deduce the final velocity.

14 A car of mass 1000 kg tows a caravan of mass 700 kg. Their acceleration up a gradient of 1 in 20 is 12 km h^{-1} per minute; estimate the force then being exerted by the tow-bar on the caravan, stating the assumptions involved in this estimate. What can you say about the magnitude of the driving force provided by the car's engine? (You need not at this stage consider the chain of mechanisms whereby the thrust of the pistons is converted into a forward force on the car's body.)

15 A lift of mass 1000 kg has an upward acceleration of 0.4 m s^{-2}. Given that its motion is opposed by frictional resistances of 500 N, deduce the magnitude of the forces exerted on the lift by the cables.

16 The fictitious economist Dr Nutt has formulated three laws, viz.
Nutt's first law: The know-how of any working group is a constant characteristic of the group.
Nutt's second law: The effectiveness of any working group is directly proportional to the number of cups of tea drunk, and also to the know-how of the group.
Nutt's third law: The wage-level of any group is a direct multiple of its effectiveness.

132 An introduction to mechanics and modelling

(i) Explain, using steps parallel to those used in §§ 4.8 and 5.1 onward for the definition of mass, how Nutt's laws may be used to provide a quantitative definition of know-how.
(ii) State whether, in your answer to (i) you made any assumption not explicitly authorized by the laws as stated.
(iii) Choose a suitably derived British unit of know-how.
(iv) How could the validity of Nutt's laws be tested 'experimentally'?

17 A proton initially has a horizontal velocity of 3×10^6 m s^{-1} *in vacuo*. Its path takes it through a uniform electric field which exerts a downward force of 9×10^{-14} dynes on it. The dyne is the c.g.s. unit of force, $= 1$ cm g s^{-2}.) The mass of the proton is 1.66×10^{-24} gm. Find the proton's horizontal and vertical components of velocity at time t seconds. If the particle emerges from the electric field after $t = 10^{-4}$ seconds, through what angle has its path been deflected? [**Note:** Newtonian mechanics is applicable, and the units given are all absolute units in the same (c.g.s.) system.]

18 Is equation 5–6 a straightforward extension of equation 5–5, or is an additional postulate required? (cf. § 6.1 onwards.)

5.3 Forces of nature

5.31 Gravity

The force on any body due only to the presence of other masses in the universe, is called the force of gravity. The conclusion drawn from many experiments, covering a much wider range of locations than those of our fell-walker in § 4.62, may be stated as a law of nature, or fundamental postulate of our scheme of mechanics. Thus, the acceleration of any body (i.e. of its centre of mass) due only to the presence of other masses in the universe, is **G** where **G** depends only on the position of the test body relative to the other masses of the universe. The force **F'** of gravity on any mass m therefore equals $m\mathbf{G}$ where **G** is a function of position.

$$\mathbf{F'} = m\mathbf{G} \qquad 5\text{–}11$$

5.311 Newton's law of gravitation

This law gives the information necessary for the calculation of **G** if the locations and magnitudes of all the masses exerting appreciable effect on the test mass are known. The law may be stated:

> Every mass M in the universe attracts every other mass m with a force whose magnitude F is proportional to the product of the two masses concerned, and inversely proportional to the square of the distance r between the masses. Thus $F = \mathrm{G}\, Mm/r^2$ where G is a scale-dependent constant (cf. § 1.5) whose value in SI units is 6.7×10^{-11}.

As in the case of Newton's second law, the statement of a relationship true for 'bodies' presents us with difficulty in interpreting the symbol r – to which point of the body does it refer? The cleanest line of approach is to postulate the truth of the law for particles only, and subsequently to equate those particles with infinitesimal elements in a new, continuous model of matter; that is, we use two different models of matter concurrently: thereby theoretical results of great utility can be derived. For two particles, the distance r is unambiguous, and the direction of F is along the line joining the particles (attraction).

If the process of integration is applied to the infinitesimal elements of a spherically symmetric distribution of mass, it is found (see sequel volume) that the law of gravitation for particles implies an identical law of gravitation, viz. $F = GMm/r^2$, between non-overlapping spherically symmetric bodies of finite size, the symbol r now signifying the distance between their centres.

Thus for a particle or spherical distribution of matter just above the surface of the earth, the gravitational pull towards the centre of the earth is of magnitude $m(GM/R^2)$, where M is the earth's mass and R its radius. The constant g introduced in Chapter 4 as a measured acceleration is therefore related to M and R by the equation

$$g = GM/R^2 \qquad 5\text{--}12$$

Relation 5–12 both illustrates the origin of the acceleration of a falling body, and enables GM to be estimated from measurements of g and R.

Because neither the distribution of density nor the shape of the earth is perfectly spherical, formula 5–12 is an approximation: actual values of g vary (by something like 1%) from point to point on the earth's surface.

Note that although in the case of a **spherical** distribution, the appropriate value of r is that measured to its centre of mass, there is no theoretical justification for attributing gravitational significance to the centre of mass (as defined in the next chapter) when the mass distribution is not spherical.

5.312 'Weighing the earth'
The constant of gravitation, G, has been measured experimentally in several different ways. Typically, a very sensitive balance is observed, (a) in the absence, and (b) in the presence of a dense mass close to one of the suspended masses.

Having determined G, and knowing $GM = R^2 \times$ known experimental average value of g, the earth's mass M may be deduced.

The colloquial phrase 'weighing an object' usually means 'determining the mass of the object'.

Exercise 2
Several of these questions are suitable for discussion in class.
1 Given a standard mass and a device for the measurement of force, show how, by making two force measurements at the same location, and using equation 5–11, the mass of an arbitrary test body can be determined.
2 Given a device for finding the ratio of two parallel forces (e.g. the yard-arm balance of FIG. 5.5) show how a single measurement allows the determination of the mass of a test body as a multiple of a standard mass (usually incorporated as part of the balance).

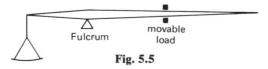

Fig. 5.5

134 An introduction to mechanics and modelling

3. Given that $G = 6.7 \times 10^{-11}$ SI, and that the mean value of the earth's circumference = 40 000 km, find the earth's mass in kg.

4. A spaceship is observed above the atmosphere, travelling directly away from the earth with speed v satisfying $v^2 = 10\,000 + C/x$ where x is the spaceship's distance from the earth's centre. Show that this observation is consistent with free flight under Newton's law of gravitation provided that C has an appropriate value.

5. 'Weighing the sun'. Model the earth's motion as a circular path round the sun, described at constant speed. The distance of the earth from the sun is 1.50×10^8 km. (This can be established by astronomical observation. In practice indirect methods are used, but in effect, the direction of the sun relative to the background of the stars is compared at different observation posts. The angle α shown in FIG 5.6 corresponding to observation posts at opposite ends of the earth's diameter has the value $\alpha = 8.8''$.

Fig. 5.6

The earth's diameter d is 1.276×10^4 km, so the sun's distance can be deduced.) Use the fact that one circuit takes $365\frac{1}{4}$ days to calculate the angular speed (ω) of the earth's centre round the sun, and hence (with the assistance of Newton's law of gravitation $F = GMm/r^2$ and the appropriate form of $\mathbf{F} = m\mathbf{a}$, namely $F = mr\omega^2$, deduce the mass M of the sun. (*Answer*: $M \simeq 2 \times 10^{30}$ kg.)

6. An earth satellite describes a circular path at height h above the earth's surface. Its (uniform) speed is such that each orbit takes 20 hours. Derive a relationship between GM, h and the earth's mean radius R. Do you think that a determination of GM based on this relation should have any advantage over the determination previously suggested in § 5.312?

7. The polar equation of any ellipse (with origin taken at a focus) may be written: $A/r = 1 + B \cos \theta$, where A is a constant related to the size of the ellipse, and B is a constant determining the ratio of the lengths of its major and minor axes. Kepler's analysis of the astronomical data compiled by Tycho Brahe led him to the conclusion that every planet describes an elliptical orbit with focus at the sun; and also that, for each planet, $r^2 \dot\theta$ = constant, whence $d\theta/dt = C/r^2$, say.

From the information given above, deduce the transverse and radial components of force on a planet of mass m (ref. equations 4–4) and verify that they are consistent with Newton's law of gravitation.

8. A planet describes an ellipse as in question 7. Show that its radial component of velocity is zero when $\theta = 0, \pi$. Given that when $\theta = 0$ the planet's speed is V (the sun being regarded as stationary), find its speed when $\theta = \pi$, giving the answer in terms of V and B.

9. (Geometry of the ellipse.) Sketch an ellipse and roughly indicate the position of a focus (a point on the major axis, not at the centre; any such point shown will serve the purposes of this question).

    (i) For what values of θ has r (where $A/r = 1 + B \cos \theta$) its greatest value? Mark the initial line $\theta = 0$ in your diagram.

    (ii) Mark the point L of the ellipse for which $\theta = \pi/2$. The line OL is called the semi-latus rectum, and its length (OL) is usually denoted l. Evaluate the constant A in terms of l.

    (iii) Given l and B, find the total length of the major axis of the ellipse. (**Hint:** the sum of the values of r corresponding to $\theta = 0, \pi$.)

    (iv) By writing $\pi - \theta' = \theta$, show that $l/r = 1 - B \cos \theta'$ represents the same ellipse as does $l/r = 1 + B \cos \theta$. Show that, with an arbitrarily chosen direction of the initial line, the same ellipse is represented by an equation of the form $l/r = 1 + B \cos(\theta + \varepsilon)$.

(v) By considering the value of $1 + B \cos \theta$ as θ increases from 0 to 2π, show that for an ellipse, $|B| < 1$. The value of $|B|$ is called the eccentricity of the ellipse, conventionally written as e. The standard form of the equation of an ellipse is then $l/r = 1 \pm e \cos \theta$.

5.313 *Weightlessness and weight*

It is now possible to enlarge a little on what was said about 'weightlessness' in § 4.3. We shall conclude that the term weightlessness is an expressive but unfruitful description of a phenomenon which can be understood in terms of ordinary Newtonian mechanics.

When a spacecraft is in free non-rotating flight with its engines cut off, the resultant force on the craft is purely gravitational in origin, and thus equal to $M\mathbf{G}$, where M is the mass of the spacecraft; the gravitational force on an object in mid-air within the craft is likewise $m\mathbf{G}$, where m is the mass of the object. Thus in the absence of other forces, m and M have equal accelerations $\mathbf{G}$, i.e. the mass m has zero acceleration relative to the spacecraft, and, in particular, if it starts at rest relative to the spacecraft, it will remain so instead of appearing to fall. The colloquial attribute of 'weightlessness' does not denote a property of the object, but a property of the frame of reference within which that object is observed.

When the spacecraft (constituting the frame of reference) is powered, there will be a relative acceleration between it and the unpowered object. Let $\mathbf{a}'$ denote this relative acceleration, i.e. the part of the spacecraft's total acceleration which is due to the thrust of its engines. Then the acceleration of the unsupported mass m relative to the spacecraft is $-\mathbf{a}'$. Seeing the mass m moving spontaneously (relative to his frame), a naive observer within the craft might interpret the direction of its relative motion as 'downwards', the magnitude of the relative acceleration as a gravitational constant $g' (= |\mathbf{a}'|)$ say, and the quantity mg' as the weight of the object.

The quantity mg' is equal and opposite to the force required to keep the object at rest relative to the spacecraft, and some textbooks of physics adopt this quantity as the definition of the 'weight' of the object in the extraordinary conditions of space travel. So defined, the weight of any body depends not only on its constitution and location, but also on the acceleration of the frame of reference of the observer. Mathematicians do not subscribe to so complicated and so contingent a definition.

The mathematician would refer to mg' as the **apparent weight** in the spaceman's frame of reference. When he refers to 'weight' without qualification, he means the quantity $m\mathbf{G}$ which can be calculated by Newton's law of gravitation, and whose magnitude is well-approximated by mg near the earth's surface. (See § 9.3 for more about frames of reference.)

Assignment

Either explain, using words and equations, why the housewife's concept of weight may be reconciled with the mathematician's concept (by a suitable renaming of units), but not with the physics textbook's concept.

Or model the ascending/descending journeys of a lift by periods of constant acceleration/velocity/deceleration. Discuss the corresponding changes of force

136 An introduction to mechanics and modelling

between the feet of a traveller and the lift floor during an upward/downward journey:

(a) by equations using variables measured in an unaccelerated frame,
(b) by equations using variables measured in the lift as reference frame.

Make your own comments.
Or envisage a man standing at the middle of a long slippery plank which is rotating with constant angular velocity ω about a vertical axis through one end, so that the plank sweeps out a horizontal circle.

(a) Use the first of equations 4–4 to show that the man will start to slip outwards from the centre of rotation.
(b) Suppose that the man has his eyes shut, and is unaware that the plank is rotating, (so to him r is an x-coordinate) and he attributes his outward motion to a pseudo-gravitational pull. Use your answer to (a) to say what value he ascribes to this outward force on himself. Note that this is his modelling of the situation: it is not wrong, but, to the mathematician, it seems more confusing than method (a).
(c) Use equations 4–4 to show that, when $\dot{r} > 0$, the man must cling to the plank in order not to be thrown off sideways. What transverse force must he exert? If he still refuses to believe that the plank is moving, what force components does he now believe to be acting on him?
(d) Discuss the advantages/disadvantages of model (a) and model (c).

Note: the pseudo-force postulated in (b) is called the **centrifugal** component, and the additional component in (c) the **Coriolis** component of the fictitious force.

5.32 Normal reaction
The forces of contact between two rigid bodies can be fully discussed only when the rotational equations of motion, as well as the translational equation $M\mathbf{a} = \mathbf{F}$, become available. The main principles, however, are exposed in the following discussion of the contact forces between a particle and a solid body: a simple extension to two bodies of finite size is given at the end of the section.

Consider a particle (FIG. 5.7a–c), stationary or in motion, in contact with a solid body, which itself may be either stationary or moving. The word 'solid' is interpreted as meaning that the particle may not penetrate the interior of the body, i.e. that the body will exert on the particle just such a normal component of force, R say, as is sufficient to prevent penetration. The interior of the body is forbidden; the direction of R is that of the **outward** normal.

A perfectly general mathematical representation of this condition is too complicated to be useful. Particular examples are far more helpful. For the fixed plane and fixed cylinder whose sections are represented in FIG. 5.7b–c, the normal force R required to maintain contact with the stationary surface, satisfies (for FIG. 5.7b)

$$R + Y = m\ddot{y} = 0 \qquad\qquad 5-13$$

where Y is the sum of the y-components of all the other forces acting on the

Mass and force; momentum; forces of nature 137

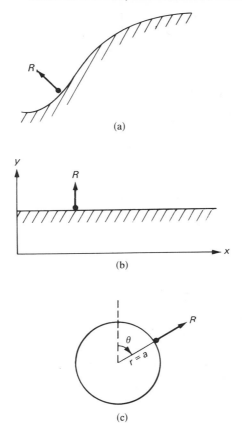

Fig. 5.7

particle, and (for FIG. 5.7c)

$$R + R' = m(\ddot{r} - r\dot{\theta}^2)$$
$$= -ma\dot{\theta}^2 \qquad 5\text{--}14$$

where R' is the algebraic sum of the radial components of all the other forces on the particle.

The value of R given by equations 5–13 or 5–14 is the value required to maintain contact.

> If the necessary value is **positive**, i.e. if a push force is required, that value will be provided by the solidity of the body; and that value of R is the actual value of the normal reaction.

But if the value of R required by equation 5–13 or 5–14 is negative, representing a pull force or attraction, it will **not** normally be provided by a surface; only if the surface is sticky or magnetic or in some other way exceptional can it resist separation. The usual conclusion is that,

> if the R value required for contact is **negative**, separation will take place, and the actual value of the normal reaction is zero.

Suppose now that FIG. 5.7b represents a particle on a plane which is itself accelerating downwards with acceleration $\ddot{y} = -20$ m s^{-2}. Then, for contact

$$R - mg = m\ddot{y} = -20m$$

i.e.

$$R = m(g-20) \simeq -10m, \quad \text{since} \quad g \simeq 10 \text{ m s}^{-2}$$

We conclude that unless the plane is sticky, contact will be lost (the plane accelerating away from the freely falling particle) and the actual value of the normal reaction will be zero.

5.33 Friction

A similar procedure is possible when the tangential component of the interaction force is under consideration. We start by finding what force F in the tangential direction would be required to ensure that the particle does not slide relative to the surface. To describe the surface as 'smooth' is to say straight away that no such force could be provided.

If the surface were described as 'perfectly rough' we should interpret this to mean that, however large the required value of F, it would be supplied by a rub force between the plane and the particle. No real surface, of course, is perfectly rough: in any particular context the description means 'you may assume that no slipping occurs'.

Real surfaces lie between the two extremes of 'smooth' and 'perfectly rough', and, to indicate that neither of those over-simplifications is acceptable in a problem under discussion, the surface is customarily referred to as 'rough', i.e. not smooth.

> A rough surface exerts on any body with which it is in contact a force whose tangential component is just sufficient to prevent sliding, **provided** that the magnitude required, F, is less than or equal to some limiting value F_{max} which itself can be estimated in terms of more readily calculable quantities.

Experiment shows that, at a given point contact with a surface of given material, the value of F_{max} increases as the particle and surface are pushed together, i.e. the value of F_{max} increases as the normal reaction force increases. The empirical relation

$$F_{max} = \mu R$$

where μ depends only on the materials of which the particle and surface are made, has proved to be a very useful one. For purposes of calculation, the uncertainties in μ values are an acceptable price to pay for so handy a means of estimating F_{max} and hence F. The quantity μ is called the 'coefficient of friction' between the materials concerned.

TABLE 5.1 Coefficients of friction

Substances in contact	Coefficient of friction
Wood/wood	0.2–0.5
Wood/metal	0.2–0.6
Metal/metal	0.15–0.3
Smooth/greased surfaces	0.03–0.1
Leather/wood	0.25–0.4
Leather/metal	0.15–0.6
Masonry components	0.5–0.7
Earth/earth	0.25–1.0

NOTE: The figures indicate the expected range.

Table 5.1 can be of use when only orders of magnitude are of interest. Estimates of greater precision may be possible if μ is determined experimentally in a situation closely resembling that to which the calculations refer.

Summary of method: to determine whether slipping occurs or to find the magnitude of the frictional component of interaction for bodies starting from relative rest, we proceed as follows.

Find the value of F required to prevent slipping; if this value lies in the range

$$-\mu R < F < \mu R \qquad 5{-}15$$

then there will be no slipping, and F is the actual tangential component of the interaction; if, however, $F > \mu R$, then slipping will take place, and in this case the actual value of the tangential component of reaction will be of magnitude F_{sliding}. Experiment shows that, like F_{max}, the quantity F_{sliding} depends primarily on the magnitude of R and the nature of the materials composing the bodies in contact. Sometimes there is considerable velocity dependence as well. For some purposes the approximation

$$F_{\text{sliding}} = \mu' R \qquad 5{-}16$$

where μ' is the coefficient of sliding friction, is adequate. In order to understand stick–slip phenomena, e.g. the sticking drawer which comes out with a rush, it should be recognized that μ' is less than μ for many materials. The difference between μ' and μ is often small, and, except for stick–slip situations, F_{sliding} may be estimated under the assumption $\mu' \simeq \mu$ (whence $F_{\text{sliding}} \simeq F_{\text{max}}$) in conjunction with Table 5.1.

If the particle is originally in motion relative to the surface, $F = F_{\text{sliding}}$ until it comes to relative rest. Thereafter inequality 5–15 applies.

The above account refers to a point particle in contact with a surface. That choice was made because the equations of motion of a particle are readily available to us, so the mathematical consequences of the modelling of friction are easy to demonstrate. The forces between two extensive bodies at a point of contact are of identical nature, but we cannot yet pursue the mathematical detail because we have not yet derived the equations of angular motion of rigid bodies.

140 An introduction to mechanics and modelling

For non-rotating bodies with a point of contact no addition is needed to the above account.

When bodies of finite size have contact over a whole area, calculations are simple if that area is plane. Experiment shows that, given plane surfaces of contact, the resultant tangential (frictional) component, F, of interaction is related to the resultant normal component R in just the same way as for a point contact, i.e. F takes the value required to prevent slipping provided the magnitude required $|F| \leq \mu R$, where the value of μ is given by Table 5.1, i.e. μ is independent of the area of the surface of contact.

5.331 Example 1
A plank is required as a ramp whereby people may get up a one metre step (FIG. 5.8). What is the least length of wood you would consider suitable for the purpose?

Unless a person can stand on the plank without slipping, he/she will certainly be unable to ascend. As a first approach, the investigator may therefore choose to seek a formula giving the greatest angle of inclination of the plank consistent with no slipping when the person is stationary. The plank will be modelled as rigid, unbending.

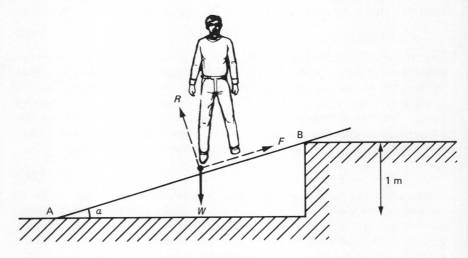

Fig. 5.8

Let T represent the point (or area) of the person's body which touches the plank (FIG. 5.8), and W his body weight. Writing α for the angle the plank makes with the horizontal, the equations of equilibrium of the contact point P are:

$$\left. \begin{array}{l} F - W \sin \alpha = 0 \\ R - W \cos \alpha = 0 \end{array} \right\} \qquad 5\text{--}17$$

Since $R = W \cos \alpha > 0$, this is the actual value of R (contact maintained). The value $F = W \sin \alpha = (\tan \alpha)\, R$ is the actual value only if $\tan \alpha < \mu$, otherwise

Mass and force; momentum; forces of nature 141

slipping occurs. Thus the required greatest angle of inclination consistent with a person being able to stand on the plank is

$$\alpha = \tan^{-1} \mu.$$

Actual values of the angle require values of μ. The author chooses to consider the least leather/wood value of Table 5.1, viz. 0.25. The corresponding estimate of α is 14°. Since the step height, 1 m, $= \sin \alpha \times$ slant length, the estimate of the length AB in FIG. 5.8 is cosec 14°, or rather more than 4 metres.

This is a starting point, but the author would not be happy to make a recommendation on this alone. The possibility of slipping at points A and B needs investigating, as does the effect of the plank-walker's acceleration from rest when he steps forward.

5.332 Example 2
(The artificiality of the statement of the following problem indicates to the reader that any real-life situation to which it may refer has already been idealized/modelled up to the point where mathematical treatment takes over.)

A particle is projected horizontally from the highest point of a sphere, with speed V. Given that the sphere's radius is a, and the coefficient of (sliding) friction between the particle and the sphere is μ, discuss the subsequent motion. The appropriate diagram is shown in FIG. 5.9. For initial maintenance of contact, $mV^2/a = mg - R$, requiring $R = mg - mV^2/a$, so $R > 0$ only if $V^2 < ag$.

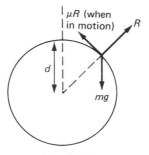

Fig. 5.9

We conclude: if $V^2 > ag$, the particle moves as a free projectile; if $V^2 < ag$ the motion proceeds on the sphere as follows. While slipping is maintained,

$$ma\dot{\theta}^2 = mg \cos \theta - R \quad \text{(radial equation of motion)}$$

$$ma\ddot{\theta} = mg \sin \theta - \mu R \quad \text{(tangential equation of motion)}$$

$$= mg \sin \theta - \mu(mg \cos \theta - ma\dot{\theta}^2)$$

whence

$$a\ddot{\theta} - \mu a\dot{\theta}^2 = g(\sin \theta - \mu \cos \theta) \qquad 5\text{-}18$$

Note that in principle m, a, μ and g are known quantities, and t is an independent variable describing the course of time, so that, in proceeding from the equations

142 An introduction to mechanics and modelling

of motion to equation 5–18, we have reduced two equations in two unknowns θ, R to one equation in one unknown. Equation 5–18 applies so long as the particle both remains in motion (for $F = \mu R$ to apply), and remains in contact with the surface (for $a\ddot{\theta}$ and $-a\dot{\theta}^2$ to represent the acceleration components); i.e., so long as both $\dot{\theta} > 0$ and $a\dot{\theta}^2 < g \cos \theta$.

Equation 5–18 can be integrated by one of the standard methods of Chapter 7 (using equation 7–1 as in the treatment of equation 7–36),

$$a\dot{\theta}^2 = \frac{V^2}{a} \exp(2\mu\theta) + \frac{2g}{4\mu^2 + 1}[(1 - 2\mu^2)(\exp(2\mu\theta) - \cos \theta) - 3\mu \sin \theta]$$

5–19

From equation 5–19 we may deduce the velocity of the particle at each position, namely $a\dot{\theta}$ in the tangential direction, so long as this mode of motion continues. A different mode of motion takes over when either of the conditions $\dot{\theta} > 0$ or $a\dot{\theta}^2 < g \cos \theta$ would be violated by continuance in the mode of equation 5–19. To find out what happens, we calculate which of $\dot{\theta} = 0$ and $a\dot{\theta}^2 = g \cos \theta$ occurs first, according to equation 5–19.

Numerical conclusions depend on the numerical values of V, μ etc. To illustrate the procedure we consider $\mu = \frac{1}{2}$, and $V^2/a = \frac{1}{8}g$. Equation 5–19 then reduces to

$$a\dot{\theta}^2 = \tfrac{1}{4}g[2.5 \exp \theta - (2 \cos \theta + 6 \sin \theta)]$$

5–20

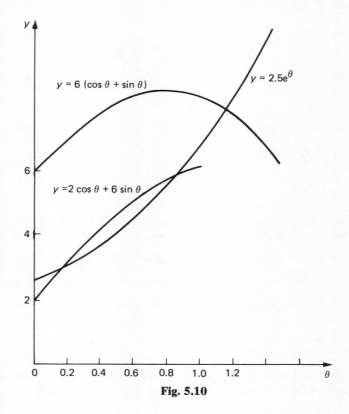

Fig. 5.10

The value of θ for which equation 5–20 gives $\dot{\theta} = 0$ may be found graphically, as in FIG. 5.10 where interest focuses on the first point of intersection of the curves $y = 2.5 \exp(\theta)$ and $y = 2 \cos \theta + 6 \sin \theta$. This first intersection occurs at $\theta = 0.16$ (radians).

The value of θ for which equation 5–20 gives $a\dot{\theta}^2 = g \cos \theta$, i.e. gives the transition point ($R = 0$) between $R > 0$ and $R < 0$, may be similarly found. When the right-hand side of equation 5–20 is equated to $g \cos \theta$, so that $R = 0$,

$$4 \cos \theta = 2.5 \exp(\theta) - (2 \cos \theta + 6 \sin \theta)$$

i.e.

$$2.5 \exp(\theta) = 6(\cos \theta + \sin \theta) \qquad 5\text{--}21$$

The point of intersection of the curves $y = 2.5 \exp(\theta)$ and $y = 6(\cos \theta + \sin \theta)$ see FIG. 5.10, occurs at $\theta \simeq 1.15$ (radians).

Since the value of θ for the cessation of motion is less than the value of θ at which the particle would leave the surface, we conclude, equation 5–20 is valid only up to $\theta = 0.16$ radians $= 9.2°$ at which angle the particle comes to rest on the sphere.

The student should note that equations (such as 5–21) for which it would be difficult to find algebraic solutions, are not therefore insoluble: graphical or numerical methods will give practical answers.

Exercise 3

1. A car of mass M crosses a hump-backed bridge at 30 km h^{-1}. Given that the radius of curvature of the bridge at its highest point is 40 m, find, as a multiple of the car's weight, the magnitude of the normal reaction R between the road and the car at that highest point. Itemize weaknesses of the model.
2. A dip in the road has radius of curvature equal to 40 m at its lowest point. Find the magnitude of the normal reaction between the road and a car (weight W) travelling at 30 km h^{-1} at the lowest point.
3. The motorist in questions 1 and 2 attempts to accelerate at the highest/lowest point. In which of the two cases are his wheels more likely to slip? (**Hint:** to avoid deep waters, restrict yourself to 4-wheel drive.)
4. A box is placed on a plane inclined at 30° to the horizontal. The coefficient of friction between the box and the plane is $\frac{1}{2}$. Show that the subsequent acceleration of the box is constant, and hence deduce the time taken for it to move 1 m, starting from rest.
5. (An elementary but useful exercise; work carefully.) In FIG. 5.11, a particle of mass m kg is in contact with a plane OX, the coefficient of friction at the contact point

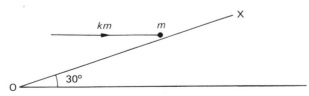

Fig. 5.11

being $\mu = \frac{1}{3}$. In addition to gravity and contact forces, the mass is acted on by the horizontal force km (newtons). Using the approximation $g = 10$, find for (i) $k = 20$, (ii) $k = 8$, (iii) $k = 3$, and (iv) $k = 1.5$, the friction force between the plane and the particle, and the position and velocity of the mass after 1 second starting from rest.

6 A particle of mass m is free to move on a rough plane inclined at angle α to the horizontal. The coefficient of friction between the particle and the plane is $\mu < \tan \alpha$. Show that in the absence of any applied forces the particle will slide down the plane. Find the least force that can be applied (i) parallel to the plane, (ii) horizontally, (iii) in any direction, to hold the particle at rest.

7 A mass M is placed on a smooth horizontal table, and connected by a light inextensible string passing over a small smooth pulley at the edge to a mass m hanging freely. Find the tension in the string and the acceleration of each particle. (**Hint:** the forces are identified in FIG. 5.12, and equations of motion are written down separately for mass M and for mass m, noting that the same magnitude $\ddot{x}$ which stands for the vertical acceleration of the hanging mass measures the horizontal acceleration of the mass M.)

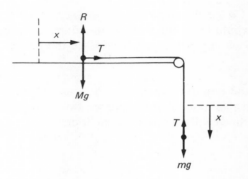

Fig. 5.12

8 The two inclined faces of a fixed wedge make acute angles α and β respectively with the horizontal third face. The inclined faces are equally rough, and on them rest two particles of masses m_1 and m_2 respectively. The particles are connected by a light inextensible string which passes over a small smooth pulley at the edge where the two inclined faces meet. Given that $\alpha > \beta$ and that the friction is insufficient to prevent motion, show that each particle has an acceleration of magnitude a where

$$(m_1 + m_2)a = (m_1 \sin \alpha - m_2 \sin \beta - \mu m_1 \cos \alpha - \mu m_2 \cos \beta)g.$$

9 Identify the modelling assumptions about the nature of friction which were needed to solve question 5 using only the information given. Discuss the modifications which would be needed
(i) if $\mu_{\text{sliding}} = \frac{1}{4}$,
(ii) if $\mu_{\text{sliding}} = \frac{1}{8}v$
where v is the particle's speed.

10 A bead of mass m is threaded on to a smooth straight wire inclined at 60° to the horizontal. The whole wire moves horizontally in the vertical plane containing its length. Given that the vertical component of the bead's acceleration is $4g/3$ downwards, deduce the magnitude of the contact force exerted by the wire on the bead. What is the magnitude of the wire's acceleration?

11 A second wire, whose inclination is the same as the wire in question 10, has coefficient of friction $\mu = \frac{1}{4}$ with the bead threaded on it. For what range of horizontal accelerations of the wire will the bead remain at rest relative to the wire?

Mass and force; momentum; forces of nature   145

12 A horizontal platform, e.g. the floor of a lift, or helicopter, accelerates downwards (i) at $2\,\text{m s}^{-2}$, (ii) at $6\,\text{m s}^{-2}$, (c) at $20\,\text{m s}^{-2}$. On this platform initially is a mass of 7 kg, which is acted on at all times by a horizontal force of magnitude 10 N. Using the approximation $g = 10$, and given that the coefficients of static and sliding friction between the mass and the floor are respectively $\frac{1}{4}$ and $\frac{1}{5}$, find the friction force exerted by the platform on the mass in each of the situations (i), (ii) and (iii).

13 The experiment described in § 5.332, Example 2, is repeated with $\mu = \frac{1}{2}$ and a different speed of projection, all other conditions being unchanged. The new speed V satisfies $V^2/a = \frac{1}{4}g$. Show that in this case the particle will leave the sphere, and determine the corresponding value of θ.

14 A small mass slides down a great circle on the surface of a fixed sphere, centre O, radius 0.2 m, starting at the highest point A. When the mass is at the point B such that $\angle AOB = 25°$, its speed is $1\,\text{m s}^{-1}$. Evaluate the normal component of reaction between particle and sphere (in terms of an unknown constant which you must identify). If the coefficient of friction is 0.05, what is the magnitude of the particle's angular acceleration $\ddot\theta$ at point B?

15 A light inextensible string AB of length 0.5 m has a particle attached to end B. The particle moves in a circle on a smooth horizontal table with the string taut and point A (i) at fixed point O of the table, (ii) held at a point 0.3 m above O. In each case the particle describes its circular path in 2 seconds at a uniform speed. Prove that the tensions in the string AB in situations (i) and (ii) are equal, but that the normal reactions between the mass and the table are (approximately) in the ratio 10:7.

With the geometry of situation (ii), what is the least number of revolutions per second needed to ensure that contact with the table is lost? Describe the motion which ensues if this number of revolutions is exceeded, substantiating your statement with applicable equations of motion.

16 Two small masses m, $2m$ slide in a smooth groove in the line of the x-axis, the mass m being to the left of the mass $2m$ ($x_1 < x_2$). Initially ($t = 0$) the two masses are in contact and a force is applied to mass m which ensures that, until such time as contact with the $2m$ mass is lost, the position of mass m is given by $x = 2 - \cos 3t$. Find the magnitude of the force applied to m and express it in terms of x. Describe the motion of the $2m$ mass after contact is lost. Given that, after loss of contact with $2m$, the force applied to m is the same function of x as formerly, write down the equation of motion appropriate to the subsequent motion of m.

17 The shaded areas, A, B and C, in FIG. 5.13 represent three flat pieces of metal lying on a smooth horizontal table. Forces which are not shown in the diagram are applied to disc A, causing it to rotate; C is held stationary, and B moves under the influence of contact forces only. No contacts are broken. On a diagram show all the unknown contact forces acting on A, B and C respectively: (i) on the supposition that A rotates, but the contacts are smooth; (ii) on the supposition that the contacts are not smooth, but that they are not rough enough at any contact to prevent slipping;

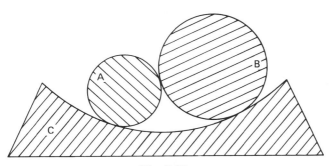

Fig. 5.13

146 An introduction to mechanics and modelling

(iii) on the supposition that A rotates, but the contacts between A, C and A, B are rough enough to prevent slipping at these points. (This is an exercise in selecting a minimum sufficient number of unknown symbols.)

In questions 18, 19 and 20, the configuration is best specified by one ordinary and one relative coordinate; then use result 2–6(b).

18 A wedge of mass M and angle α is placed on a rough horizontal plane, the coefficient of friction being μ. A smooth particle of mass m is placed carefully on the inclined face of the edge. Show that, if the wedge moves, its acceleration will be

$$g(m \cos \alpha \sin \alpha - \mu m \cos^2 \alpha - \mu M)/(m \sin^2 \alpha - \mu m \sin \alpha \cos \alpha + M)$$

(Two equations of motion are available for the wedge, and two for the particle.)

19 A smooth plane is inclined at angle α to the horizontal. On it is placed a wedge of angle α and mass M, in such a way that the upper face of the wedge is horizontal. A smooth mass m is placed on that horizontal upper face. Prove that the acceleration of the particle is $(M+m)g \sin^2 \alpha/(M+m \sin^2 \alpha)$, and find the normal reaction between the wedge and the plane.

20 A particle P of mass m rests on a rough horizontal table whose coefficient of friction with the mass is μ. A light inextensible string is attached at one end to the mass m; it then passes over a smooth small fixed pulley at the edge of the table, under a smooth small movable pulley of mass m, over a second fixed, small, smooth pulley, and finally passes vertically down to its point of attachment to a second particle of mass m, hanging freely. All portions of string not in contact with any pulley are vertical, except the first horizontal section. Prove that if $\mu > \tfrac{3}{5}$, P will not move, and that if $\mu < \tfrac{3}{5}$, the hanging particle has an acceleration of $(3+\mu)g/6$.

21 'A man can walk more easily over a snowfield if he is wearing snow-shoes.' Discuss the laws of friction in this context. (**Note:** you are on your own; there is no orthodox opinion to refer to.)

5.34 Coefficient of restitution

An empirical law which is sometimes of use states: the velocity of separation of two bodies after an impact is a multiple e of their velocity of approach immediately before impact. By the 'velocity of separation' is meant the magnitude of the normal component of the relative velocity; thus, in FIG. 5.14a, which shows

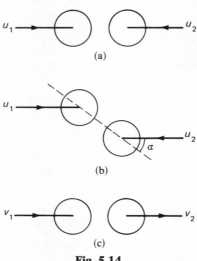

Fig. 5.14

Mass and force; momentum; forces of nature

the direct approach of two spheres, the velocity of approach is $u_1 + u_2$; but in FIG. 5.14b, the relative velocity makes an angle α with the normal at the point of contact, so the velocity of approach is $(u_1 + u_2) \cos \alpha$. If FIG. 5.14c represents the system before impact, the velocity of approach is $v_1 - v_2$; but if it represents the system after impact, it gives the velocity of separation as $v_2 - v_1$.

The quantity e is called the 'coefficient of restitution'. It is approximately constant for all impacts of two given bodies of regular shape, but, since it depends on the shape and other factors as well as the materials of the impinging bodies, useful tables of experimental values are not readily available. Sample values have, however, been given by D. P. Thomas in *Mathematics Applied to Mechanics* (see Table 5.2).

TABLE 5.2 Coefficients of restitution

Substances in contact	Coefficient of restitution
'Cool' squash ball and racquet	0.25
'Warm' squash ball and racquet	0.32
Two wooden balls	0.50
Two steel balls	0.55
Golf ball and club (professional golfer)	0.74
Golf ball and club (handicap golfer)	0.79
Two ivory balls	0.89
Two synthetic rubber balls	0.92

It should be noted that the golf ball and club figures apply to high speed drives; quite different values would apply to putting shots. The value given for 'two steel balls' of unspecified steel and dimensions, appears rather low for consistency with the motion observed in a popular executive toy consisting of a row of contiguous steel balls suspended by parallel threads, wherein the impact of one of the end balls on the remaining system sets up a repetitive motion whose amplitude diminishes more slowly than the value $e = 0.55$ would suggest. (Dr Thomas himself now believes this value 0.55 to be misleading, although he was quoting from a reputable Russian source.)

All of which amounts to a recognition that the law

$$\text{velocity of separation} = e \times \text{velocity of approach} \qquad 5\text{--}22$$

(in which e is constant) is an approximation useful only in a restricted range. The appropriate restrictions vary from case to case.

Collisions for which the approximation $e = 0$ is acceptable are called inelastic. When the approximation $e = 1$ is acceptable, the collision is called perfectly elastic.

5.4 Empirical laws, accuracy and approximation

The traditional modelling of the forces exerted by springs and elastic strings will be described in § 7.23, and the modelling of the forces of air and water resistance

148 An introduction to mechanics and modelling

will be discussed in section 7.4. As in the modelling of friction by $F = \mu R$, linear laws of force, viz. $F = \lambda x$, $F = kv$, are proposed, and have their use in limited ranges of x, v values respectively. Outside the appropriate ranges – which can be determined only by experiment, not by mathematical logic – the approximations are unsatisfactory. Sometimes simple alternative laws are available, e.g. $F = Kv^2$ for fluid resistance at higher velocities: only reference to tables of physical data can settle the issue.

> The student in search of physical data may start by consulting any one of the following, in the stated or any other edition,
> Kaye, G. W. C. and Laby, T. H., *Tables of Physical and Chemical Constants* (Longmans 1973);
> ed. Weast, R. C., *CRC Handbook of Chemistry and Physics* (CRC Press, Florida, USA 1979);
> ed. Washburn, E. W., *NRC, International Critical Tables* (McGraw-Hill 1926–33);
> ed. Bolz, R. E. and Tuve, G. L., *CRC Handbook for Applied Engineering Science* (CRC, Ohio, USA 1973).

The usefulness of many standard types of approximation is well attested, provided the proper scrutiny is made of the range of validity. In the first place the reader may allow himself to be guided by tradition.

Apart from inadequacies of the model (just discussed), error may arise from other sources, viz.

(1) approximations occurring during the course of the mathematical analysis;
(2) inaccuracies resulting from graphical or numerical evaluation of a correctly derived final mathematical expression (as in Example 2 of § 5.332);
(3) mistakes.

Of these three sources, errors from (1) may be catastrophic. By approximating at too early a stage of calculation, errors may snowball and sometimes even swamp the effect which is being investigated. Errors springing from source (2) are usually less objectionable; their effect is relatively easy to estimate, and the risk of drawing false conclusions is slight. Mistakes will occur sooner or later in anyone's work. If undetected, the work may be worse than useless. The student must acquire sufficient self-knowledge to appreciate the amount of re-checking that each piece of work needs before it is likely to be mistake-free.

The reader is advised to look critically at every completed piece of work (except, perhaps the more trivial practice exercises), first to make sure that no avoidable source (1) or source (3) errors persist, and secondly, to estimate – however roughly – the error from all remaining sources. The author regrets that space does not allow the latter part of this prescription to be set out in detail for problems described in the pages of this book.

Summary

Mass is essentially a ratio of two velocity magnitudes, with the standard mass as unit.

Force = mass × acceleration

Momentum ($\Sigma m\mathbf{v}$) is conserved in the absence of external forces.

Newton's law of gravitation states $F = GMm/r^2$

Normal reaction R is of just the magnitude needed to prevent penetration, friction F is of just the value required to prevent slipping, i.e. relative tangential motion at the point of contact provided this value is less than μR. Otherwise $F = \mu' R$. The attribution of constant values to μ, μ' represents a crude approximation.

Assignment
Either discuss the following definition of mass: 'Mass is the quantity of matter in a body, and its magnitude is determined by use of a chemical balance'.
Or find an empirical relation between breathing rate and speed of walking/running. (Time yourself indoors or outdoors or both, as convenient.) Can you propose a simple relation in any range? Can you widen the range of applicability, or propose a different relation to accommodate a different range of speeds?

6
A system-of-particles model; centre of mass; motion of a rigid body (in two dimensions)

6.1 Modelling a body as a system of particles

So far, our problem-solving has involved only the motion of a single point of any body under consideration, namely its centre of mass. This is sometimes rather misleadingly referred to as **single-particle** modelling or particle mechanics. To take into account the body's extension, we now introduce a new, **system-of-particles** model. The theoretical development of the new model followed through in this chapter should be regarded, not as hard dogma, but as a journey of exploration which will bring handsome rewards.

The rewards of the journey are: (i) a new appreciation of some already known results (e.g. the relation between $m\mathbf{a} = \mathbf{F}$ for a particle and $M\mathbf{a} = \mathbf{F}$ for a body of finite extent); (ii) the derivation of quantitative formulae, (especially those for the position of the centre of mass and for the angular motion of a body); and (iii) consolidation of understanding of the nature of mechanical interactions in complex systems. The simple $m\mathbf{a} = \mathbf{F}$ model will have been much enriched, and the reader should gain the sense that he/she understands mechanics up to the current level. It would be a mistake to assume that any model – even this rather good one – mirrors all aspects of mechanical behaviour. The use of different models leads to different, albeit related, branches of mechanics, for example continuum mechanics or quantum mechanics.

6.11 Point particle model of a rigid body; internal and external forces

We conceive the body as consisting of a finite number of point particles of finite mass $m_1, m_2, m_3, \ldots$, etc. as in FIG. 6.1.

The particles of such a system would fall away from one another like grains of dry sand did we not postulate the existence of mutual forces between them. (Note how the model, by creating artificial gaps, has pointed unequivocally to the need to recognize forces other than those applied by external means.)

Let particle m_2 exert a force $\mathbf{P}_{1(2)}$ on particle m_1. We note that $\mathbf{P}_{1(2)}$ is intrinsically unmeasurable, for its magnitude could be deduced only by observing the forced motion of m_1 under its action; but such untrammelled motion is

A system-of-particles model 151

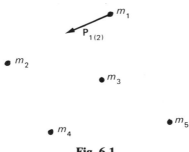

Fig. 6.1

forbidden so long as m_1 is a particle of a rigid body. In order that we shall not be landed with an indefinitely large number of unrelated and therefore uneliminable unknowns, some physical relation must be postulated between the mutual forces in our model. Newton's third law – though not directly verifiable for forces deep in the interior of rigid bodies – provides such a relation. We assume its validity for our model, i.e. we postulate that if particle m_2 exerts force $\mathbf{P}_{1(2)}$ on particle m_1 then particle m_1 exerts force $\mathbf{P}_{2(1)} = -\mathbf{P}_{1(2)}$ on particle m_2. In general let particle m_j exert force $\mathbf{P}_{i(j)}$ on particle m_i: then all forces of form $\mathbf{P}_{i(j)}$ are called **internal** forces between the particles of the body; and

> the vector sum of all the internal forces on the body is zero,

the forces cancelling in pairs, because

$$\mathbf{P}_{i(j)} + \mathbf{P}_{j(i)} = \mathbf{P}_{i(j)} - \mathbf{P}_{i(j)} = \mathbf{0}.$$

Forces which are not internal forces (e.g. a force $\mathbf{F}_{1(K)}$ exerted on m_1 by means of body K external to the rigid body under consideration) are called **external** forces or **applied** forces. As well as pushes, pulls and rubs applied to the surface of a rigid body the term 'external force' embraces gravitational or electromagnetic forces exerted on the rigid body by reason of an external mass or apparatus.

If an **external** force (FIG. 6.2) is applied to a rigid body the equal and opposite reaction will be applied not to a particle of the body but to the external object or apparatus: it will not be included in any summation of forces acting on particles of the body, and there will be no automatic cancellation of external forces in summations of this kind.

6.12 Equations of motion of the particles of a rigid body
Newton's second law makes available to us an equation of the form

'mass × acceleration = total force on the particle'

for each constituent particle of the body.

Let m_s be the mass and $\mathbf{r}_s$ the position vector of the sth particle of the body $(s = 1, 2, 3, \ldots, s, \ldots)$, and let $\mathbf{F}_s$ be the resultant of all the external forces

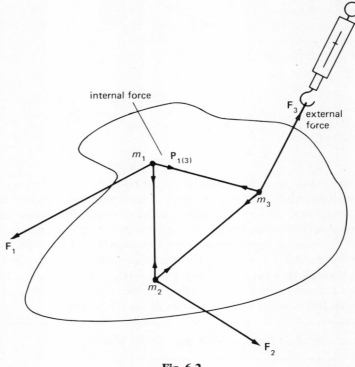

Fig. 6.2

applied to mass m_s. Then we have the system of equations

$$m_1\ddot{\mathbf{r}}_1 = \mathbf{F}_1 + \text{internal forces acting on particle 1}$$
$$m_2\ddot{\mathbf{r}}_2 = \mathbf{F}_2 + \text{internal forces acting on particle 2}$$
$$m_s\ddot{\mathbf{r}}_s = \mathbf{F}_s + \text{internal forces acting on particle } s$$

6–1

and so on.

Given equations 6–1, every additional assumption which we make about the nature of internal forces opens up an opportunity to manipulate and combine these equations in such a way as to yield a corresponding equation about motion and applied forces, from which reference to unknowable internal forces has been eliminated. Thus the assumption of the validity of Newton's third law for internal forces tells us that the sum of all the internal forces on all the right-hand sides of equations 6–1 is zero and hence by adding

$$m_1\ddot{\mathbf{r}}_1 + m_2\ddot{\mathbf{r}}_2 + m_3\ddot{\mathbf{r}}_3 + \cdots = \mathbf{F}_1 + \mathbf{F}_2 + \mathbf{F}_3 \cdots$$
$$= \mathbf{F} \text{ say}$$

6–2

where for convenience, **the symbol F is used for the vector sum of all the external forces applied to the body regardless of their points of application.** Equation 6–2 would apply equally to any group of particles, whether held together rigidly or not. We may anticipate that when it is further known that the particles move together in a rigid framework, the left-hand side of equation 6–2 will provide information about the motion of that framework.

We note that though the particles move, their masses remain constant, so

$$m_1\ddot{\mathbf{r}}_1 + m_2\ddot{\mathbf{r}}_2 + m_3\ddot{\mathbf{r}}_3 + \cdots = \frac{d^2}{dt^2}(m_1\mathbf{r}_1 + m_2\mathbf{r}_2 + m_3\mathbf{r}_3 \cdots) = \mathbf{F} \quad \text{6–2(a)}$$

6.2 Interpretation of $\Sigma m_i r_i$; centre of mass

Our aim is to replace reference to numerous position vectors $\mathbf{r}_1, \mathbf{r}_2, \mathbf{r}_3 \ldots$ in equation 6–2(a) by a reference to the position vector of a single point of the body.

Let A be any point of the body. Let $\mathbf{r}_A$ denote the position vector of A, and write

$$\mathbf{r}_1 = \mathbf{r}_A + \mathbf{R}_1$$
$$\mathbf{r}_2 = \mathbf{r}_A + \mathbf{R}_2$$
\quad 6–3

i.e. $\mathbf{R}_1, \mathbf{R}_2, \ldots$ are the position vectors of the particles $m_1, m_2, \ldots$ relative to the point A of the body. We note that, unlike $\mathbf{r}_1, \mathbf{r}_2, \ldots$ the *lengths* of $\mathbf{R}_1, \mathbf{R}_2, \ldots$ remain unchanged during any motion of the rigid body. Moreover, though the orientations of $\mathbf{R}_1, \mathbf{R}_2$ alter during the motion, the magnitude of the angle between the directions of $\mathbf{R}_1, \mathbf{R}_2$ (FIG. 6.3a) remains unaltered, for the particles m_1, m_2 and the point A are connected in a rigid triangle.

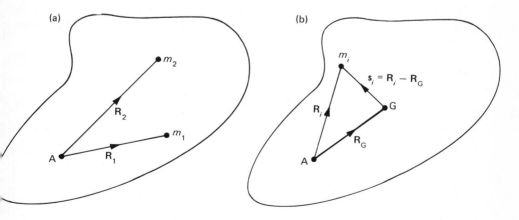

Fig. 6.3

From the above it follows also that the lengths of the vectors $m_1\mathbf{R}_1$ and $m_2\mathbf{R}_2$ and also the magnitude of the angle between them remain unchanged whatever the motion of the rigid body. Hence the magnitude of the vector sum $m_1\mathbf{R}_1 + m_2\mathbf{R}_2$ remains constant during the motion, and its direction bears a constant relationship to the directions of $\mathbf{R}_1$ and $\mathbf{R}_2$. Thus any point whose position vector **relative to A** is a constant multiple of $m_1\mathbf{R}_1 + m_2\mathbf{R}_2$, is a point moving with the body as though it were a fixed point in that body. (cf also questions 12 and 13 of Exercise 1 below). By the same argument it follows that

$$\sum m_i \mathbf{R}_i \equiv m_1\mathbf{R}_1 + m_2\mathbf{R}_2 + m_3\mathbf{R}_3 + \cdots \qquad 6\text{–}4$$

is a vector of constant magnitude, rotating with the body. Since the magnitude of $\sum m_i \mathbf{R}_i$ has units of mass times length, this constant vector may conveniently be written as $M\bar{\mathbf{R}}$, where M is a mass of any, as yet, unspecified magnitude, and

$$\bar{\mathbf{R}} = \frac{1}{M} \sum m_i \mathbf{R}_i \qquad 6\text{–}5$$

is a vector of constant length, rotating with the body. $\bar{\mathbf{R}}$ can be represented as $\overrightarrow{AG}$ where G, like A, is a fixed point in the body (Fig. 6.3b). Using equations 6–3,

$$\sum m_i \mathbf{r}_i = \sum m_i (\mathbf{r}_A + \mathbf{R}_i)$$
$$= (\sum m_i)\mathbf{r}_A + \sum m_i \mathbf{R}_i$$

To simplify, the mass M is identified with the total mass of the body, $M = \sum m_i$,

$$\sum m_i \mathbf{r}_i = M\mathbf{r}_A + M\bar{\mathbf{R}}$$
$$= M(\mathbf{r}_A + \bar{\mathbf{R}})$$
$$= M\bar{\mathbf{r}} \quad \text{say} \qquad 6\text{–}6$$

where $\bar{\mathbf{r}}$ is the position vector of the fixed point G of the body. (Fixed, that is, in the body, not fixed in space.)

The position of G in the body may be determined once and for all, by placing the body in any convenient position, and locating G by use either of the formula $\overrightarrow{AG} = (\sum m_i \mathbf{R}_i)/M$, or the formula $\overrightarrow{OG} = (\sum m_i \mathbf{r}_i)/M$, whichever is the more convenient.

Equation 6–6 holds, for this point G of the body, at all times, so 6–6 may be differentiated with respect to time,

$$\frac{d^2}{dt^2} \sum m_i \mathbf{r}_i = M\ddot{\bar{\mathbf{r}}} \qquad 6\text{–}7$$

Equation 6–2(a) may therefore be expressed in the very convenient form,

$$M\ddot{\bar{\mathbf{r}}} = \mathbf{F} \qquad 6\text{–}8$$

where

$$\bar{\mathbf{r}} = \bar{\mathbf{r}}(t) = \frac{1}{M} \sum m_i \mathbf{r}_i \qquad 6\text{–}6(a)$$

is the position vector of a point G of the rigid body, called the 'centre of mass' or 'mass-centre'.

This is a key result. It is because of equation 6–8 that $\mathbf{F}$, the vector sum of all the forces acting on a rigid body, is of such significance for mechanics.

A system-of-particles model

To the extent that more loosely connected systems of particles move together, equation 6–8 retains some significance, though the point G defined by equation 6–6(a) then migrates; it does not coincide with any single particle, but maintains a central position at all times.

6.3 Summary of the principal algebraic properties of the centre of mass

I   If $\mathbf{r}_i$ denotes position vectors from an origin fixed in space

$$m_1\mathbf{r}_1 + m_2\mathbf{r}_2 + \cdots = M\bar{\mathbf{r}} \qquad 6\text{–}9$$

where $M = m_1 + m_2 + \cdots =$ total mass under consideration.

II   If $\mathbf{s}_i$ denotes position vectors relative to the centre of mass FIG. 6.3b, (so $\bar{\mathbf{s}} = \overline{GG} = \mathbf{0}$)

$$m_1\mathbf{s}_1 + m_2\mathbf{s}_2 + \cdots = \mathbf{0} \qquad 6\text{–}10$$

Relations 6–9 and 6–10 are valid at all times, and hence the process of differentiation with respect to time may be carried through, giving

III $$m_1\dot{\mathbf{r}}_1 + m_2\dot{\mathbf{r}}_2 + \cdots = M\dot{\bar{\mathbf{r}}} \qquad 6\text{–}9(a)$$

and

$$m_1\dot{\mathbf{s}}_1 + m_2\dot{\mathbf{s}}_2 + \cdots = \mathbf{0} \qquad 6\text{–}10(a)$$

and, on further differentiation with respect to time,

IV $$m_1\ddot{\mathbf{r}}_1 + m_2\ddot{\mathbf{r}}_2 + \cdots = M\ddot{\bar{\mathbf{r}}} \qquad 6\text{–}9(b)$$

$$m_1\ddot{\mathbf{s}}_1 + m_2\ddot{\mathbf{s}}_2 + \cdots = \mathbf{0} \qquad 6\text{–}10(b)$$

If the name 'linear momentum' is given to the product 'mass × velocity' of a particle, relation 6–9(a) may be expressed thus.

> The sum of the linear momenta of the constituent particles is equal to the linear momentum of the whole mass (M) moving with the velocity of the centre of mass.

Equations 6–10(a) and 6–10(b) are of particular use in the derivation of further equations of motion of a rigid body in § 6.72.

From equation 6–8, known as the equation of translation of the rigid body, the motion of the point G of the body can be deduced immediately from the vector sum of all the external forces on the body regardless of their points of application, in just the same way as the motion of a point particle is determined by its equation of motion $m\ddot{\mathbf{r}} = \mathbf{F}$.

When cartesian coordinates are used, e.g. $\mathbf{r}_i = x_i\mathbf{i} + y_i\mathbf{j} + z_i\mathbf{k}$, or $\mathbf{s}_i = \mathsf{x}_i\mathbf{i} + \mathsf{y}_i\mathbf{j} + \mathsf{z}_i\mathbf{k}$ the x-components of equation 6–?, 6–9(a) and 6–9(b) are:

$$\left. \begin{array}{l} m_1 x_1 + m_2 x_2 + \cdots = M\bar{x} \\ m_1 \dot{x}_1 + m_2 \dot{x}_2 + \cdots = M\dot{\bar{x}} \\ m_1 \ddot{x}_1 + m_2 \ddot{x}_2 + \cdots = M\ddot{\bar{x}} \end{array} \right\} \qquad 6\text{–}9(c)$$

the y-, z-component equations following the same pattern. Similarly the x-components of equations 6–10, 6–10(a) and 6–10(b) are:

$$\left. \begin{array}{l} m_1 x_1 + m_2 x_2 + \cdots = 0 \\ m_1 \dot{x}_1 + m_2 \dot{x}_2 + \cdots = 0 \\ m_1 \ddot{x}_1 + m_2 \ddot{x}_2 + \cdots = 0 \end{array} \right\} \qquad 6\text{--}10(c)$$

with corresponding equations in y_i and z_i and their time derivatives.

6.4 Identification of additive sets of forces

Given any set of particles, equation 6–9 defines their centre of mass; the quantity $\bar{\mathbf{r}}$ is a weighted mean of the position vectors $\mathbf{r}_1, \mathbf{r}_2, \ldots$ and since all the m_i's are necessarily positive, this locates G within, though not necessarily totally enclosed by the group of particles – see questions 12–14 of Exercise 1 – in a position biased towards the greatest concentrations of mass. When the particles are disconnected or non-rigidly connected, any motion of the particles relative to one another is reflected in a smaller migration of the centre of mass within the group of particles. Not only for a rigid body, but for any system of particles,

$$M \ddot{\bar{\mathbf{r}}} = m_1 \ddot{\mathbf{r}}_1 + m_2 \ddot{\mathbf{r}}_2 + m_3 \ddot{\mathbf{r}}_3 + \cdots$$
$$= \mathbf{F}_1 + \mathbf{F}_2 + \mathbf{F}_3 + \cdots$$
$$= \mathbf{F}, \text{ say,}$$

i.e. the value of **F** determines the acceleration of the mass centre of the system of particles. It is this which invests the vector sum of the set of forces acting on a prescribed set of particles with significance.

The sum of all the external forces acting on a wriggling cat is significant because it determines the motion of the centre of mass of that cat. The vector sum of the forces on three grains of rice in France and four grains of rice in Australia is a nonsense except to the fanatic who wishes to discuss the motion of the centre of mass of those seven grains.

6.5 Examples using centre of mass relations

Example 1
Find the position of the centre of mass of 2 kg at point $(1, 5, -3)$, 3 kg at $(4, 2, 2)$, 5 kg at $(1, -6, 0)$.

$M = 2 + 3 + 5 = 10;$ and the component equations of equation 6–9 are thus:

$10\bar{x} = (2 \times 1) + (3 \times 4) + (5 \times 1) = 19,$ whence $\bar{x} = 1.9$

$10\bar{y} = 10 + 6 - 30 = -14 \qquad \bar{y} = -1.4$

$10\bar{z} = -6 + 6 + 0 = 0 \qquad \bar{z} = 0$

The required position is $(1.9, -1.4, 0)$.

A system-of-particles model

Example 2
Find the velocity of the centre of mass of three particles, 2 kg travelling north-west at 10 m s^{-1}, 500 g travelling due east at 6 m s^{-1} and 300 g travelling vertically upwards at 20 m s^{-1}.

We use the component equations of equation 6–9(a), making sure the mass units are consistent. Then $M = 2.8$ kg, and taking east, north, and vertical as x-, y-, z-directions,

$$2.8\dot{x} = -2 \times \frac{10}{\sqrt{2}} + (0.5 \times 6)$$

$$2.8\dot{y} = 2 \times \frac{10}{\sqrt{2}}$$

$$2.8\dot{z} = 0.3 \times 20$$

so the required velocity is $(-3.98, 5.05, 2.14)$.

Example 3
Two equal masses are joined by a light inextensible string which is initially horizontal. One mass is projected vertically with speed 30 m s^{-1}. Find the motion of the mid-point of the string during the interval of time for which the string remains taut and neither particle hits the ground.

So long as the string remains taut, its mid-point is the centre of mass of the two particles [for, from this mid-point, $m_1\mathbf{s}_1 + m_2\mathbf{s}_2 = m\mathbf{s} + m(-\mathbf{s}) = 0$]. The initial velocity of the mid-point is therefore given by equation 6–9(a).

$$2mv = (m \times 30) + (m \times 0) \text{(upwards)}, \text{ i.e. } v = 15 \text{ upwards}.$$

The forces due to string tension on the two particles are equal and opposite 'internal' forces of the system. The only external forces are those of gravity, whose sum is equal to

$$mg + mg = 2mg \text{ vertically downwards}.$$

Writing $\bar{x}$, $\bar{y}$, $\bar{z}$ for the coordinates of the mid-point of the string, the component equations corresponding to equation 6–8 using 6–9(b) are

$$2m\ddot{\bar{x}} = 0 \quad \text{where initially} \quad \dot{\bar{x}} = 0$$
$$2m\ddot{\bar{y}} = 0 \quad \quad \quad \quad \quad \quad \quad \dot{\bar{y}} = 0$$
$$2m\ddot{\bar{z}} = -2mg \quad \quad \quad \quad \quad \dot{\bar{z}} = 15.$$

Integrating these equations with respect to time t, we deduce that, in the prescribed interval,

$$\dot{\bar{x}} = \text{constant} = 0; \quad \dot{\bar{y}} = \text{constant} = 0; \quad \dot{\bar{z}} = -gt + \text{constant} = -gt + 15.$$

These are the required components of velocity of the mid-point of the string so long as it remains taut.

158 An introduction to mechanics and modelling

Exercise 1

Some important results are derived in this exercise.

1 Find the centre of mass of four particles, of masses 2, 3, 5 and 8 units respectively, which are situated at the corners, taken in order, of a square OXPY. (Take OX and OY as axes of coordinates, and let the side of the square be of length b units.)
2 Find the centre of mass of six particles, of masses, 1, 2, 3, 4, 5 and 6 kg situated at the vertices of a regular hexagon of side 1 m.
3 Find the centre of mass of $4n$ particles of masses $m_1, m_2, m_3, \cdots, m_n$; $m_1, m_2, m_3, \ldots, m_n$; $m_1, m_2, m_3, \ldots m_n$; $m_1, m_2, m_3, \ldots, m_n$; situated in a plane at points $(x_1, y_1), (x_2, y_2), (x_3, y_3), \ldots, (x_n, y_n)$; $(-x_1, y_1), (-x_2, y_2), \ldots, (-x_n, y_n)$; $(x_1, -y_1), (x_2, -y_2), \ldots, (x_n, -y_n)$; $(-x_1, -y_1), (-x_2, -y_2), \ldots, (-x_n, -y_n)$; respectively.
4 Prove that the centre of mass of the first two sets of masses of question 3 lies on the y-axis; that of the second and third sets lies at the origin; and that of the first and third sets on the x-axis.
5 Using the point sets of questions 3 and 4 as models, discuss the position of the centres of mass of symmetrical bodies, and complete the formulation of laws beginning as follows:
    (i)   The centre of mass of a plane body which has an axis of symmetry lies ...
    (ii) The centre of mass of a solid body which has a plane of symmetry lies ...
    (iii) If a solid body has two planes of symmetry, its centre of mass lies ...
    (iv) If a solid body has three planes of symmetry meeting at a point P, its centre of mass lies ...
6 Use the laws of question 5 to deduce the position of the centre of mass of:
    (i)   a uniform rectangular block;
    (ii) a uniform straight rod;
    (iii) a uniform circular disc;
    (iv) a uniform sphere.
7 Use the laws of question 5 to partially locate the position of the centre of mass of:
    (i)   a hemispherical shell;
    (ii) a sector of a circle;
    (iii) a right pyramid whose base is a regular polygon.
8 A grand piano is pushed over the edge of a cliff, and thereafter moves freely under gravity. Write down equations from which the horizontal and vertical motion of the piano's centre of mass during descent may be determined. (Air resistance may be neglected.)
9 Find the position of the centre of mass, G, of two particles, each of mass m. Given that the position vectors of the particles at time t are $\mathbf{r}_1, \mathbf{r}_2$ deduce from question 5(iv) the position vector of G and verify that your result is in accordance with equation 6–9.
10 Two particles, each of mass 1 kg are joined by a light inextensible string. Initially the string is taut, and the particles rotate freely about the mid-point of the string, the whole system being supported on a large smooth horizontal table. A force of 4 N directed horizontally eastwards is applied to one of the particles. Given that the string remains taut, find the distance which the mid-point of the string has moved after 3 seconds.
11 A railway train consisting of an engine and several carriages, of total mass 400 tons runs away down a track whose gradient is 1/200. The resistance due to friction, etc. on the various parts of the train is equivalent to a retarding force of 1.5 ton-force. What principle enables you to estimate the acceleration of the train? What reservations are needed about the interpretation of this value as the acceleration of a particular carriage especially in the first few seconds of motion? Find the speed of a carriage 2 minutes after starting from rest, subject to this reservation. (**Hint**: you may write 1 ton = K kg. The constant K will subsequently cancel out.)

> Note: 1 ton-force = weight of 1 ton (mass)
> 1 ton = 1016 kg = 1.016 tonne
> By 'gradient' mathematicians mean tan θ, where θ is the angle with the horizontal.

12 Use the method of Example 3 of § 2.4 (with $\overrightarrow{OA} = \mathbf{r}_1$ and $\overrightarrow{OB} = \mathbf{r}_2$ to establish the result

$$m_1\mathbf{r}_1 + m_2\mathbf{r}_2 = (m_1 + m_2)\bar{\mathbf{r}} \qquad (A)$$

where $\bar{\mathbf{r}}$ is the position vector of the point C which divides AB in the ratio m_2/m_1. If A and B are points of a rigid body, deduce that the time variation of the position vector $\bar{\mathbf{r}} = \bar{\mathbf{r}}(t)$ describes the motion of a point fixed in the body.

13 Given that point masses m_1, m_2, m_3 are rigidly connected together, and that at time t their position vectors are $\mathbf{r}_1(t), \mathbf{r}_2(t), \mathbf{r}_3(t)$ respectively, extend the result of question 12 above to verify that the point $\bar{\mathbf{r}}$ defined by the equation

$$(m_1 + m_2 + m_3)\bar{\mathbf{r}} = m_1\mathbf{r}_1 + m_2\mathbf{r}_2 + m_3\mathbf{r}_3$$

is fixed in the body.

14 Using the method of questions 12 and 13, prove by induction that the point $\bar{\mathbf{r}}$ defined by

$$(m_1 + m_2 + \cdots + m_n)\bar{\mathbf{r}} = m_1\mathbf{r}_1 + m_2\mathbf{r}_2 + \cdots + m_n\mathbf{r}_n$$

(where the masses m_i constitute a rigid body) is a point fixed in the body. This is an alternative way of establishing the main result of § 6.2.

6.6 Methods of finding the position of the centre of mass of simple rigid bodies (without integration)

6.61 Symmetry properties as an aid to the determination of the centre of mass

> (i)   The centre of mass of a body which has an axis of symmetry lies on that axis (FIG. 6.4a).
> (ii) The centre of mass of a solid body which has a plane of symmetry lies on that plane (FIG. 6.4b).
> (iii) If a solid body has two planes of symmetry, its centre of mass lies on the line of intersection of those planes (FIG. 6.4c).
> (iv) If a solid body has three planes of symmetry meeting at a point P, its centre of mass is at P.

The reader who has worked through Exercise 1, questions 3, 4 and 5, will know these results already; the reader who has not, is advised to work questions 4 and 5 before proceeding further. From these symmetry results we deduce that the centre of mass of uniform spheres, rods, rectangular boxes, etc. coincide with their geometrical centres.

The calculation of the position of the centre of mass of a body of irregular shape or non-uniform material must await the reader's introduction to continuum modelling.

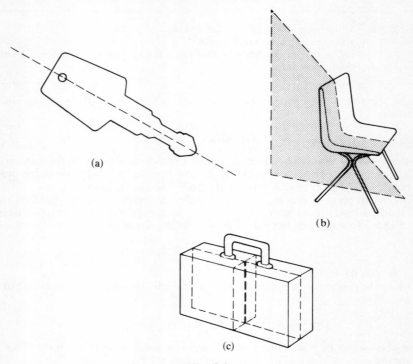

Fig. 6.4

6.62 Composite body theorem

If the shape of a body is conveniently described in terms of simpler shapes – as, for example, the door in FIG. 6.5a, might be described as a semicircle on top of a square or the ashtray in FIG. 6.5b might be modelled as a cylinder with a

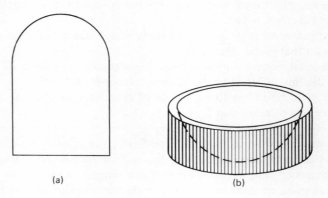

Fig. 6.5

portion in the shape of a spherical cap removed – then the following corollary, proved in § 6.63 is often useful.

> **Composite body theorem.** If the centre of mass of mass M_1, e.g. the square part of the door, is at $\bar{\mathbf{r}}_1$, and the centre of mass of M_2 is at $\bar{\mathbf{r}}_2$, then the centre of mass of the composite body of mass $(M_1 + M_2)$ is at the position $\bar{\mathbf{r}}$ satisfying
>
> $$(M_1 + M_2)\bar{\mathbf{r}} = M_1 \bar{\mathbf{r}}_1 + M_2 \bar{\mathbf{r}}_2 \qquad 6\text{--}11$$

Example
To apply result 6–11 to the door of FIG. 6.5 we first note that G must lie on the axis of symmetry. Next we choose a convenient x-axis, e.g. the diameter of the semicircle, and write the y-component of equation 6–11 as

$$(M_1 + M_2)\bar{y} = M_1 \bar{y}_1 + M_2 \bar{y}_2$$

where M_1 = mass of square; $\bar{y}_1 = -a$ (writing $2a$ = side of square = diameter of circle); M_2 = mass of semicircle; and $\bar{y}_2$ may be found from a table such as that given in § 6.94. There we find for a sector of half-angle α, that $\bar{y} = \frac{2}{3}a \sin \alpha/\alpha$, so in the present case, with $\alpha = \pi/2$, $\bar{y}_2 = 4a/3\pi$. If we further write $M_1 = 4a^2\sigma$; $M_2 = \frac{1}{2}(\pi a^2)\sigma$ we find $\bar{y} = -20a/(3\pi + 24)$. This completes the determination of G.

Example
To apply formula 6–11 to the ashtray, the procedure would be slightly modified, since it is the centre of mass of the part, not of the complete cylinder, whose calculation concerns us. Here we choose to assign the symbols

$$M_1 = \text{mass of ashtray}$$
$$M_2 = \text{mass of removed spherical cap}$$

so that

$$(M_1 + M_2) = \text{mass of cylinder before its excavation,}$$

and

$$\bar{\mathbf{r}} = \text{position of centre of mass of the entire cylinder}$$
$$= \text{mid-point of the axis of the cylinder.}$$

Assuming the mass centre ($\bar{\mathbf{r}}_2$) of a spherical cap to be known (see § 6.94), the position of the mass centre of the ashtray ($\bar{\mathbf{r}}_1$) may be derived from equation 6–11 rewritten

$$M_1 \bar{\mathbf{r}}_1 = (M_1 + M_2)\bar{\mathbf{r}} - M_2 \bar{\mathbf{r}}_2 \qquad 6\text{--}12$$

6.63 Composite body theorem derived from particle model

We enumerate the particles of the first part of the body $1, 2, 3, \ldots, s$ say, and the particles of the second part $s+1, s+2, \ldots, v$ say. Then

$$m_1 + m_2 + \cdots + m_s = M_1; \quad \text{centre of mass of } M_1 \text{ denoted } \bar{\mathbf{r}}_1$$
$$m_{s+1} + m_{s+2} + \cdots + m_v = M_2; \quad \text{centre of mass of } M_2 \text{ denoted } \bar{\mathbf{r}}_2$$
$$m_1 + m_2 + \cdots + m_v = M_1 + M_2; \quad \text{centre of mass of } (M_1 + M_2) \text{ denoted } \bar{\mathbf{r}}$$

162 An introduction to mechanics and modelling

By definition

$$(M_1 + M_2)\bar{\mathbf{r}} = m_1\mathbf{r}_1 + m_2\mathbf{r}_2 + \cdots + m_s\mathbf{r}_s + m_{s+1}\mathbf{r}_{s+1} + \cdots + m_v\mathbf{r}_v$$
$$= (m_1\mathbf{r}_1 + m_2\mathbf{r}_2 + \cdots + m_s\mathbf{r}_s) + (m_{s+1}\mathbf{r}_{s+1} + \cdots + m_v\mathbf{r}_v)$$
$$= M_1\bar{\mathbf{r}}_1 + M_2\bar{\mathbf{r}}_2$$

The result may be extended to any number of constituent parts by only a trivial modification of the proof, thus

$$(M_1 + M_2 + M_3 + \cdots + M_n)\bar{\mathbf{r}} = M_1\bar{\mathbf{r}}_1 + M_2\bar{\mathbf{r}}_2 + \cdots + M_n\bar{\mathbf{r}}_n \qquad 6\text{-}13$$

Exercise 2
1 Model a uniform triangular lamina (thin sheet) as a system of uniform rods parallel to one side of the triangle. By taking the opposite vertex as origin, use the composite body theorem to show that the centre of mass of the lamina lies on a median of the triangle.
2 Use the result of question 1 to show that

> the centre of mass of any uniform triangular lamina is at the point of intersection of two of its medians.

Deduce that the three medians are concurrent. (**Note** that $\bar{\mathbf{r}}$ is defined uniquely by equation 6–6.)

3 Find the centre of mass of each of the uniform laminae shown in FIG. 6.6a–d.

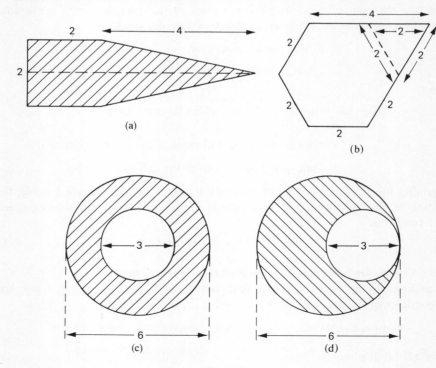

Fig. 6.6

6.7 Motion of rotation of a rigid body

6.71 Property of equal and opposite moments for pairs of internal forces

Sections 6.1 and 6.2 showed how the simple 'equal and opposite' postulate about internal forces, in conjunction with the point particle model of matter, led to a most informative result, viz. that the acceleration of the centre of mass G of a body (or indeed any set of bodies) is determined by the external forces through the relation

$$\ddot{\mathbf{r}} = \frac{1}{M} \mathbf{F}.$$

If the starting position and motion of the body are known, then the velocity and position of the point G of the body at all subsequent times can, in principle, be deduced from its acceleration (more about this in Chapter 7).

> If the body is rigid and, as well as the motion of G, the rotation of the body about G is known, the specification of its motion is complete.

It is the theory of this motion of rotation which we now pursue, again starting from the many-particle model.

Information about the behaviour of the particles as a group can be deduced from the equations of motion of the constituent particles (viz. $m_1 \mathbf{r}_1 = \mathbf{F}_1 +$ internal forces on particle 1 etc.) by use of an additional postulate/hypothesis about the internal forces. A simple hypothesis which leads to experimentally verifiable conclusions about the rotational motion of the body as a whole is the following.

> Forces between the particles of the model are **either** forces of contact so that the equal and opposite forces act at the same point, **or**, if the particles are situated at separated points, their interaction is one of pure attraction or pure repulsion (see FIG. 6.7).

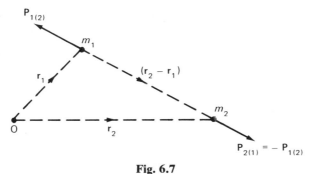

Fig. 6.7

Using the notation of § 6.11, in which $\mathbf{P}_{1(2)}$ is the force exerted on the mass m_1 situated at $\mathbf{r}_1$ by the mass m_2, our new postulate may be expressed algebraically, **either** $\mathbf{r}_1 = \mathbf{r}_2$ **or** $\mathbf{P}_{1(2)}$ is parallel or antiparallel to $(\mathbf{r}_2 - \mathbf{r}_1)$ (see FIG. 6.7).

Now consider the vector product $(\mathbf{r}_2 - \mathbf{r}_1) \wedge \mathbf{P}_{1(2)}$; the first alternative may be restated $\mathbf{r}_2 - \mathbf{r}_1 = 0$, and hence the vector product is zero; the second alternative also ensures that the vector product is zero since the vector product of two parallel vectors is zero; so our hypothesis ensures that $(\mathbf{r}_2 - \mathbf{r}_1) \wedge \mathbf{P}_{1(2)} = \mathbf{0}$ in either case.

When

$$(\mathbf{r}_2 - \mathbf{r}_1) \wedge \mathbf{P}_{1(2)} = \mathbf{0}, \qquad \mathbf{r}_2 \wedge \mathbf{P}_{1(2)} - \mathbf{r}_1 \wedge \mathbf{P}_{1(2)} = \mathbf{0} \qquad 6\text{--}14$$

i.e. since in our model $\mathbf{P}_{2(1)} = -\mathbf{P}_{1(2)}$,

$$\mathbf{r}_2 \wedge \mathbf{P}_{2(1)} + \mathbf{r}_1 \wedge \mathbf{P}_{1(2)} = \mathbf{0}. \qquad 6\text{--}15$$

This last form shows that our new postulate is equivalent to saying

> the sum of the moments, about any point, of any **pair** of internal forces is zero.

Hence, by summing over all particles of the body, the sum of the moments of all internal forces is zero.

$$\sum_i [\mathbf{r}_i \wedge (\text{\bf sum of internal forces on mass } m_i)] = \mathbf{0} \quad (i = 1, 2, \ldots, n) \quad 6\text{--}16$$

Although position vectors $\mathbf{r}_i$ are usually measured from a fixed origin, nothing in the above argument requires the reference point to be stationary; for example, if $\mathbf{s}_i$ denotes the position vector of mass m_i relative to the moving point G (FIG. 6.3) then at each specified time t,

$$\sum_i [\mathbf{s}_i \wedge (\text{\bf sum of internal forces on mass } m_i)] = \mathbf{0} \qquad .6\text{--}17$$

6.711 *d'Alembert's principle*

The two postulates (i) the (vector) sum of all internal forces is zero, and (ii) the (vector) sum of the moments of internal forces is zero, are together equivalent to d'Alembert's principle. d'Alembert expressed his postulate in language coloured by the somewhat hybrid force model of his day. His statement may be translated 'the system of effective forces on the particles of the body is equivalent at each instant to the external force system'.

The concept of equivalent systems of force needs to be defined before d'Alembert's statement can be interpreted. Equivalent systems of force for rigid bodies will be discussed in Chapter 8. Our needs are met by equations 6–16 and 6–17.

6.72 Rate of change of angular momentum

We recall equations 6–1, viz.

$$m_1 \ddot{\mathbf{r}}_1 = \mathbf{F}_1 + \text{internal forces acting on particle 1} \qquad 6\text{--}1(a)$$

$$m_2 \ddot{\mathbf{r}}_2 = \mathbf{F}_2 + \text{internal forces acting on particle 2} \qquad 6\text{--}1(b)$$

and so on.

To use equation 6–16 in conjunction with equations 6–1, we first form the vector product of $\mathbf{r}_1$ with the left-hand side of 6–1(a), the vector product of $\mathbf{r}_2$ with the left-hand side of 6–1(b), and so on, then add the resulting equations. On the right-hand side, as well as the sum $\sum_i \mathbf{r}_i \wedge \mathbf{F}_i$, we get $\sum (\mathbf{r}_i \wedge \text{\bf sum of internal forces}$

A system-of-particles model 165

acting on particle i) which is zero because of equation 6–16. Hence,

$$\mathbf{r}_1 \wedge m\ddot{\mathbf{r}}_1 + \mathbf{r}_2 \wedge m\ddot{\mathbf{r}}_2 + \cdots = \mathbf{r}_1 \wedge \mathbf{F}_1 + \mathbf{r}_2 \wedge \mathbf{F}_2 + \cdots$$

= vector sum of the moments of external forces on the rigid body about the origin O, equals $\mathbf{M}_0$ say.   6–18

The right-hand side of equation 6–18 may be referred to as the total moment of external forces about O. N.B. Symbols M and **M** denote moments, whereas italic M denotes the mass of a body.

We note that

$$\frac{d}{dt}(\mathbf{r}_i \wedge m_i \dot{\mathbf{r}}_i) = \dot{\mathbf{r}}_i \wedge m_i \dot{\mathbf{r}}_i + \mathbf{r}_i \wedge m_i \ddot{\mathbf{r}}_i \quad \text{(cf. qn. 15, Ex. 3, Chap. 4)}$$

$$= \mathbf{r}_i \wedge m_i \ddot{\mathbf{r}}_i$$

Hence equation 6–18 may be written

$$\frac{d}{dt}(\mathbf{r}_1 \wedge m_1 \dot{\mathbf{r}}_1 + \mathbf{r}_2 \wedge m_2 \dot{\mathbf{r}}_2 + \cdots) = \text{moment of external forces about O}$$

$$= \mathbf{M}_0 \qquad 6\text{–}19$$

The occurrence of the sum

$$\sum_i \mathbf{r}_i \wedge m_i \dot{\mathbf{r}}_i$$

on the left-hand side of equation 6–19 – which is an equation referring to an arbitrary system of particles – indicates that

$$\sum \mathbf{r}_i \wedge m_i \dot{\mathbf{r}}_i$$

is a quantity of significance, worthy of a name; it is called the **angular momentum** of the system of particles about the origin O (or, in older books, the moment of momentum); it is not expected that it will at first be seen by the reader as anything more than a nasty mathematical expression. As a step to better appreciation of its nature, let us consider the special case of a plane body (e.g. a flywheel) rotating with angular velocity $\dot{\theta}$ about a fixed axis perpendicular to its plane (FIG. 6.8).

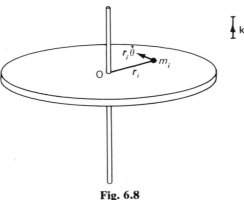

Fig. 6.8

Flywheel example

We take as origin the point where the axis intersects the plane of the rotating body. Each mass m_i describes a circle of constant radius r_i about this origin; each velocity $\dot{\mathbf{r}}_i$ is of magnitude $r_i\dot{\theta}$ in a direction at right angles to $\mathbf{r}_i$ (FIG. 6.8). So,

$$\mathbf{r}_i \wedge m_i\dot{\mathbf{r}}_i = m_i r_i^2 \dot{\theta}\mathbf{k}$$

where $\mathbf{k}$ is unit vector in the direction of the axis of rotation. Consequently

$$\sum \mathbf{r}_i \wedge m_i\dot{\mathbf{r}}_i = \dot{\theta}(\sum m_i r_i^2)\mathbf{k}$$

i.e. the angular momentum of the system is a vector whose magnitude in this special case, is proportional to the angular velocity $\dot{\theta}$ and whose direction is the direction $\mathbf{k}$ of the axis of rotation. The multiple

$$\sum m_i r_i^2$$

is big when large masses are situated at large distances from the axis of rotation – as in a large, heavy flywheel. Equation 6–19 says that the angular momentum can be changed by applying a moment about O proportional to the desired rate of change, $M_O = \text{constant} \times \ddot{\theta}$. The greater the initial angular momentum (heavy flywheel rotating fast), the more difficult will it be to stop the rotation.

Let us now return to the general case. Equation 6–19 may be expressed as a theorem.

> The rate of change of angular momentum of any system about a fixed point is equal to the vector sum of the moments of all external forces acting on the system, about that point.

When no point of the system is fixed, the evaluation of neither side of equation 6–19 is easy. In such a case it is more convenient to make measurements relative to a point moving with the system; and there is one such point, namely G whose motion is known to us (through $M\ddot{\bar{\mathbf{r}}} = \mathbf{F}$). Equation 6–19 can be translated into terms of these more convenient quantities by writing (FIG. 6.9),

$$\mathbf{r}_1 = \bar{\mathbf{r}} + \mathbf{s}_1$$
$$\mathbf{r}_2 = \bar{\mathbf{r}} + \mathbf{s}_2$$

6–20

and so on.

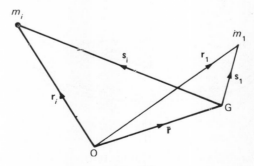

Fig. 6.9

Using equation 6–17, viz.

$$\sum \mathbf{s}_i \wedge (\text{sum of internal forces on } m_i) = \mathbf{0},$$

in conjunction with equations 6–1 (the equations of motion of the separate particles), we find

$$\mathbf{s}_1 \wedge m_1\ddot{\mathbf{r}}_1 + \mathbf{s}_2 \wedge m_2\ddot{\mathbf{r}}_2 + \cdots = \mathbf{s}_1 \wedge \mathbf{F}_1 + \mathbf{s}_2 \wedge \mathbf{F}_2 + \cdots \qquad \text{6–21}$$

i.e.

$$m_1 \mathbf{s}_1 \wedge (\ddot{\mathbf{r}} + \ddot{\mathbf{s}}_1) + m_2 \mathbf{s}_2 \wedge (\ddot{\mathbf{r}} + \ddot{\mathbf{s}}_2) + \cdots = \sum_i \mathbf{s}_i \wedge \mathbf{F}_i$$

whence

$$(m_1 \mathbf{s}_1 + m_2 \mathbf{s}_2 + \cdots) \wedge \ddot{\mathbf{r}} + \mathbf{s}_1 \wedge m_1 \ddot{\mathbf{s}}_1 + \mathbf{s}_2 \wedge m_2 \ddot{\mathbf{s}}_2 + \cdots = \sum_i \mathbf{s}_i \wedge \mathbf{F}_i$$

i.e. $(\sum m_i \mathbf{s}_i) \wedge \ddot{\mathbf{r}} + \sum (\mathbf{s}_i \wedge m_i \ddot{\mathbf{s}}_i) = \sum \mathbf{s}_i \wedge \mathbf{F}_i$
But, from equation 6–10 $\sum m_i \mathbf{s}_i = 0$, so

$$\sum (\mathbf{s}_i \wedge m_i \ddot{\mathbf{s}}_i) = \sum \mathbf{s}_i \wedge \mathbf{F}_i \qquad \text{6–22(a)}$$

i.e.

$$\frac{d}{dt} \sum (\mathbf{s}_i \wedge m_i \dot{\mathbf{s}}_i) = \sum \mathbf{s}_i \wedge \mathbf{F}_i \qquad \text{6–22(b)}$$

which may be expressed,

> **Theorem:** the rate of change of angular momentum of any system relative to its centre of mass G is equal to the sum of the moments of external forces about G. 6–22(c)

The significance of the angular momentum theorems becomes clearer when we restrict our attention to the constituent particles of a rigid body, for then the left-hand side of equation 6–22(a) can be simplified. The simplification is particularly neat when the motion of the body is two-dimensional, i.e. when

> each particle of the body moves in a plane, the planes of motion of different particles all being parallel (definition of two-dimensional motion)

as, for example, when a piece of furniture on castors is moved across the floor in any way (the planes are all horizontal), or when a hoop flies through the air rotating in the vertical plane which contains the path of its centre.

6.73 Interpretation of $\sum (\mathbf{s}_i \wedge m_i \ddot{\mathbf{s}}_i)$ for a rigid body moving in two dimensions

6.731 Preliminaries
Our definition of two-dimensional motion gives us two items of information about the way each of the $\mathbf{s}_i$'s varies with time, viz.

(a) the mass m_i at point A_i (FIG. 6.10) moves in a plane parallel to the plane of motion of G,
(b) the points A_i and G are rigidly connected, so the distance of A_i from G is constant.

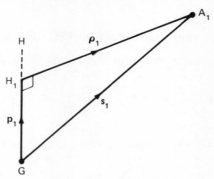

Fig. 6.10

It follows that A_i lies on the curve of intersection of its own (constant) plane of motion referred to in (a), and the moving sphere of points distance A_iG from G. The motion of A_i relative to G is thus a circle, in a plane parallel to the plane of motion of G. (You may think of yourself located at G and moving with G: the motion relative to G is what you observe.)

Let GH_1 be the perpendicular from G on to the plane of motion of the point A_1; write $\overrightarrow{GH_1} = \mathbf{p}_1$ (FIG. 6.10), then

$$\mathbf{s}_1 = \mathbf{p}_1 + \boldsymbol{\rho}_1$$

where $\mathbf{p}_1$ is a constant vector, and $\boldsymbol{\rho}_1$ is a vector which sweeps out a circle in the plane of motion, which is perpendicular to $\mathbf{p}$.

Similarly for any point A_i;

$$\mathbf{s}_i = \mathbf{p}_i + \boldsymbol{\rho}_i$$

where all the $\mathbf{p}_i$ are parallel to each other, and all the $\boldsymbol{\rho}_i$ sweep out circles in parallel planes.

6.732 Condensed derivation of two-dimensional angular equation of motion
Since the full argument given in §§ 6.733 and 6.74 for the derivation of the equation $I_G\ddot{\theta} = M_G$ is rather long, we here give a condensed preview. We start from

$$\sum \mathbf{s}_i \wedge m_i\ddot{\mathbf{s}}_i = \sum \mathbf{s}_i \wedge \mathbf{F}_i \qquad 6\text{-}22(a)$$

For two-dimensional motion the variables, $\mathbf{s}_i$, $\ddot{\mathbf{s}}_i$, $\mathbf{F}_i$ may be expressed in terms of $\boldsymbol{\rho}_i$, $\ddot{\boldsymbol{\rho}}_i$ and the component vector $\mathbf{P}$ (=component of $\mathbf{F}$ in the plane of motion). It is then found that the original equation is equivalent to

$$\sum \boldsymbol{\rho}_i \wedge m_i\ddot{\boldsymbol{\rho}}_i = \sum \boldsymbol{\rho}_i \wedge \mathbf{P}_i$$

The transverse component of $m\ddot{\boldsymbol{\rho}}$ in polar coordinates is $m\rho\ddot{\theta}$ so

$$\sum \boldsymbol{\rho}_i \wedge m_i\ddot{\boldsymbol{\rho}}_i \to \sum m_i\rho^2\ddot{\theta} = \ddot{\theta}\sum m_i\rho_i^2$$
$$= I_G\ddot{\theta}$$

whence

$$I_G\ddot{\theta} = M_G, \text{ where by definition } I_G = \sum m_i\rho_i^2 \text{ (constant)}.$$

6.733 Expanded discussion

In FIG. 6.10 $\boldsymbol{\rho}_1$ is of constant length ρ_1 say, but variable direction,

$$\boldsymbol{\rho}_1 = \rho_1 \mathbf{I} \qquad 6\text{-}23$$

where $\mathbf{I}$ is a unit vector which rotates in the plane of motion. We note that $\mathbf{p}_1$ is constant, so $\ddot{\mathbf{p}}_1 = 0$, and

$$\ddot{\mathbf{s}}_1 = \ddot{\boldsymbol{\rho}}_1 = -\rho_1 \dot{\theta}_1^2 \mathbf{I} + \rho_1 \ddot{\theta}_1 \mathbf{J} \qquad 6\text{-}24$$

Equation 6–24 is the algebraic expression of the fact that the relative motion (namely motion in a circle) has acceleration components $\rho_1 \dot{\theta}^2$ towards the centre and $\rho_1 \ddot{\theta}_1$ tangentially.

The motion of the triangle A_1GH_1 in FIG. 6.10 determines that of the whole rigid body, which can therefore be described as a rotation about the moving axis GH (whose direction is constant and whose location at any instant is determined by the current value of $\bar{\mathbf{r}}$). The angle θ_1 between the initial and the current position of $\boldsymbol{\rho}_1$ is the angle between the initial and current positions of plane A_1GH_1, and is therefore equal to the angle θ through which the body as a whole has rotated about axis GH. It follows that the angles $\theta_2, \theta_3, \ldots$, etc. (which are defined as the angles between initial and current positions of the $\boldsymbol{\rho}$ vectors of the masses $m_2, m_3, \ldots$) are all equal,

$$\theta_1 = \theta_2 = \theta_3 \cdots = \theta. \qquad 6\text{-}25$$

Equation 6–22a is a vector equation, equivalent to three component equations. The simple result which we are seeking arises from the third of these component equations, viz. the one which involves only the components of $\mathbf{s}_i$, $\ddot{\mathbf{s}}_i$, $\mathbf{F}_i$ in the plane of motion. We draw the reader's attention to a handy property of the vector product $\mathbf{a} \wedge \mathbf{b}$: the component of $\mathbf{a} \wedge \mathbf{b}$ in any direction is independent of the components of $\mathbf{a}$ and $\mathbf{b}$ in that direction, e.g. $(\mathbf{a} \wedge \mathbf{b}) \cdot \mathbf{k} = a_1 b_2 - a_2 b_1$ is independent of the $\mathbf{k}$-components a_3 and b_3 of $\mathbf{a}$ and $\mathbf{b}$.

Now let $\mathbf{I}$ = the (presumably variable) unit vector parallel to $\overrightarrow{H_1 A_1}$; $\mathbf{J}$ = the (presumably variable) unit vector perpendicular to $\overrightarrow{H_1 A_1}$ in the plane of motion; $\mathbf{K}$ = the constant unit vector perpendicular to the plane of motion.

By definition, $\mathbf{s}_1 = \mathbf{p}_1 + \boldsymbol{\rho}_1$ where $\mathbf{p}_1$ is parallel to $\mathbf{K}$ so $\mathbf{s}_1$ and $\boldsymbol{\rho}_1$ differ only in their third or $\mathbf{K}$-component: the third, i.e. $\mathbf{K}$-components of $\mathbf{s}_1 \wedge \ddot{\mathbf{s}}_1$ and of $\boldsymbol{\rho}_1 \wedge \ddot{\mathbf{s}}_1$ are therefore equal. Using suffix K to identify these third components,

$$[\mathbf{s}_1 \wedge \ddot{\mathbf{s}}_1]_K = [\boldsymbol{\rho}_1 \wedge \ddot{\mathbf{s}}_1]_K$$
$$= \rho_1^2 \ddot{\theta}$$

because of equation 6–24 and 6–25. The required quantity

$$\left[\sum_i \mathbf{s}_i \wedge m_i \ddot{\mathbf{s}}_i \right]_K$$

takes the form

$$\left[\sum_i \mathbf{s}_i \wedge m_i \ddot{\mathbf{s}}_i \right]_K = \sum (m_i \rho_i^2 \ddot{\theta}) \text{ where } \ddot{\theta} \text{ is the same for all } m_i$$
$$= \left(\sum m_i \rho_i^2 \right) \ddot{\theta}$$
$$= I_G \ddot{\theta} \quad \text{say}, \qquad 6\text{-}26$$

170 An introduction to mechanics and modelling

> where $I_G (\equiv \sum m_i \rho_i^2)$ depends only on the distances of the constituent masses from the **K**-axis through G. I_G is independent of time – it is determined by the constitution of the body, and is called the 'moment of inertia' of the body about the specified axis through G.

6.74 Use of the angular equation of motion of a rigid body moving in two dimensions

From equations 6–22 and 6–26 we derive the key formula,

> $$I_G \ddot{\theta} = \left[\sum \mathbf{s}_i \wedge \mathbf{F}_i\right]_K \qquad \text{6–27(a)}$$
> $$= M_G \quad \text{say,} \qquad \text{6–27(b)}$$
>
> where M_G is the **K**-component of the total moment of external forces, and $I_G = \sum_i (m_i \rho_i^2)$.

With this equation, the student can solve problems like examples 1 and 2 below and question 5 of Exercise 3. If **s** and **F** both lie in the plane of motion, the evaluation of $\mathbf{s} \wedge \mathbf{F}$ is made easy by reference to FIG. 6.11. The magnitude of $\mathbf{s} \wedge \mathbf{F}$ is

$$sF \sin \phi$$

and this may be evaluated either as sS, where S is the component of **F** at right angles to **s**, or as pF, where p is the perpendicular distance from G on to **F**.

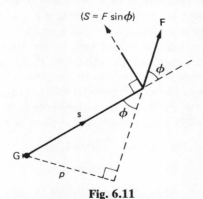

Fig. 6.11

Even if **s** and **F** are not both in the plane of motion, similar ideas can be used.

> If **P** is the component vector of **F** in the plane of motion, (just as **ρ** is the component vector of **s** in the plane of motion),
>
> $$I_G \ddot{\theta} = \sum_i \pm |\boldsymbol{\rho}_i \wedge \mathbf{P}_i| \qquad \text{6–27(c)}$$
>
> where each $|\boldsymbol{\rho} \wedge \mathbf{P}|$ may be evaluated as ρS or as pP; $S = P \sin \phi$, $p = \rho \sin \phi$.

Equation 6–27, in any of the forms (a), (b) or (c) is called the **angular equation of the rigid body**: it determines the angular acceleration $\ddot{\theta}$ of the body from the moment M_G of external forces. It is as important in the discussion of the rotation of a body as Newton's second law is for discussion of the motion of the centre of mass.

Example 1
A bicycle wheel is modelled as consisting of a heavy line distribution of mass round the rim, and 'light' spokes. Given that its radius is 0.3 m and that its total mass is 2 kg, find its moment of inertia about the axle. Find the constant moment which must be applied (as a result of all external causes) about this axle in order to produce a rotation rate of 120 r.p.m. at the end of 10 seconds, starting from rest.

Every particle of the 2 kg mass is at distance $\rho_i = 0.3$ from the axis through G perpendicular to the plane of rotation of the wheel. Hence

$$I_G = \sum_i m_i (0.3)^2 = 0.09 \sum m_i$$
$$= 0.09 \times 2$$
$$= 0.18 \quad \text{(SI units)}$$

The angular equation of motion is thus

$$0.18 \ddot{\theta} = M_G \text{ where } M_G \text{ is constant.}$$

Integrating with respect to time (t), using the fact that M_G is constant,

$$0.18 \dot{\theta} = M_G t + \text{constant } C \qquad \text{6–28(a)}$$

We are given that $\dot{\theta} = 0$ when $t = 0$, so we deduce $C = 0$. Moreover when $t = 10$ we require

$$\dot{\theta} = 120 \times 2\pi \text{ radians per minute}$$
$$= 4\pi \text{ radians per sec (SI units)} \qquad \text{6–28(b)}$$

From equations 6–28(a) and (b),

$$0.18 \times 4\pi = 10 M_G, \quad \text{so} \quad M_G = 0.226 \text{ (SI units)}, \quad \text{i.e. N m.}$$

The result is valid whether the axle is stationary or progressing.

Example 2
A uniform rod AB of mass M and length $2l$ units rests on a smooth horizontal table, and rotates with angular velocity Ω about a vertical axis through the end A. The connection at A suddenly fails, and the rod subsequently moves freely across the table. Find the position of end B at any time t during the subsequent motion.

The original motion of steady rotation about end A is given; equations of motion for the interval of time prior to release are therefore unnecessary. The instantaneous release of the connection at A is interpreted as a process for which no infinite accelerations, i.e. no instantaneous velocity changes occur, so the

172 An introduction to mechanics and modelling

horizontally unconstrained motion starts with the mid-point of the rod moving with velocity $l\Omega$ (perpendicular to the rod) and the rod rotating with angular velocity Ω.

Taking x-, y-axes in the plane of the table, parallel and perpendicular to the initial position of the rod (FIG. 6.12) the equations of motion are

$$M\ddot{\bar{x}} = 0; \quad \dot{\bar{x}} = \text{constant} = 0,$$

$$M\ddot{\bar{y}} = 0; \quad \dot{\bar{y}} = \text{constant} = l\Omega,$$

$$I_G \ddot{\theta} = 0; \quad \dot{\theta} = \text{constant} = \Omega.$$

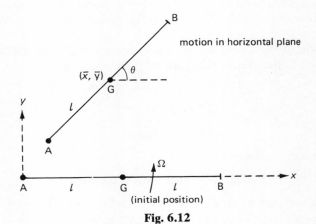

Fig. 6.12

Choosing the origin and initial line of O such that at time $t = 0$, $\bar{x} = \bar{y} = \theta = 0$, we deduce that at any time t,

$$\bar{x} = 0$$

$$\bar{y} = (l\Omega)t$$

$$\theta = \Omega t$$

We note that the centre of mass is the mid-point of the rod, so the coordinates of point B are

$$x = \bar{x} + l \cos \theta = l \cos \Omega t$$

$$y = \bar{y} + l \sin \theta = l\Omega t + l \sin \Omega t.$$

Exercise 3

1 (i) Given that

$$\mathbf{r} = x\mathbf{i} + y\mathbf{j} + z\mathbf{k}$$

$$\mathbf{F} = X\mathbf{i} + Y\mathbf{j} + Z\mathbf{k}$$

evaluate the third (i.e. $\mathbf{k}$-) component of (a) $\mathbf{r} \wedge \ddot{\mathbf{r}}$ (b) $\mathbf{r} \wedge \mathbf{F}$.

(ii) Given that $m\ddot{x} = X$, $m\ddot{y} = Y$, verify that your result for (b) is m times your answer for (a).

2 Using the cartesian method of question 1, and including internal forces (X', Y') as well as external forces (X, Y) prove that, for a system of particles moving in a plane,

$$\frac{d}{dt}\sum m_i(x_i\dot{y}_i - \dot{x}_iy_i) = \sum m_i(x_i\ddot{y}_i - \ddot{x}_iy_i)$$

$$= \sum (x_iY_i - y_iX_i)$$

3 Using vector notation, prove from memory that, for any system of particles, the rate of change of angular momentum about the origin, i.e. of $\sum \mathbf{r}_i \wedge m\dot{\mathbf{r}}_i$, is equal to the total moment of external forces about the origin.

4 A rigid body rotates about a fixed axis with angular velocity $\dot{\theta}$. Using cylindrical polar coordinates (r, θ, z), with the z-axis coinciding with the axis of rotation, write down the components of acceleration of any constituent particle mass m_i, coordinates r_i, θ_i, z_i, in each of the r, θ, z directions. Deduce that

$$\sum \mathbf{r}_i \wedge m_i\ddot{\mathbf{r}}_i = I_0\ddot{\theta}\mathbf{k} \text{ where } I_0 = \sum m_ir_i^2$$

(I_0 is called the moment of inertia of the body about the fixed axis, $\mathbf{k}$ is unit vector parallel to the z-axis).

5 A uniform hoop of radius a and mass M is free to move in a horizontal plane across the surface of a smooth horizontal table. Initially point A of its circumference is stationary and the diameter through A has angular velocity Ω. Find:
(i)   the initial velocity of the hoop's centre of mass,
(ii) the velocity of the centre of mass after 4 seconds,
(iii) the position of the centre of mass after 4 seconds,
(iv) the angular acceleration and angular velocity of the hoop at any instant,
(v)   the position and velocity of point A of the hoop after 4 seconds.

6 Example 2 in the text (§ 6.74) is modified as follows: in the motion after release, a time-dependent force of magnitude $3t$ is applied to the rod at its centre G. The direction of the time-dependent force is at all times parallel to the initial velocity of G. Find the position of B at time t under the new conditions.

7 Find the moment of inertia I_G of the system of four equal masses m shown in FIG. 6.13, the direction of the axis through the centre of mass being:
(i)   parallel to the dotted line of symmetry,
(ii) in the plane of the paper, perpendicular to the dotted line,
(iii) perpendicular to the plane of the paper.

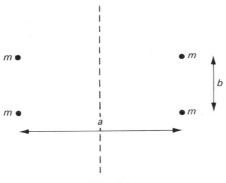

Fig. 6.13

8 Find the moment of inertia I_0 (as defined in question 4) about the z-axis of the system of masses: 2 kg at point $(1, 1, 1)$; 3 kg at point $(0, 2, 1)$; 4 kg at point $(-1, 0, 5)$.

9 Find the moment of inertia of the system of particles given in question 8, (i) about the x-axis, (ii) about the y-axis.

174 An introduction to mechanics and modelling

10 (Non-vectorial proof of $I_G\ddot{\theta} = M_G$.)

A typical mass m_i (one of a system of particles $i = 1, 2, \ldots$) has coordinates $x_i = \bar{x} + \mathsf{x}_i$, $y_i = \bar{y} + \mathsf{y}_i$, z_i = unchanging during the motion; and, in addition to the 'internal' forces exerted on it by other members of the system, the mass m_i is subject to the applied force X_i, Y_i, Z_i.

(i) Write down the x-component equation of motion for the mass m_i, $(m_i \ddot{x}_i = \cdots)$ in terms of derivatives of $\bar{x}$, x_i.
(ii) Multiply both sides of your equation by y_i.
(iii) Write down the component y-equation in terms of $\bar{y}$, y_i.
(iv) Multiply both sides of the y-equation by x_i.
(v) From the result of (iv) subtract the result of (ii).
(vi) In the result of (v), which terms represent the z-component of the moment about the centre of mass of forces applied by other particles of the system? (refer to equation 3–32).
(vii) Following § 6.71, postulate that the sum of the moments of each pair of equal and opposite internal forces (about the centre of mass G) is zero. Use 6–10c and deduce from (v) that

$$\sum m_i(\mathsf{x}_i \ddot{\mathsf{y}}_i - \mathsf{y}_i \ddot{\mathsf{x}}_i) = \sum (\mathsf{x}_i Y_i - \mathsf{y}_i X_i).$$

(viii) Show that if the particles are constituents of a rigid body, the left-hand side of the result of (vii) may be written

$$\frac{d}{dt} \sum m_i(\mathsf{x}_i \dot{\mathsf{y}}_i - \mathsf{y}_i \dot{\mathsf{x}}_i)$$

where $\mathsf{x}_i = \rho_i \cos\theta$ and $\mathsf{y}_i = \rho_i \sin\theta$ and ρ_i does not change with time. Deduce that the left-hand side may be written as

$$\frac{d}{dt} \sum m_i \rho_i^2 \dot{\theta}.$$

(ix) Express the result of (viii) as $I_G \ddot{\theta} = M_G$.

11 Use the particle model to justify the statement that if a rigid body is composed of two parts whose moments of inertia about a given axis are I_1, I_2 respectively, then the moment of inertia of the entire composite body about the axis is $I_1 + I_2$.

6.75 Two-dimensional equations of motion as ordinary differential equations

> The equations of motion of any rigid body moving in two dimensions are (for the motion of G)
>
> $$M\ddot{\mathbf{r}} = \mathbf{F} \qquad \qquad 6\text{--}8$$
>
> (for motion relative to G)
>
> $$I_G \ddot{\theta} = M_G \qquad \qquad 6\text{--}27$$
>
> Together these two equations allow a complete description to be given of the subsequent motion of the body, provided its initial position, velocity and angular velocity are specified.

Equations 6–8 and 6–27 are both called equations of motion. To distinguish them, the former is called the equation of **translation** of the body, the latter being the equation of rotation, or **angular equation of motion**.

A system-of-particles model 175

Equation 6–8 may be expressed as three cartesian equations, viz.

$$M\ddot{x} = X; \quad M\ddot{y} = Y; \quad M\ddot{z} = Z$$

Each of these, like $I_G\ddot{\theta} = M_G$, is of the mathematical form $c\ddot{x} = f(\text{variables}$ affecting the force system) where x stands for the position variable, c is a constant and f is a supposedly known function, usually depending on position, velocity and time. The mathematical form

$$c\ddot{x} = f(x, \dot{x}, t) \qquad 6\text{–}29$$

embraces most of the common soluble equations of translation or rotation. The procedures which allow us to deduce $\dot{x}$ and x from equations of form 6–29 are discussed at length in Chapter 7.

The reader's confidence in dealing with the angular motion of rigid bodies in two dimensions may be consolidated by observing the closeness of the parallel between motion in a straight line and rotation about an axis (either fixed or through G), as seen in the Table 6.1.

TABLE 6.1 Parallels between motion in a straight line and rotation about a fixed axis.

	Motion in a straight line	Angular motion
Coordinate	x	θ
Velocity/angular velocity	$\dot{x}$	$\dot{\theta}$
Acceleration/angular acceleration	$\ddot{x}$	$\ddot{\theta}$
Force/moment of forces	X	M
(Inertial) mass/moment of inertia	m	I
Equation of motion	$m\ddot{x} = X$	$I\ddot{\theta} = M$
When $X = 0$/when $M = 0$	$m\dot{x} = $ constant	$I\dot{\theta} = $ constant
Linear momentum/angular momentum	$m\dot{x}$	$I\dot{\theta}$
Integral of equation of motion with respect to coordinate	$[\tfrac{1}{2}m\dot{x}^2] = \int X\,dx$	$[\tfrac{1}{2}I\dot{\theta}^2] = \int M\,d\theta$
'Kinetic energy'	$\tfrac{1}{2}m\dot{x}^2$	$\tfrac{1}{2}I\dot{\theta}^2$
'Work done'	$\int X\,dx$	$\int M\,d\theta$
When force/moment is constant	$\ddot{x} = $ constant	$\ddot{\theta} = $ constant
	$= a$, say	$= \alpha$, say
If also at $t = 0$,	$\dot{x} = u$	$\dot{\theta} = \omega$
	and $x = x_0$	and $\theta = \theta_0$
then at any t	$\dot{x} = u + at$	$\dot{\theta} = \omega + \alpha t$
	$\dot{x}^2 = u^2 + 2ax$	$\dot{\theta}^2 = \omega^2 + 2\alpha\theta$
	$x - x_0 = ut + \tfrac{1}{2}at^2$	$\theta - \theta_0 = \omega t + \tfrac{1}{2}\alpha t^2$

6.76 Rigid body with three-dimensional motion

When the restrictive two-dimensional hypothesis (a) of § 6.73 is not satisfied, it is still possible to simplify $\sum \mathbf{s}_i \wedge m_i \ddot{\mathbf{s}}_i$ by use of the rigid body postulate (b), and to conduct three-dimensional mechanics in a roughly similar way. However, the processes are much heavier, and the student will do best first to master the use of the two-dimensional equations. (Experiment abundantly confirms that the equations of motion derived from the particle model, as above, give reliable predictions.)

176 An introduction to mechanics and modelling

The reader should recognize however, that some phenomena cannot be understood without introducing the third dimension: in particular, the theory of gyroscopic effects is essentially three-dimensional, and no two-dimensional analogy will 'explain' them.

6.8 Rigid body rotation about a fixed axis

When a rigid body rotates about a fixed axis, the equations $M\ddot{\mathbf{r}} = \mathbf{F}$ and $I_G\ddot{\theta} = M_G$ are still true, even though they are complicated by the contributions to $\mathbf{F}$ and M_G from unknown forces exerted by the axis. It might seem that the presence of these unknown forces would make the equations of motion useless, but this is not so.

> The very constraint which causes an unknown force to arise imposes a corresponding geometrical condition. There is different, but still sufficient information for the solution of the problem.

The fixed axis ensures that, instead of three coordinates $\bar{x}$, $\bar{y}$, θ, only one, namely θ, is needed to determine the position of the body. Instead of finding $\bar{x}$ from the equation of motion $M\ddot{\bar{x}} = X$, it can be determined by a geometrical condition of the form $\bar{x} = d\cos\theta$, where d is the distance of the centre of mass from the axis of rotation. Similarly, instead of $M\ddot{\bar{y}} = Y$, a relation $y = d\sin\theta$ is available. Moreover, knowing the forms of x and y, the unknown reaction contributions in X and Y can be deduced, and this provides enough information to insert in the equation $I_G\ddot{\theta} = M_G$ for the determination of θ.

The general equations of motion are not, however, the neatest tool available when there is a fixed axis. We can conveniently apply equation 6–18 to any rigid body rotating about a fixed axis, extending to a body of any shape the discussion previously restricted to a plane body (flywheel) rotating about an axis perpendicular to its plane.

With an origin on the fixed axis of rotation, the acceleration of any particle of the rigid body may be written

$$\ddot{\mathbf{r}}_i = -R_i\dot{\theta}^2\mathbf{I} + R_i\ddot{\theta}\mathbf{J}$$

where R_i is the distance of mass m_i from the (fixed) axis of rotation, and θ is the angle through which the body has rotated measured from some standard position. Paralleling the derivation of equations 6–26 and 6–27, we find

$$[\sum \mathbf{r}_i \wedge m_i\ddot{\mathbf{r}}_i]_K = \sum m_iR_i^2\ddot{\theta}$$
$$= I_0\ddot{\theta}$$

where

$$I_0 (\equiv \sum m_iR_i^2)$$

does not vary during the motion, because the R_i's do not change when the body rotates about the prescribed axis. I_0 is called the moment of inertia of the body about the axis.

It follows from equation 6–18 that

> $I_0\ddot{\theta}$ = the third, i.e. **K**-component of the total moment $\mathbf{M}_0$ of external forces about 0
>
> $$I_0\ddot{\theta} = \mathbf{M}_0 \text{ say} \qquad 6\text{–}30$$
>
> wherein $\mathbf{M}_0$ may conveniently be evaluated as
>
> $$\mathbf{M}_0 = \sum (x_i Y_i - y_i X_i) \qquad 6\text{–}31$$

Equation 6–30 is the **equation of angular motion of the rigid body about the given fixed axis**.

6.81 Parallel axis theorem

In FIG. 6.14 we show the fixed axis through O (pointing upwards) and a parallel axis through the centre of mass G, **d** represents the displacement of the axis through G from the fixed axis (**d** is perpendicular to the axis)

$$\mathbf{R}_i = \mathbf{d} + \boldsymbol{\rho}_i$$

where $\boldsymbol{\rho}_i$ has the same significance as in § 6.73 and FIG. 6.10. Consequently,

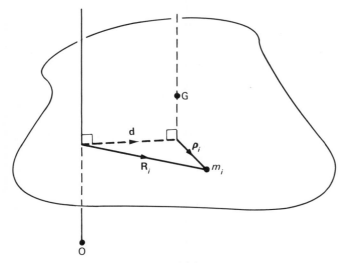

Fig. 6.14

$$I_0 = \sum m_i R_i^2 = \sum m_i \mathbf{R}_i \cdot \mathbf{R}_i = \sum m_i (\mathbf{d} + \boldsymbol{\rho}_i) \cdot (\mathbf{d} + \boldsymbol{\rho}_i)$$
$$= (\sum m_i)\mathbf{d} \cdot \mathbf{d} + 2(\sum m_i \boldsymbol{\rho}_i) \cdot \mathbf{d} + \sum (m_i \boldsymbol{\rho}_i \cdot \boldsymbol{\rho}_i)$$

We write $\sum m_i = M$. $\sum m_i \boldsymbol{\rho}_i$ is the component in the plane of motion of the quantity $\sum m_i \mathbf{s}_i$ (cf. equation 6–23, in which **p** is perpendicular to the plane of motion), and, from equation 6–10, $\sum m_i \mathbf{s}_i = 0$. Hence

$$I_0 = Md^2 + \sum m_i \rho_i^2$$
$$= Md^2 + I_G \qquad 6\text{–}32$$

178 An introduction to mechanics and modelling

The moment of inertia (I_0) about any axis is therefore greater than that about the parallel axis through the centre of mass by amount Md^2 where d is the perpendicular distance between those two axes.

6.82 Examples

Example 1
A light-topped tea trolley on castors is set rotating about one of its four vertical legs, the castor of this leg being trapped on the floor. Given that the friction effects at the castors are adequately represented by supposing a coefficient of friction of 0.1 between each trolley leg and the floor, estimate the rate at which the trolley must be set rotating in order to come to rest in 1 second.

All equations of motion will involve the trolley's mass. This is easy to discover by weighting, so we suppose that M is known.

Some estimate of the trolley's moment of inertia about the axis of rotation is needed. In the absence of relevant experimental or theoretical information, we propose a crude estimate as follows: let the trolley top be of length a and width b (both easily measurable); we shall assume all the mass of the trolley to be in its four equal legs. Thus in FIG. 6.15, O represents the axis of rotation and masses $\tfrac{1}{4}M, \tfrac{1}{4}M, \tfrac{1}{4}M$ are situated at distance a, $\sqrt{(a^2+b^2)}$, b from O. Our estimate of I_0 is consequently

$$I_0 \simeq \tfrac{1}{4}M[a^2+(a^2+b^2)+b^2]$$
$$= \tfrac{1}{2}M(a^2+b^2) \qquad\qquad 6\text{--}33$$

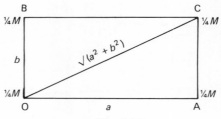

Fig. 6.15

Next consider the forces acting on the trolley, particularly in respect of their contribution to the moment M_0 about the vertical axis of rotation (through O). We remind ourselves of the three useful categories of force, viz.

(i)   forces at a distance (especially gravity);
(ii) known applied forces;
(iii) contact forces not already accounted for.

(i)   Gravitational forces are vertical, and therefore contribute nothing to the moment M_0.
(ii) During the period of rotation between the initial and final positions, there are no known applied forces except that at the foot of the 'fixed' leg (i.e.

through the axis of rotation) which is known only in the sense that it is just sufficient to prevent the leg moving. Its contribution to M_0 is zero.

(iii) The horizontal components of the other contact forces at the feet are – by our stated assumption – calculable as friction forces provided the normal reactions are known. Supposing that the weight of the table is supported equally by its four legs, the normal reaction at each is $\frac{1}{4}Mg$ (equation of motion $4R - Mg = 0$). The horizontal, frictional component of the contact force at each of the three moving feet will therefore be represented by $0.025\,Mg$ acting in a direction opposite to the direction of motion; thus the friction force at A acts perpendicular to OA and contributes a moment $0.025\,Mga$, that at C contributes a moment $0.025\,Mg\sqrt{(a^2+b^2)}$ and that at B a moment $0.025\,Mgb$, all the moments being in the sense which opposes rotation.

The equation of angular motion about the fixed axis therefore takes the form

$$\tfrac{1}{2}M(a^2+b^2)\ddot{\theta} \simeq -0.025Mg[a+\sqrt{(a^2+b^2)}+b]$$

whence, integrating with respect to time,

$$\tfrac{1}{2}M(a^2+b^2)\dot{\theta} = -0.025Mg[a+b+\sqrt{a^2+b^2}]t+C \qquad 6\text{--}34$$

We are given that $\dot{\theta} = 0$ when $t = 1$, so $C = 0.025\,Mg[a+b+\sqrt{(a^2+b^2)}]$, and using equation 6–34 with this value for C, we find that the initial angular velocity (the value of $\dot{\theta}$ at time $t = 0$) is

$$\dot{\theta}_0 = 0.05g[a+b+\sqrt{(a^2+b^2)}]/(a^2+b^2) \qquad 6\text{--}35$$

Comment: we should be a bit sceptical about two of the modelling assumptions, namely, the mass concentrated in the legs only, and the equivalence of castor friction effects to ordinary sliding friction. The author nevertheless thinks that equation 6–35 might be of some use to give an 'order of magnitude' estimate. For a $1 \times 4/3$ m trolley (diagonal length 5/3), the estimated value of $\dot{\theta}_0 = 0.7°$ radian/sec. For more accurate predictions, preliminary experiments to provide better values for I_0 and the castor friction effects would have to be undertaken.

Example 2
(Impulsive moments and angular momentum). What is the least impulse which could produce the angular velocity $\dot{\theta}_0$ in the tea trolley in question '1, starting from rest?

Just as in § 5.2 we converted the equation

$$m\mathbf{a}\left[= \frac{d}{dt}(m\mathbf{v}) \right] = \mathbf{F}$$

into

$$m\mathbf{v}(t_2) - m\mathbf{v}(t_1) = \int_{t_1}^{t_2} \mathbf{F}\,dt \qquad 6\text{--}36(a)$$

and gave the left-hand side the name 'change of momentum' and the right-hand

180 An introduction to mechanics and modelling

side the name 'impulse of the force **F**', so from the equation

$$I_0\ddot{\theta}\left[\simeq \frac{d}{dt}(I_0\dot{\theta})\right] = M_0$$

we deduce

$$I_0\dot{\theta}(t_2) - I_0\dot{\theta}(t_1) = \int_{t_1}^{t_2} M_0 \, dt \qquad 6\text{-}36(b)$$

To the left-hand side we give the name 'change of angular momentum about the axis at O', and to the right-hand side the name 'impulsive moment' (about that axis).

The component parallel to the axis of rotation, of any force, contributes nothing to the moment M_0; so the least force required to produce a given M_0 lies in the plane perpendicular to the axis of rotation, and is of magnitude F such that

$$pF = M_0 \quad \text{i.e.} \quad F = M_0/p,$$

where p, the perpendicular distance of F from the axis has the greatest magnitude consistent with the dimensions of the body – in this case $p = \sqrt{(a^2+b^2)}$.

Since $M_0 = pF$ and p is constant for the duration of F

$$\int_{t_1}^{t_2} M_0 \, dt = p \int_{t_1}^{t_2} F \, dt$$
$$= \sqrt{(a^2+b^2)} \times (\text{the impulse of force } \mathbf{F})$$

In the present case we put $\dot{\theta}(t_2) = \dot{\theta}_0$ and $\dot{\theta}(t_1) = 0$ in equation 6-36(b). We deduce that the required least impulse is applied at the end of the diagonal OC in the plane of the table top, at right angles to OC, and is of magnitude

$$I_0\dot{\theta}_0/p = 0.025 Mg[a+b+\sqrt{(a^2+b^2)}]/\sqrt{(a^2+b^2)}.$$

6.821 Footnote on impulses

An impulse signifies a very large force of magnitude F acting for a very short time $t_2 - t_1$ during which the direction of the force may be assumed to remain unchanged; the magnitude of its impulse

$$J \equiv \int_{t_1}^{t_2} F \, dt \leqslant (t_2 - t_1) \times \text{maximum value of } F.$$

Consider a body subjected to a blow whose duration is so short that the distance moved by the body during the blow is regarded as negligible. The modelling assumption 'distance moved $\simeq 0$' may also be written '(mean speed) $\times (t_2 - t_1) \simeq 0$'. Consistency requires that any finite multiple of $(t_2 - t_1)$ is considered to be negligible. In particular, the impulse of any **finite** force such as the weight Mg of the body is negligible. By contrast, for the blow under consideration, F is conceived to take such large values during its interval of action that

$$J = \int_{t_1}^{t_2} F \, dt$$

has a non-zero value.

A system-of-particles model   181

A direction is associated with an impulse, $J = \int_{t_1}^{t_2} F \, dt$, and a vectorial definition of impulse may be given, namely $\mathbf{J} = \int_{t_1}^{t_2} \mathbf{F} \, dt$, but the evaluation of integrals of vectors will not be discussed in this volume; consideration of components will serve our purposes. These matters will be clarified in section 9.2.

If a friction force acts on the body, it may be either finite (a zero impulse) or impulsive, according as the corresponding normal reaction R is finite or impulsive. From equation 6–36(a) we see that any force which causes an effectively instantaneous change of velocity is impulsive – it constitutes an impulse whose magnitude is equal to the (instantaneous) change of momentum.

Exercise 4

1. A rigid body rotates about a fixed axis under the action of forces which produce a time dependent moment M_0 about this axis. The body's moment of inertia I_0 (about the fixed axis) is of magnitude 6 SI units. Find its angular velocity after 2 seconds in each of the following situations:
   (i)   Body starts from rest; $M_0 = 3t + 2$ (SI units).
   (ii) Body starts with $\dot{\theta} = 5$; $M_0 = 3t - 2$.
   (iii) Body starts from rest, $M_0 = 5 \cos t$.
   (iv) Body starts with $\dot{\theta} = -3$, $M_0 = \sin 5t$.

2. A system of particles in the x-y plane consists of 1 kg at $(2, 1, 0)$; 2 kg at $(-1, -1, 0)$; 3 kg at $(3, 0, 0)$. Write down the distance of each particle from (a) the x-axis, (b) the z-axis. Denoting the moments of inertia of the system about the x-axis by I_x, about the y-axis by I_y, and about the z-axis by I_z, find I_x, I_y and I_z and verify that $I_z = I_x + I_y$.

3. **(Perpendicular axis theorem for a plane distribution of mass.)** In the notation of question 2, show that $I_x + I_y = I_z$ for any system of masses in the x-y plane. Explain why it is not true for a three-dimensional distribution of mass.

4. A mass m is at distance a from the origin O. Show that whatever its direction from O, $I_x + I_y + I_z = 2ma^2$.

5. Model a uniform spherical shell of mass M as a system of particles m_i each at distance a from the centre O. Deduce that for the shell, $I_x + I_y + I_z = 2Ma^2$.

6. By writing $I_x = I_y = I_z$ in the result of question 5, deduce the moment of inertia of a uniform spherical shell about any diameter.

7. By use of the result of question 6, and the parallel axis theorem of § 6.81, deduce that the moment of inertia of a uniform spherical shell of radius a about any tangent line is of magnitude $5Ma^2/3$.

8. A netball/basketball post consists of a straight hollow metal pole of mass 20 kg, supporting a hoop of mass 2 kg, whose diameter is 0.3 m and from which hangs a net. The foot of the pole is well sunk below the surface of the ground, so that the pole always remains vertical, but the subterranean portion has worked loose, so that the pole can rotate in its socket.
   (i)   Find the moment of inertia of the hoop about a vertical axis through its centre.
   (ii) Find the moment of inertia of the hoop about the axis of the pole (use the parallel axis theorem).
   (iii) The pole's inner diameter is not known; its outer diameter is about 5 cm. What estimate would you suggest for the moment of inertia of the complete netball post about its axis of rotation.

9. The netball/basketball post of question 8 receives an impulsive blow on the rim of its hoop, the line of action of the blow being horizontal, at a perpendicular distance of 0.1 m from the pole and 0.05 m from the hoop's centre. The magnitude of this impulse is 4 N s. Assuming that the resisting moment exerted by the socket on the foot of the pole is constant, and that the rotation of the pole continues for 2 seconds, estimate the magnitude of this resisting moment.

182 An introduction to mechanics and modelling

10 Assuming that the impulse in question 9 was delivered by a ball of mass $\frac{1}{2}$ kg flying horizontally, what change would have been observed in the ball's velocity?

11 Enumerate the defects you can see in the modelling proposed in questions 8, 9 and 10 above.

12 Accepting that measurable forces can be applied to a given body, describe (i.e. invent) a series of experiments whereby the validity of $I_0 \ddot{\theta} = M_0$ as a 'law of nature' could be tested.

6.83 Rigid body rotation about a moving axis (two-dimensional motion)

The advanced student, or any practitioner of two-dimensional mechanics who is much concerned with the behaviour of unfastened doors on moving vehicles, may be interested in the form taken by the angular equation of motion when **an axis which does not pass through the centre of mass** is considered. For brevity we consider all particles in the plane of motion, the normal to which is the unit vector **k**.

Taking an axis, perpendicular to the plane of motion, through the point Á of § 6.2, FIG. 6.3, we write

$$\mathbf{r}_i = \mathbf{r}_A + \mathbf{R}_i, \qquad \mathbf{M}_A = \sum \mathbf{R}_i \wedge \mathbf{F}_i = \sum \mathbf{R}_i \wedge m_i \ddot{\mathbf{r}}_i, \quad \text{and} \quad M_A = \mathbf{M}_A \cdot \mathbf{k}$$

Then

$$\mathbf{M}_A = \sum \mathbf{R}_i \wedge m_i (\ddot{\mathbf{r}}_A + \ddot{\mathbf{R}}_i)$$
$$= \sum \mathbf{R}_i \wedge m_i \ddot{\mathbf{r}}_A + \sum \mathbf{R}_i \wedge m_i \ddot{\mathbf{R}}_i$$
$$= \sum (m_i \mathbf{R}_i) \wedge \ddot{\mathbf{r}}_A + I_A \ddot{\theta} \mathbf{k}$$

or

$$M_A = (M \mathbf{R}_G \wedge \ddot{\mathbf{r}}_A) \cdot \mathbf{k} + I_A \ddot{\theta}$$

This formula is particularly useful when the acceleration $\ddot{\mathbf{r}}_A$ of the point A is known. $\mathbf{R}_G$ is the position vector of the centre of mass relative to the point A; the reader should consider the minor modifications necessary when the points of the body move, not in the same plane, but in parallel planes.

Exercise

A train door has mass 150 kg, width 0.8 m and moment of inertia 30 kg m² about its hinge. The hinge is on the leading edge of the door, and the door is partly open when the train starts. Given that the train starts from rest with uniform acceleration 0.2 m s⁻², show that the acute angle θ between the door and the side of the train initially satisfies the equation

$$\ddot{\theta} = -0.4 \sin \theta.$$

When does this equation cease to be valid?

6.9 Summary of formulae valid for any body of matter; tables of centres of mass and moments of inertia of rigid bodies of standard shapes

6.91 Systems of particles which do not constitute a single rigid body

Although in this chapter we have directed our thoughts mainly to the description of the motion of a single rigid body, all the work on systems of particles prior to the assumption of a rigid connection is applicable to any identifiable body of matter (perhaps a volume of fluid, perhaps a finite number of loosely associated rigid bodies, perhaps a composite of rigid and deformable elements).

We summarize below the general results which may be applied in all these situations. In applications to non-rigid bodies of matter, the expression 'centre of mass' signifies a point which, while central to the system of objects considered, changes its distance from the component parts of the system as those constituents themselves move relative to one another.

6.92 General properties relating to centres of mass

I $M = \sum m_i$ (definition of M)
   $M\bar{\mathbf{r}} = \sum m_i \mathbf{r}_i$ (definition of $\bar{\mathbf{r}}$)
   $M\ddot{\bar{\mathbf{r}}} = \mathbf{F} \equiv$ vector sum of all external forces on the system (i.e. $\bar{\mathbf{r}}$ is the position vector of the point to which Newton's second law applies).

II $M\dot{\bar{\mathbf{r}}} = \sum m_i \dot{\mathbf{r}}_i \equiv$ the total linear momentum of the system. The linear momentum of any system $(\sum m_i \dot{\mathbf{r}}_i)$ is equal to that of the whole mass M moving with the velocity of the centre of mass $(M\dot{\bar{\mathbf{r}}})$.
   When the external force is zero, $\mathbf{F} = 0$, $M\ddot{\bar{\mathbf{r}}} = 0$, $M\dot{\bar{\mathbf{r}}} =$ constant, i.e. **in the absence of external forces, linear momentum is conserved**.

III   Composite body theorem:

$$M\bar{\mathbf{r}} = M_1 \mathbf{r}_1 + M_2 \mathbf{r}_2 + \cdots$$

where the right-hand side includes every subdivision of the system in question.

6.93 Angular momentum properties of general systems

I   The total angular moment of a system about the origin is defined as

$$\sum (\mathbf{r}_i \wedge m_i \dot{\mathbf{r}}_i).$$

The total angular momentum of the system relative to its centre of mass is

$$\sum (\mathbf{s}_i \wedge m_i \dot{\mathbf{s}}_i) \qquad (1)$$

The total angular momentum about the centre of mass is defined as

$$\sum (\mathbf{s}_i \wedge m_i \dot{\mathbf{r}}_i) = \sum [\mathbf{s}_i \wedge m_i (\dot{\bar{\mathbf{r}}} + \dot{\mathbf{s}}_i)] \qquad (2)$$
$$= (\sum m_i \mathbf{s}_i) \wedge \dot{\bar{\mathbf{r}}} + \sum (\mathbf{s}_i \wedge m_i \dot{\mathbf{s}}_i)$$
$$= \sum (\mathbf{s}_i \wedge m_i \dot{\mathbf{s}}_i).$$

So there is no need to distinguish between the quantities labelled (1) and (2) above.

II $\dfrac{d}{dt} \sum (\mathbf{r}_i \wedge m_i \dot{\mathbf{r}}_i) = \mathbf{M}_0$

i.e. **the rate of change of the angular momentum of any system about a fixed point** (which may be taken as origin) **is equal to the vector sum of the moments of external forces about that point**.

184 An introduction to mechanics and modelling

III $\quad \dfrac{\mathrm{d}}{\mathrm{d}t}\sum(\mathbf{s}_i \wedge m_i\dot{\mathbf{s}}_i) = \mathbf{M}_G$

i.e. **the rate of change of the angular momentum of the system about its centre of mass is equal to the total moment of external forces about the centre of mass.**

IV $\quad \sum \mathbf{r}_i \wedge m_i\dot{\mathbf{r}}_i = \sum(\bar{\mathbf{r}}+\mathbf{s}_i) \wedge m_i(\dot{\bar{\mathbf{r}}}+\dot{\mathbf{s}}_i)$

$\qquad = \bar{\mathbf{r}} \wedge \sum m_i\dot{\bar{\mathbf{r}}} + \left(\sum m_i\mathbf{s}_i\right) \wedge \dot{\bar{\mathbf{r}}} + \bar{\mathbf{r}} \wedge \sum m_i\dot{\mathbf{s}}_i + \sum(\mathbf{s}_i \wedge m_i\dot{\mathbf{s}}_i)$

$\qquad = \bar{\mathbf{r}} \wedge M\dot{\bar{\mathbf{r}}} + \sum(\mathbf{s}_i \wedge m_i\dot{\mathbf{s}}_i)$

The angular moment about the origin is equal to the angular momentum of the whole mass M moving with the centre of mass together with the angular momentum of the system relative to the centre of mass.

V   (a) If the moment $\mathbf{M}_0$ of external forces is zero, angular momentum about the origin is conserved.
(b) If the moment $\mathbf{M}_G$ of external forces is zero, angular momentum about the centre of mass is conserved.

VI For two-dimensional motion of a rigid body (axis of rotation parallel to $\mathbf{k}$), Angular momentum

$$\sum \mathbf{s}_i \wedge m_i\dot{\mathbf{s}}_i = I_G\dot{\theta}\mathbf{k};$$

Equation of angular motion

$$I_G\ddot{\theta} = \mathbf{M}_G;$$

If an axis in the body is fixed and 0 is a point on it

$$I_0\ddot{\theta} = \mathbf{M}_0.$$

Assignment

(i) Summarize the line of thought whereby in Chapters 4, 5 and 6 the equations

$$M\ddot{\mathbf{r}} = \mathbf{F}; \qquad I_G\ddot{\theta} = \mathbf{M}_G$$

for the two-dimensional motion of a rigid body have been derived from the equation of motion $m\ddot{\mathbf{r}} = \mathbf{F}$ for a point particle.

(ii) Newton's law was originally formulated with real bodies in mind; so what insights do you think have been provided by the theoretical 'derivation' of $M\ddot{\mathbf{r}} = \mathbf{F}$?

6.94 Centres of mass of standard bodies (uniform density)

	Body considered	Position of centre of mass
(i)	Body with centre of symmetry	Centre of symmetry
(ii)	Body with axis of symmetry	On axis of symmetry
(iii)	Circular arc, radius a, subtending angle 2α at centre O	$OG = (a \sin \alpha)/\alpha$
(iv)	Circular sector, radius a, angle 2α at centre O	$OG = (\tfrac{2}{3}a \sin \alpha)/\alpha$
(v)	Hemisphere, radius a, centre O	$OG = \tfrac{3}{8}a$

A system-of-particles model 185

	Body considered	Position of centre of mass
(vi)	Solid spherical segment height h, near face at distance $a-h$ from centre O	$OG = 3(2a-h)^2/4(3a-h)$
(vii)	Section of spherical shell, radius a, bounded by parallel planes distances d_1, d_2 from centre O	$OG = \frac{1}{2}(d_1+d_2)$ (d_1, d_2 same side) $= \frac{1}{2}|d_2-d_1|$ (d_1, d_2 opposite sides of O)
(viii)	Triangle	Point of intersection of medians, i.e. point $\frac{1}{3}$ of way from base to vertex
(ix)	Pyramid on plane base (including tetrahedron and similarly for cone)	$\frac{1}{4}$ of the way up the line joining the centres of mass of sections parallel to the base

The centres of mass of bodies of the following shapes may be derived from (i)–(ix) by use of the composite body theorem:

(x)	Quadrilateral (plane)	Use (viii)
(xi)	Segment of circle	Use (iv) and (viii)
(xii)	Parallel-faced slice of cone or pyramid	Use (ix)
(xiii)	Parallel-faced slice of sphere	Use (vi)

6.95 Moments of inertia of standard bodies (uniform density)

Body considered (mass M)	Location of axis		Moment of inertia
Straight rod, length $2a$ ($=l$)	(i)	Through centre, at right angles to length	$\frac{1}{3}Ma^2$ ($=Ml^2/12$)
	(ii)	Through end at right angles to length	$\frac{4}{3}Ma^2$ ($=\frac{1}{3}Ml^2$)
Uniform circular disc, radius a	(i)	Through centre, at right angles to plane	$\frac{1}{2}Ma^2$
	(ii)	Diameter	$\frac{1}{4}Ma^2$
Rectangle, sides $2a$ and $2b$	(i)	Through centre, parallel to side $2b$	$\frac{1}{3}Ma^2$
	(ii)	Through centre, at right angles to plane	$\frac{1}{3}M(a^2+b^2)$
Spherical shell, radius a	Diameter		$\frac{2}{3}Ma^2$
Uniform sphere, radius a	Diameter		$\frac{2}{5}Ma^2$
Solid circular cylinder radius a, length l	(i)	Axis of cylinder	$\frac{1}{2}Ma^2$
	(ii)	Diameter of one end	$M(\frac{1}{4}a^2+\frac{1}{3}l^2)$
Solid circular cone, radius a, height h	(i)	Axis of cone	$3Ma^2/10$
	(ii)	Diameter of base	$3M(\frac{1}{2}a^2+\frac{1}{3}h^2)/10$
Solid anchor ring, radius a, cross-section radius b	Axis of ring		$M(a^2+\frac{3}{4}b^2)$
Triangle, sides a, b and c	Through mass centre, at right angles to plane		$M(a^2+b^2+c^2)/36$

Radius of gyration (definition): the radius of gyration (k) of a body about a specified axis is that length k for which $Mk^2 = I$, where I denotes the moment of inertia of the body about the specified axis.

Parallel axis theorem: $I_O = Md^2 + I_G$
Perpendicular axis theorem:
$$(2D) \quad I_x + I_y = I_z$$
$$(3D) \quad I_x + I_y + I_z = 2\sum m_i r_i^2$$

7

Motion and differential equations

It will be assumed that the reader has already a good knowledge of calculus and some acquaintance with simple differential equations such as $dy/dx = ky$. By expressing the basic equation of mechanics, $M\ddot{\mathbf{r}} = \mathbf{F}$, in terms of components, that vector equation is replaced by scalar equations, e.g. $M\ddot{x} = X$, etc. The right-hand side (X) is commonly expressible as a function of position, time, velocity, or a combination of those; solution of problems concerning the motion of objects therefore often involves discussion of relations of the forms $dv/dt = f(v, t)$ or $\ddot{x} = g(\dot{x}, x, t)$. The former is a differential equation of the first order in v, t; the latter is a second-order differential equation with t as independent and x as dependent variable.

Some results from the calculus which the student will need when solving differential equations of mechanics are summarized in § 7.1 before starting our discussion of methods of solution.

7.1 Some basic tools of calculus

(i) The two statements

$$\dot{u} = v(t) \text{ and } u = \int v(t)\, dt$$

are equivalent. If $\dot{u} = v(t)$ and

$$\int v(t)\, dt = w(t) + C$$

in a range of t which includes t_1 and t_2, then

$$[u(t)]_{t_1}^{t_2} \equiv u(t_2) - u(t_1) = w(t_2) - w(t_1) = \int_{t_1}^{t_2} v(q)\, dq \text{ (any symbol } q\text{)}$$

(ii) If $u = u(s)$ and $s = s(t)$, then

$$\frac{du}{dt} = \frac{du}{ds} \cdot \frac{ds}{dt}.$$

In particular

$$\frac{d\dot{s}}{dt} = \frac{d\dot{s}}{ds} \cdot \dot{s} = \frac{d}{ds}\left(\frac{\dot{s}^2}{2}\right),$$

so

$$\ddot{s} = \frac{d}{ds}\left(\frac{\dot{s}^2}{2}\right)$$

whence

$$\int \ddot{s}\,ds = \frac{\dot{s}^2}{2} + C \qquad 7\text{–}1$$

After $M\ddot{\mathbf{r}} = \mathbf{F}$, equation 7–1 is perhaps the most useful formula in this book.

(iii) Integration by parts. If u and v are functions of t,

$$\int uv\,dt = u\int v\,dt - \int \left(\int v\,dt\right)\frac{du}{dt}\cdot dt \qquad 7\text{–}2(a)$$

The statement of this formula in words, learnt parrot-fashion, is undoubtedly more speedy and convenient in use than is substitution in formula 7–2(a). The following form of words may be used.

> The integral of a product is equal to the first function times the integral of the second, minus the integral of (the integral of the second, times the derivative of the first).

Which of the two factors u, v is regarded as the 'first' is a matter of choice; the factor which is less readily integrated is usually the better choice.

Note that when definite integrals are to be evaluated, the appropriate interpretation of equation 7–2(a) is

$$\int_{t_1}^{t_2} uv\,dt = \left[u\int v\,dt\right]_{t_1}^{t_2} - \int_{t_1}^{t_2}\left(\int v\,dt\right)\frac{du}{dt}\,dt \qquad 7\text{–}2(b)$$

The positions of t_1, t_2 on the right-hand side should be noted with care. The integral for which no limits are inserted is an **indefinite** integral, a function of t, in which it is immaterial whether an arbitrary constant is retained or jettisoned – the outcome is the same in either case.

(iv) Since derivation of the formulae

$$\frac{d}{d\theta}(\sin\theta) = \cos\theta, \qquad \frac{d}{d\theta}(\cos\theta) = -\sin\theta \text{ etc,}$$

involves use of the relation

$$\lim_{\delta\theta\to 0}\frac{\sin\delta\theta}{\delta\theta} = 1,$$

it is essential that $\delta\theta$ and therefore θ shall be measured in **radians** whenever trigonometrical functions are involved in the solution of differential equations.

188 An introduction to mechanics and modelling

Exercise 1

1 Write down:

(a) $\dfrac{d}{dx}(\tfrac{1}{2}\dot{x}^2)$, (b) $\dfrac{d}{dt}(\tfrac{1}{2}\dot{x}^2)$.

2 Integrate each of the following equations, performing one integration with respect to whichever variable seems convenient.

$\ddot{s} = 2s + 3$; (ii) $\ddot{s} = 2t + 3$; (iii) $\ddot{\theta} = -4\theta$; (iv) $\ddot{p} = 3\sin p$; (v) $\ddot{y} = \tfrac{1}{2}\dot{y}^2$.

3 Evaluate

$$\int_2^3 \ln x \, dx \left(\equiv \int_2^3 1 \cdot \ln x \, dx \right).$$

4 A stone is dropped from a third-floor window of a building. Its equation of motion through the air may be written $\ddot{y} = -g$, where $g \simeq 10 \text{ m s}^{-2}$. Explain, in the context of (i) above, why its position after 10 seconds is not 1000 metres below its starting point.

7.2 Special significance of the exponential function in discussions of differential equations

(i) If

$$s = A \exp mt$$

where m and A are constants, then

$$\dot{s} = mA \exp mt$$

and

$$\ddot{s} = m^2 A \exp mt$$

from which it follows that

$$\dot{s} - ms = 0; \quad \ddot{s} - 2m\dot{s} + m^2 s = 0; \quad \ddot{s} - m^2 s = 0; \text{ etc.}$$

Between any selection of derivatives of s, and s itself, there exists at least one linear relationship with constant coefficients.

Conversely, if a linear relationship with constant coefficients is known to hold between s and derivatives of s, say

$$\ddot{s} + a\dot{s} + bs = 0 \qquad\qquad 7\text{--}3$$

where a and b are constants, it will be reasonable to look for a solution related to the form

$$s = A \exp mt.$$

Exercise: find the value or values of m for which the suggested form $s = A \exp mt$ satisfies the given linear relationship $\ddot{s} + a\dot{s} + bs = 0$ for all t.

(ii) If

$$s = t \exp mt$$

then

$$\dot{s} = (1 + mt) \exp mt$$

Motion and differential equations

and
$$\ddot{s} = (2m + m^2 t) \exp mt$$

As well as linear relations with variable coefficients – for example $(1+mt)s = t\dot{s} - s$, $\dot{s}$, $\ddot{s}$ satisfy a linear relation with **constant** coefficients, namely

$$\ddot{s} - 2m\dot{s} + m^2 s = 0 \qquad 7\text{-}4$$

When looking for functions $s(t)$ whose derivatives satisfy linear relationships involving $\ddot{s}$, this property of $t \exp mt$ is often useful. When higher derivatives $\dddot{s}$, $\ddddot{s}$ are involved, $t^2 \exp mt$, $t^3 \exp mt$ have similar relevance.

A relationship between a function and its derivatives is called a 'differential equation'. The process of finding a function, preferably the most general function, which satisfies the relation is called 'solving' (or integrating) the differential equation. A linear relationship between s and its derivatives is called a linear differential equation with s as dependent variable.

Example
Find solutions of (i) $\ddot{s} - 6\dot{s} - 7s = 0$ (ii) $\ddot{s} - 6\dot{s} + 9s = 0$.

(i) Substitution of $s = A \exp mt$ yields
$$A(m^2 - 6m - 7) \exp mt = 0.$$

This is satisfied for non-zero A and all t if and only if $m^2 - 6m - 7 = 0$, i.e. $(m-7)(m+1) = 0$. We deduce that $s = A_1 \exp 7t$ is a solution, and that $s = A_2 \exp(-t)$ is also a solution, for arbitrary values of A_1 and A_2.

(ii) $s = A \exp(mt)$ satisfies the second given equation for all t provided
$$m^2 - 6m + 9 = 0; \text{ i.e. } (m-3)^2 = 0$$

We deduce that $s = A \exp(3t)$ is a solution 'twice'. The repeated root gives notice that this is a special case; further inspection shows us that the given equation is of the same form as equation 7-4, with $m = 3$. We conclude that, as well as $s = A \exp(3t)$, $s = Bt \exp(3t)$ is a solution, in which any numerical values may be given to A and B. In particular, $\exp(3t)$ and $t \exp(3t)$ are solutions.

Exercise 2
Find two solutions of each of the following differential equations;
1 $\ddot{s} - 3\dot{s} + 2s = 0$.   2 $\ddot{s} - 5\dot{s} + 6s = 0$.
3 $\ddot{s} - \dot{s} - 6s = 0$.   4 $\ddot{s} - 4\dot{s} + 4s = 0$.

7.21 Superposition of solutions and a use of complex numbers
If $s_1(t)$, $s_2(t)$ are two functions, each of which satisfies a linear differential equation, for example equation 7-3,

$$\ddot{s} + a\dot{s} + bs = 0,$$

then

$$\ddot{s}_1 + a\dot{s}_1 + bs_1 = 0$$

and
$$\ddot{s}_2 + a\dot{s}_2 + bs_2 = 0$$

Multiplying the s_1-equation by k_1, the s_2-equation by k_2, and adding the resulting equations,
$$(k_1\ddot{s}_1 + k_2\ddot{s}_2) + (ak_1\dot{s}_1 + ak_2\dot{s}_2) + (bk_1s_1 + bk_2s_2) = 0,$$
i.e.
$$\frac{d^2}{dt^2}(k_1s_1 + k_2s_2) + a\frac{d}{dt}(k_1s_1 + k_2s_2) + b(k_1s_1 + k_2s_2) = 0,$$
i.e.

> if each of $s_1(t)$ and $s_2(t)$ satisfies equation 7-3, then so also does the function $s(t) = k_1 s_1 + k_2 s_2$ (for any values of the constants k_1 and k_2).   7-5

So, from any two particular solutions, a wide class of solutions incorporating two arbitrary constants can be derived. The property 7-5 is a statement of the 'principle of superposition' in a form relevant to equation 7-3. The reader will see for himself that the principle is equally applicable to equations involving higher-order derivatives and coefficients a and b which are functions of t, provided only that those equations are **linear** differential equations in s, $\dot{s}$ and $\ddot{s}$.

Example
(i) $\ddot{s} - 6\dot{s} - 7s = 0$, (ii) $\ddot{s} - 6\dot{s} + 9s = 0$, (iii) $\dddot{s} + 3\ddot{s} + 3\dot{s} + s = 0$.

For (i), $s_1 = \exp(7t)$ and $s_2 = \exp(-t)$ are known to be solutions, so, by the principle of superposition
$$s = k_1 \exp(7t) + k_2 \exp(-t)$$
is a more general solution, with two arbitrary constants.
For (ii)
$$s = A \exp(3t) + Bt \exp(3t)$$
is a solution (with two arbitrary constants A and B).
For (iii)
$$s = C_1 \exp(-t) + C_2 t \exp(-t) + C_3 t^2 \exp(-t)$$
is a solution (with three arbitrary constants).

Exercise 3
Apply the principle of superposition to the equations of Exercise 2, writing down for each equation a solution containing two arbitrary constants.

7.211 Terminology
Equation 7-3 is called a second-order, linear, homogeneous differential equation. It is **second order** because the highest-order derivative occurring in it is a second

derivative, $\ddot{s}$. It is **linear** because s and its derivatives $\dot{s}$ and $\ddot{s}$ are to be found only in separate terms, multiplied by constants or functions of independent variables: the form of the most general linear second-order differential equation in s and t is

$$f_0(t)\ddot{s} + f_1(t)\dot{s} + f_2(t)s = f_3(t).$$

This equation is **inhomogeneous** because of the presence of the term $f_3(t)$. Equation 7–3 is **homogeneous** because it contains no terms except multiples of $\ddot{s}$, $\dot{s}$ and s (the coefficients being free from reference to s).

7.212 *An important application of the principle of superposition*

The following application of the principle of superposition to homogeneous second-order linear differential equations with constant coefficients is of special importance.

Given the equation

$$\ddot{s} - 2a\dot{s} + bs = 0, \qquad 7\text{–}6$$

(i.e. the general homogeneous second-order differential equation with constant coefficients, with the coefficient of $\dot{s}$ written as $-2a$ in order to avoid clumsy fractions and to neaten the signs in subsequent working), $s = \exp(mt)$ is a solution provided

$$m^2 - 2am + b = 0$$

i.e. if m has either of the values

$$m = a \pm \sqrt{(a^2 - b)}$$

When $a^2 - b$ is negative, $\sqrt{(a^2 - b)} = ic$ where c is real;

$$s = s_1 = \exp[(a + ic)t]$$
$$= \exp(at) \cdot \exp(ict)$$

is therefore a solution to equation 7–6, and

$$s = s_2 = \exp[(a - ic)t]$$
$$= \exp(at) \cdot \exp(-ict)$$

is also a solution. Neither s_1 nor s_2 separately can represent a real coordinate, but, using the principle of superposition, we deduce that

$$s = \exp(at)[k_1 \exp(ict) + k_2 \exp(-ict)] \qquad 7\text{–}7$$

is a solution of the given equation for any values, real or complex, of the constants k_1 and k_2. If (and only if) k_1 and k_2 are conjugate complex numbers, the expression for s (equation 7–7 with a, c both real) is real for all values of t. For, since

$$\exp(ict) = \cos ct + i \sin ct; \qquad \exp(-ict) = \cos ct - i \sin ct;$$

from equation 7–7,

$$s = \exp(at)[(k_1 + k_2)\cos ct + i(k_1 - k_2)\sin ct],$$

and if k_1 and k_2 are conjugate complex, $k_1 + k_2$ and $i(k_1 - k_2)$ are both real, $=A$ and B say, respectively. Since any two values of $k_1 + k_2$ and $k_1 - k_2$ are achievable by a suitable choice of k_1 and k_2, A and B may be chosen arbitrarily.

$$s = \exp(at)(A \cos ct + B \sin ct) \qquad 7\text{–}8$$

is equivalent to equation 7–7.

If a and c are real, any real solution of equation 7–6 may be written in the form 7–8, with real values of A and B.

(See also § 7.22 below.)

Example

$$\ddot{s} + 9s = 0$$

$s = \exp(mt)$ is a solution provided $m^2 + 9 = 0$, i.e. $m^2 = -9$, $m = \pm 3i$. Hence, using the foregoing result, and bearing in mind that $\exp 0 = 1$,

$$s = A \cos 3t + B \sin 3t$$

is a real solution of the given equation for any real values of the constants A and B.

Exercise 4
Derive (real) solutions, containing two arbitrary constants, for each of the following second-order linear differential equations with constant coefficients:

1. $\ddot{s} - 9s = 0$;
2. $\ddot{y} + 4y = 0$;
3. $\ddot{x} - 5\dot{x} + 6x = 0$;
4. $\ddot{x} + 5\dot{x} + 6x = 0$;
5. $\ddot{s} - \dot{s} - 2s = 0$;
6. $\ddot{\theta} - 3\dot{\theta} + 2\theta = 0$;
7. $d^2y/dx^2 + 3\, dy/dx - 4y = 0$;
8. $\ddot{z} - 2\dot{z} + 5z = 0$;
9. $\ddot{p} + 5p = 0$.

7.213 Simple harmonic motion

The differential equation $\ddot{x} = -Kx$ (i.e. $\ddot{x} + Kx = 0$) where K is positive, often occurs in the modelling of oscillatory systems. To display that K is positive, it is customary to write $K = \omega^2$, whereon the differential equation takes the form

$$\ddot{x} = -\omega^2 x \qquad 7\text{–}9$$

Any equation of this form is known as an equation of **simple harmonic motion**. By § 7.212 it has the solution

$$x = A \cos \omega t + B \sin \omega t$$

where A and B are arbitrary constants.

A device to which equation 7–9 applies is called a **harmonic oscillator**.

Example

Oscillations of a mass suspended by a spring (modelled by Hooke's law). With the coordinate z measured downwards from the position in which the

Motion and differential equations

spring (stiffness k) has its natural length, the equation of motion of the mass m shown in the idealized representation of FIG. 7.1 is

$$m\ddot{z} = mg - kz \qquad 7\text{--}10$$

Fig. 7.1

This equation is not homogeneous like the equations of Exercise 4, because of the constant mg. However, equation 7–10 may be rewritten

$$\ddot{z} = -(k/m)(z - mg/k).$$

This is of the form 7–9 with $z - mg/k = x$ and $k/m = \omega^2$. The position of the mass m at any time t is therefore given by

$$z - mg/k = A \cos \sqrt{(k/m)}t + B \sin \sqrt{(k/m)}t \qquad 7\text{--}11$$

and its velocity is

$$\dot{z} = -\sqrt{(k/m)}A \sin \sqrt{(k/m)}t + \sqrt{(k/m)}B \cos \sqrt{(k/m)}t \qquad 7\text{--}12$$

If two pieces of information are given about the motion, e.g. the initial position and velocity, the unknowns A and B can be determined. Thus, if the mass starts from rest ($\dot{z} = 0$) with the spring at its natural length ($z = 0$), substitution in 7–11 and 7–12 shows that A and B must satisfy $-mg/k = A$; $0 = \sqrt{(k/m)} \cdot B$; whence

$$A = -mg/k$$
$$B = 0.$$

The function

$$z = \frac{mg}{k}\left[1 - \cos \sqrt{\left(\frac{k}{m}\right)}t\right] \qquad 7\text{--}13$$

therefore both satisfies the equation of motion 7–10 and corresponds to the specified starting conditions.

194 An introduction to mechanics and modelling

The crucial question is, may we therefore deduce that 7–13 faithfully describes the motion of mass m which will actually ensue? Something must happen, so if no different solution of 7–13 exists satisfying the given starting conditions, we shall be able to conclude that 7–13 does represent the actual state of affairs. What is needed is an assurance that there are not two different solutions satisfying the same set of prescribed conditions, i.e. what is needed is a **uniqueness** theorem. Such theorems will be alluded to again in § 7.22, but we may state straight away that the student is unlikely to meet any situations in elementary mechanics where there is any difficulty about uniqueness.

> If a general solution, as defined in § 7.22 below, has been found for an equation of motion appropriate to the mechanical situation, and if the arbitrary constants in that general solution have been determined by the requirement that coordinates or velocities have correct values at one or more stated times, then insertion of the determined values in the general solution produces a solution which correctly describes the actual course of motion so long as the original equation of motion continues to apply.

According to our model of the problem under discussion, expression 7–13 therefore represents the **predicted variation** of coordinate z. The oscillatory nature of the predicted motion is apparent from the graph of 7–13, shown in FIG. 7.2. Any simple harmonic motion may be represented by a similar graph, only changes of origin and/or scale being required to cover all cases.

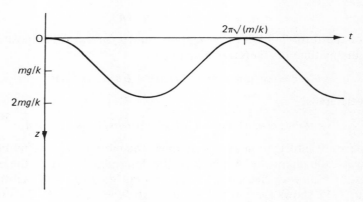

Fig. 7.2

Exercise: Verify that the variable $y = A \cos(pt + \varepsilon)$ satisfies an equation of simple harmonic motion whatever values are assigned to A and ε.

7.22 General solutions and uniqueness

One integration of an equation $\dot{s} = f_1(t)$ gives rise to one arbitrary constant in the resulting expression for s. Two integrations of an equation $\ddot{s} = f_2(t)$ give rise to two independent arbitrary constants in the expression for s.

Motion and differential equations

> Any solution s, of a first-order differential equation
> $$\dot{s} = F_1(t, s), \qquad 7\text{--}14$$
> which contains one arbitrary constant, is called a (or the) **general** solution of that equation.
>
> Any solution s, of a second-order differential equation
> $$\ddot{s} = F_2(t, \dot{s}, s), \qquad 7\text{--}15$$
> which contains two independent arbitrary constants, is called a (or the) **general** solution of that equation.

In dynamical applications, the variable s in equation 7–14 commonly represents a velocity, so that $\dot{s}$ is an acceleration; the form 7–15 arises when s is a coordinate, so that $\ddot{s}$ is an acceleration.

There will be no attempt here to **prove** the relevant uniqueness theorems. Indeed the equations of form 7–14 which the student will meet can usually be integrated directly, so recourse to a uniqueness theorem is unnecessary. For second-order equations, however, solution is seldom by direct integration. Methods based on trial forms (e.g. exp mt in the equations of Exercise 2) are customary, and a uniqueness theorem is desirable. Sufficient conditions for uniqueness are stated as follows.

Theorem: given that in some domain D of the variables t, $\dot{s}$, and s, $F_2(t, \dot{s}, s)$ is a continuous function of the three variables t, $\dot{s}$ and s; given that bounded partial derivatives $\partial F_2/\partial \dot{s}$ and $\partial F_2/\partial s$ exist in D; and given that when $t = t_0$, $\dot{s} = \dot{s}_0$ and $s = s_0$ where $(t_0, \dot{s}_0, s_0)$ lies inside domain D: then any solution of 7–15 which satisfies the given conditions at $t = t_0$ is the only function (within D) satisfying the prescribed conditions.

What is the significance of the theorem for us? It contains one reassurance and two cautions. The reassurance is that when F_2 is a function with continuous (and therefore finite) derivatives throughout the predicted course of motion, the solution derived by fitting a general solution is the solution of the physical problem under discussion.

The warnings are (a) if $F_2(t, \dot{s}, s)$ contains a surd, e.g. $\sqrt{f(t, \dot{s}, s)}$, logarithm, ln f or negative power, e.g. $1/f(t, \dot{s}, s)$, of some function f of the variables, then the uniqueness theorem does not apply at or beyond any point on the solution for which $f(t, \dot{s}, s) = 0$; and (b), when there are spatial or time limits to the domain D within which the differential equation itself is known to be valid (e.g. the equation $\ddot{y} = -g$ for a falling object ceases to be valid after that body has touched the earth's surface) one must be careful not to apply the solution outside the space or time domain to which it applies. Provided due notice is taken of what is actually happening, the student is unlikely to stumble into the pitfalls so conscientiously guarded against by the pure mathematician who formulates uniqueness theorems!

We can apply the uniqueness theorem to problems of motion governed by equations of the form 7–6 (namely $\ddot{s} - 2a\dot{s} + bs = 0$) to make specific statements as follows.

196 An introduction to mechanics and modelling

(i) Case $a^2 - b < 0$: the motion which will actually take place is of the form $s = \exp(at)(A \cos ct + B \sin ct)$ with A and B real.

(ii) Case $a^2 = b$: The motion which will actually take place is of the form $s = k_1 \exp(at) + k_2 t \exp(at)$ with k_1 and k_2 real.

(iii) Case $a^2 - b > 0$: Write $a^2 - b = \lambda^2$, then,

$$s = \exp(at)[k_1 \exp(\lambda t) + k_2 \exp(-\lambda t)]$$

with appropriate real values of k_1 and k_2 describes the motion which will take place.

7.23 Examples in mechanics – damped simple harmonic motion

Example 1

A small object of mass 0.2 kg is suspended horizontally between a spring on one side and a damping device, called a dashpot, on the other (as in FIG. 7.3). A smooth horizontal table provides the necessary vertical support. The dashpot is essentially a cylinder of fluid which delays the motion of a piston attached to, and conceived as forming part of, the moving mass. The stiffness of the spring is 0.25 SI units; the resisting force due to the dashpot is $0.45 \times$ speed in m s^{-1}.

Starting with the spring extended by 8 cm, the mass is projected with speed 16 cm sec^{-1} towards the point of attachment of the spring. Discuss the subsequent motion.

Choice of coordinates. Starting from rest, the only horizontal force on the mass is a pull along the length of the spring, so the initial motion is along this line. Hooke's elastic force is always antiparallel to the displacement from the unstretched position, and the resistive or damping force is antiparallel to the velocity, so there is no force causing the mass to depart from the line of the spring's length: a single coordinate is sufficient to specify its position. Let x be the displacement of the mass (in this line) measured **from** the position of zero stretch, in the direction of increasing spring length, FIG. 7.3.

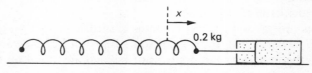

Fig. 7.3

Forces. Gravity and the normal reaction of the table act at right angles to the motion (incidentally they just balance each other): they are of no direct importance in our problem. Moreover, the table is smooth, so there is no friction. The only significant forces are those exerted by the spring and dashpot. (**Note**: we have considered (i) forces at a distance, namely gravity, (ii) known applied forces (none), and (iii) forces arising from all other contacts, enumerated and commented on above.) The forces exerted by the spring and dashpot are indicated in the statement of the question by the words 'stiffness of the spring is 0.25 SI units' and 'damping force of $0.45 \times$ speed'.

(i) **Damping force** signifies a force opposing the motion. Thus, if the velocity of the mass is $\dot{x} > 0$, the force in the x-direction due to damping is

$$-0.45|\dot{x}| = -0.45\dot{x};$$

while if the velocity of the mass is $\dot{x} < 0$, the force due to damping acts in the positive direction of x,

i.e. $\qquad +0.45|\dot{x}| = -0.45\dot{x}$ again.

No distinction need be made in the algebraic form of the forces contribution made during the outgoing and return motions: in either case $-0.45\dot{x}$ correctly represents the force in the positive direction of x.

(ii) **Force due to elasticity.** Hooke's law (§ 4.4) states that a spring with stiffness k and extension x exerts a **tension** kx. Real springs or strings do not necessarily obey this law without a fictitious interpretation about what is meant by natural length (see Question 3 of Exercise 10 of Chapter 7) and stringent restrictions on the range of x. The law is, nevertheless, extremely useful because of its simplicity, and the mere fact that a value of k is given suggests that its application is considered acceptable for the problem under discussion. A spring also resists compression, a displacement $-x$ resulting in a **thrust** kx. Whether x be positive or negative, the force in the positive sense of the x-axis is thus $-kx$.

Equation of motion. Interpreting the terminology, we find

$$0.2\ddot{x} = -0.45\dot{x} - 0.25x$$

i.e.

$$4\ddot{x} + 9\dot{x} + 5x = 0.$$

This is satisfied by $x = \exp(mt)$, provided $(4m+5)(m+1) = 0$, i.e. $m = -1$ or $-1\frac{1}{4}$. The general solution is therefore

$$x = A' \exp(-t) + B' \exp(-5t/4) \quad \text{and} \quad \dot{x} = -A' \exp(-t) - 1.25 B' \exp(-5t/4).$$

$$7\text{–}16$$

where A' and B' are arbitrary constants.

Progress of motion corresponding to the given starting conditions. Having deduced equations 7–16 in SI units, we may rewrite them in any other length units, e.g. cm, as

$$x = A \exp(-t) + B \exp(-5t/4): \qquad \dot{x} = -A \exp(-t) - 1.25 B \exp(-5t/4)$$

Given that at time $t = 0$, $x = 8$; $\dot{x} = 16$; A and B must satisfy

$$8 = A + B \quad \text{and} \quad -16 = -A - 5B/4$$

whence $A = -24$ and $B = 32$.

The position, and velocity at any time t after release are given in cm and cm s^{-1} by

$$x = 32 \exp(-5t/4) - 24 \exp(-t) \qquad \text{7–16(a)}$$

$$\dot{x} = -40 \exp(-5t/4) + 24 \exp(-t) \qquad \text{7–16(b)}$$

The graph of x is shown in FIG 7.4.

198 An introduction to mechanics and modelling

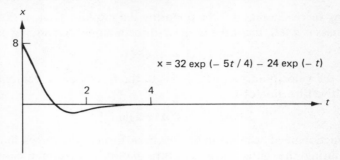

Fig. 7.4

Example 2
Discuss the solution of the same problem when, by a modification of the dashpot, its resistance has been reduced to $0.1 \times$ speed.

The equation of motion is now

$$0.2\ddot{x} = -0.1\dot{x} - 0.25x \qquad \text{i.e. } 4\ddot{x} + 2\dot{x} + 5x = 0.$$

$x = \exp(mt)$ satisfies this equation if $m = -\frac{1}{4} \pm 1.09i$. The general solution may therefore be written

$$x = \exp(-\tfrac{1}{4}t)(A \cos 1.09t + B \sin 1.09t) \qquad \qquad \text{7-17(a)}$$

The corresponding velocity is

$$\dot{x} = -\tfrac{1}{4}\exp(-t/4)(A \cos 1.09t + B \sin 1.09t)$$
$$\quad + \exp(-t/4)(-1.09A \sin 1.09t + 1.09B \cos 1.09t)$$
$$= \exp(-t/4)(D \cos 1.09t + E \sin 1.09t) \qquad \qquad \text{7-17(b)}$$

where

$$D = -\tfrac{1}{4}A + 1.09B, \quad \text{and} \quad E = -\tfrac{1}{4}B - 1.09A.$$

Given that at time $t = 0$, $x = 8$, $\dot{x} = -16$, the values of A and B (corresponding to cm units) are $A = 8$, $B = -12.84$ (since $-\tfrac{1}{4}A + 1.09B = -16$).

The variation of x with time is shown in FIG. 7.5; this may be compared with the previous solution, FIG. 7.4. The physical similarity at the beginning is clear, as is also the significance of the radical difference in mathematical form – the real exponential terms of Example 1 signifying a steady monotonic decay of x after the initial swing; the sine/cosine factors in the terms of equation 7-17(a) causing x to oscillate perpetually while it decays. Note how the oscillatory solution of FIG 7.5 always lies between the two dotted curves representing $\pm 15.1 \exp(-\tfrac{1}{4}t)$. The relation of the coefficient 15.1 to the values of A and B will be clarified in the next subsection. There too, other results will be stated which will assist the interpretation of expressions such as 7-17(a) and (b).

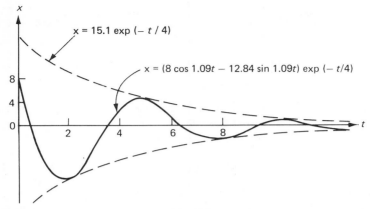

Fig. 7.5

7.24 Mathematical properties of expressions of form $A \cos \omega t + B \sin \omega t$

We first show that

> for any real numbers A and B, there is just one angle α in the range $-\pi/2 < \alpha \leq 3\pi/2$ (or its equivalent $0 < \alpha < 2\pi$) for which
>
> $$\left.\begin{array}{l}\sin \alpha = A/\sqrt{(A^2+B^2)} \\ \cos \alpha = B/\sqrt{(A^2+B^2)}\end{array}\right\} \qquad 7\text{--}18$$
>
> and

Whatever the values of A and B, positive or negative, $-1 \leq A/\sqrt{(A^2+B^2)} \leq 1$, and it is therefore possible to find an angle α_1 in the range $-\pi/2 < \alpha_1 \leq \pi/2$ such that $\sin \alpha_1 = A/\sqrt{(A^2+B^2)}$. This angle necessarily has a positive cosine (FIG. 7.6), of magnitude $\sqrt{(1-\sin^2 \alpha_1)} = |B|/\sqrt{(A^2+B^2)}$.

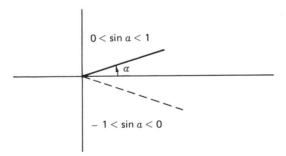

Fig. 7.6

In the same way, for the same values of A and B it is possible to find an angle α_2 in the range $\pi/2 < \alpha_2 \leq 3\pi/2$ for which the sine has the same value as before, but for which the cosine is negative, i.e. has the value $-|B|/\sqrt{(A^2+B^2)}$. Now B either equals $|B|$ or else equals $-|B|$. It follows that either α_1 or α_2 (but not both) satisfies the pair of relations postulated in 7–18.

> Hence there is a unique angle (in any range of 2π) for which both relations of 7–18 are satisfied, and for which therefore
>
> $$A \cos \omega t + B \sin \omega t = \sqrt{(A^2+B^2)}[\sin \alpha \cos \omega t + \cos \alpha \sin \omega t]$$
> $$= \sqrt{(A^2+B^2)} \sin (\omega t + \alpha) \qquad 7\text{–}19$$
>
> The graph of $x = A \cos \omega t + B \sin \omega t$ is therefore a sine wave of amplitude $\sqrt{(A^2+B^2)}$, crossing the t-axis where $\omega t + \alpha = \pi, 2\pi, \ldots$, etc. i.e. zero values of x occur at $t = (\pi - \alpha)/\omega, (2\pi - \alpha)/\omega, \ldots$, etc.

Given A and B, the simplest way of determining α is:

(i) find the acute angle α_0 for which $\tan \alpha_0 = |A/B|$;
(ii) decide, from the signs of A and B (i.e. of $\sin \alpha, \cos \alpha$) in which quadrant the required angle α must lie;
(iii) draw a diagram (e.g. FIG. 7.7 when $A > 0$ and $B < 0$) showing a vector in the required quadrant, and mark the acute angle α_0 between this vector and the line of the x-axis;

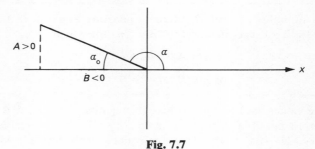

Fig. 7.7

(iv) use this diagram to attribute the correct numerical value to the angle α between the positive sense of the x-axis and the positive sense of the vector.

Exercise: the reader should verify that $A \cos \omega t + B \sin \omega t$ may also be written in the form $C \cos (\omega t + \beta)$. What can be said about the value of β if A and B are both negative?

7.25 More about simple harmonic motion

> The equation
> $$\ddot{x} = -\omega^2 x \qquad 7\text{–}20$$
> whose general solution is
> $$x = A \cos \omega t + B \sin \omega t \; [\equiv C \sin (\omega t + \alpha) \text{ where } C = \sqrt{(A^2+B^2)}]$$
> is the equation of simple harmonic motion. An equation
> $$\ddot{y} = -\omega^2 y + c \; [\equiv -\omega^2(y - c/\omega^2)] \qquad 7\text{–}21$$
> may be transformed into the standard equation of simple harmonic motion by the substitution $y - c/\omega^2 = x$; for then $\ddot{y} = \ddot{x}$ so $\ddot{x} = -\omega^2 x$.

Motion and differential equations 201

Since the values of both x and $\dot{x}$ given by the general solution are unchanged by an increase of 2π in the value of ωt (any t), the motion repeats exactly after each change of t by $2\pi/\omega$. This interval between repetitions is called the period of the simple harmonic motion

$$\text{Period } T = 2\pi/\omega.$$

The point for which $x = 0$ is called the centre of simple harmonic motion, so, in the case of equation 7–21, $y = c/\omega^2$ is the centre of the simple harmonic motion.

7.26 Damped simple harmonic motion: properties of oscillatory solutions

The equation

$$\ddot{x} = -\kappa\dot{x} - \omega^2 x + c \qquad 7\text{–}22$$

can, by the substitution $z = x - c/\omega^2$, be transformed into the standard equation of damped simple harmonic motion in the new variable z, thus

$$\ddot{z} = -\kappa\dot{z} - \omega^2 z.$$

The equation for the exponent factor m is $m^2 + \kappa m + \omega^2 = 0$, which may be expressed $(m + \kappa/2)^2 = \kappa^2/4 - \omega^2$. When $\kappa^2 < 4\omega^2$, the general solution may be written

$$z = \exp(-\kappa t/2)(A \cos nt + B \sin nt) \qquad 7\text{–}23$$

where $n^2 = \omega^2 - \kappa^2/4$, many of whose properties closely parallel those of ordinary simple harmonic motions. Note, however, the following differences.

---

1 When nt changes by 2π, i.e. when t changes by $2\pi/n$, the exponential factor $\exp(-\kappa t/2)$ changes to

$$\exp[-\kappa(t + 2\pi/n)/2] = \exp(-\kappa t/2) \cdot \exp(-\pi\kappa/n).$$

Consequently, after each 'period' of $2\pi/n$, there is a repeat with all values of z reduced by the factor $\exp(-\pi\kappa/n)$.

2 In the same way, the velocity $\dot{z}$ reduces by factor $\exp(-\pi\kappa/n)$ in each repeat period.

3 Although

$$z = \exp(-\kappa t/2)(A \cos nt + B \sin nt)$$

may be written as

$$z = \exp(-\kappa t/2) \cdot C \sin(nt + \alpha),$$

where $C = \sqrt{(A^2 + B^2)}$, the maximum of z does not occur at $\sin(nt + \alpha) = 1$.

For, the maxima or minima of z occur when $\dot{z} = 0$, i.e. when

$$C \exp(-\kappa t/2)[n \cos(nt + \alpha) - \tfrac{1}{2}\kappa \sin(nt + \alpha)] = 0,$$

i.e. correspond **not** to $\cos(nt + \alpha) = 0$, $\sin(nt + \alpha) = \pm 1$, but to $n \cos(nt + \alpha) = \tfrac{1}{2}\kappa \sin(nt + \alpha)$, $\tan(nt + \alpha) = 2n/\kappa$, of which equation there are two solutions in every increase of nt by 2π, one being a value of t for which z is a maximum, the other corresponding to a minimum of z.

The zero values of z **do** correspond to $\sin(nt+\alpha)=0$, i.e. to $nt+\alpha=0$, π, $2\pi, \ldots$ etc. We see that the maximum and minimum values of z are staggered with respect to the zero values – they do not come at the half-way points.

From the algebraic expression for z it follows that the graph of z achieves the value $\pm C \exp(-\kappa t/2)$ when $nt + \alpha = \pi/2, 3\pi/2, \ldots$; and at all other values of t lies between the two bounding curves $z = \pm C \exp(-\kappa t/2)$ as in FIG. 7.5 where $\kappa = \frac{1}{2}$ and $C = \sqrt{(8^2 + 12.84^2)} = 15.1$.

Exercise 5

1. State the amplitudes of the simple harmonic motions corresponding to the motions in a straight line whose positions or velocities are:
   (i) $x = 3\cos 2t - 4\sin 2t$;
   (ii) $x = 4\sin 11t + 8\cos 11t$;
   (iii) $y = 5 - 2\sin t$;
   (iv) $y = \cos 7t - 3\sin 7t - 4$;
   (v) $\dot{x} = 3\cos 2t + 5\sin 2t$;
   (vi) $\dot{y} = 5\cos \frac{1}{2}t - 12\sin \frac{1}{2}t$.

2. Find the value of the coordinate (x or y) at time $t = 2$ for each of the motions (i)–(iv) in question 1. (**Note**: (iv) of § 7.1.)

3. The one-dimensional motion of a damped harmonic oscillator is governed by the differential equation
$$\frac{d^2x}{dt^2} + \frac{1}{t_0}\frac{dx}{dt} + \Omega^2 x = 0.$$
Show that, if $1/2t_0 < \Omega$ the motion is oscillatory with a decreasing amplitude; and, given that $x(0) = x_0$ and $v(0) = 0$, find the position x and velocity v of the oscillator as functions of time.

4. When undamped, a given linear oscillator has period $2\pi/3$ seconds. A device is inserted which provides a damping force whose magnitude in SI units is twice the product of the velocity and mass (in SI units) of the oscillating body. Given that the mass starts from rest from a position at distance 0.5 m from the centre of the motion, find expressions for its velocity and position at any subsequent time t, and sketch graphs of velocity and position in the first 10 seconds of motion.

5. The damping device in question 4 is replaced by another providing heavier damping, namely a force whose magnitude, in newtons, is 10 times the product of velocity and mass. Find the general solution, and sketch the variation of the position coordinate, x say, (a) when the system starts with zero velocity from a point 0.5 m from the centre, and (b) when the system starts with velocity 0.5 m s^{-1} from the centre of the motion.

6. By what factor does the amplitude of the oscillation decrease in each oscillation in the motion of question 4, and what is the 'period' of oscillation? Are analogous statements possible in the situations of question 5?

7. A particle of mass m is suspended from a fixed support by a spring of stiffness k. At time $t = 0$ the particle is at rest in its equilibrium position. A constant force of magnitude P is applied to the particle in the downward direction, and maintained from time $t = 0$ to time $t = t_1$, at which time it is removed. Show that at time t_1 the extension of the spring is
$$g/\omega^2 + (P/m\omega^2)(1 - \cos \omega t_1) \text{ where } \omega^2 = k/m.$$
Find also the speed of the particle at time $t = t_1$. Hence prove that at time $t = 2t_1$ the extension of the spring is
$$g/\omega^2 + (P/m\omega^2)(\cos \omega t_1 - \cos 2\omega t_1).$$

8. The times at which a given damped harmonic oscillator passes through two successive zeros and the intervening position of maximum displacement are recorded. The zeros occur at $t = 0$, 2.5 seconds and the maximum at 0.6 seconds. Deduce the values of k, ω for the apparatus. (Notation of § 7.26 with variable z.)

9   A galvanometer coil (equation of motion $I\ddot{\theta} = M$) oscillates with a periodic time of 5.00 seconds and, successive maximum displacements are observed to be 76, 34.2, 15.3 and 6.9 scale divisions. The moment of inertia, I of the suspended system is 4.86×10^{-7} kg m². Are the readings consistent with the presence of a damping moment proportional to the angular velocity in addition to a restoring moment proportional to θ? If so, calculate the damping moment at unit angular velocity, and the restoring moment per radian. (Scale divisions assumed to be proportional, not equal, to angular displacements.)

7.3 Forced oscillations and inhomogeneous linear differential equations

Many physical systems are well modelled as harmonic oscillators (either damped or undamped) driven by an external periodic force, $F = F_0 \cos pt$ say. We are led to discuss an equation of motion of form

$$\ddot{x} + k\dot{x} + \omega^2 x = (F/m) \cos pt \qquad 7\text{--}24$$

This differs from the homogeneous linear equations we have so far discussed because the term on the right-hand side is not a multiple of $\ddot{x}$, $\dot{x}$ or x, nor can it be removed (as was the case of § 7.26, with a constant on the right-hand side) by a simple substitution. The discussion of non-homogeneous linear differential equations falls naturally into three parts.

I   From the uniqueness theorem of § 7.22 we know that any general solution of equation 7–24 will allow us to derive the desired solution of the physical problem.

II If y is a solution, not of equation 7–24 but of the **associated homogeneous equation**, obtained by deleting the right-hand side, i.e. if y satisfies

$$\frac{d^2}{dt^2} y + k \frac{d}{dt} y + \omega^2 y = 0 \qquad 7\text{--}25$$

and if x_1 is any known solution of 7–24 (maybe corresponding to quite different initial conditions from those of the problem in hand), so that we have

$$\frac{d^2}{dt^2} x_1 + k \frac{d}{dt} x_1 + \omega^2 x_1 = \frac{F_0}{m} \cos pt \qquad 7\text{--}26$$

Then, by adding equations 7–25 and 7–26,

$$\frac{d^2}{dt^2}(y + x_1) + k \frac{d}{dt}(y + x_1) + \omega^2(y + x_1) = \frac{F}{m} \cos pt$$

i.e. $x = y + x_1$ is a solution of equation 7–24. Moreover,

> if y contains two arbitrary constants, then so also will x; in other words $x = y + x_1$ is the general solution of equation 7–24.

204 An introduction to mechanics and modelling

A function y so related to a given inhomogeneous equation is called the 'complementary function' of the given equation; the function x_1 is a particular integral of the equation. So the stated result may be re-expressed:

> General solution = complementary function + particular integral

III We have already described in § 7.212 onwards how the general solution y of a **homogeneous linear differential equation** may be found. To find a particular solution x_1 of the given non-homogenous equation 7–24 – and any solution will do – we may again resort to the trial solution method introduced at the beginning of the chapter. Functions such as $\cos pt$ and $\sin pt$ came to notice because of the equivalence of the forms $k_1 \exp(ipt) + k_2 \exp(-ipt)$ and $A \cos pt + B \sin pt$. An equation

$$\ddot{x} + k\dot{x} + \omega^2 x = k_1 \exp(ipt)$$

is satisfied by $x = K_1 \exp(ipt)$ provided

$$K_1(-p^2 + ikp + \omega^2) \exp(ipt) = k_1 \exp(ipt)$$

i.e. provided

$$K_1 = k_1/(\omega^2 - p^2 + ikp) \qquad 7\text{–}27$$

The precise form of K_1 in equation 7–27 need not detain us. The point is that a solution of the form $K_1 \exp(ipt)$ exists. In the same way, an equation of the form

$$\ddot{x} + k\dot{x} + \omega^2 x = k_1 \exp(ipt) + k_2 \exp(-ipt)$$

has a solution of the form

$$x = K_1 \exp(ipt) + K_2 \exp(-ipt)$$

the values of K_1 and K_2 being determined by k_1, k_2, p, k and ω. We restate this finding as follows:

> Any equation of the form
>
> $$\ddot{x} + k\dot{x} + \omega^2 x = A \cos pt + B \sin pt$$
>
> has a (particular) solution of the form
>
> $$x = \alpha \cos pt + \beta \sin pt$$
>
> the constants α and β being determined by p, k, ω, A and B.

To find a particular solution x_1 of equation 7–24 we therefore only need to write $x_1 = \alpha \cos pt + \beta \sin pt$ and, by substitution in equation 7–24, find the values of α and β needed to satisfy that equation for all t, i.e. to ensure an **identity** between left-hand and right-hand sides. We require

$$(-\alpha p^2 \cos pt - \beta p^2 \sin pt) + k(-\alpha p \sin pt + \beta p \cos pt) + \omega^2(\alpha \cos pt + \beta \sin pt)$$
$$\equiv F_0 \cos pt/m$$

i.e.
$$-\alpha p^2 + k\beta p + \alpha\omega^2 = F_0/m$$
$$-\beta p^2 - k\alpha p + \beta\omega^2 = 0$$

i.e.
$$(\omega^2 - p^2)\alpha + kp\beta = F_0/m$$
$$-kp\alpha + (\omega^2 - p^2)\beta = 0$$

from which pair of linear equations for α and β their values can be found,

$$\left.\begin{aligned}\alpha &= \frac{(\omega^2 - p^2)}{(\omega^2 - p^2)^2 + k^2 p^2} \cdot \frac{F_0}{m} \\ \beta &= \frac{pk}{(\omega^2 - p^2)^2 + k^2 p^2} \cdot \frac{F_0}{m}\end{aligned}\right\} \quad 7\text{--}28$$

The general solution $x = y + x_1$ of the given equation 7–24 is therefore available.

7.31 Interpretation of particular integrals and complementary functions

The particular solution or integral x_1 is a sustained oscillation of amplitude

$$\sqrt{(\alpha^2 + \beta^2)} = \frac{1}{\sqrt{[(\omega^2 - p^2)^2 + k^2 p^2]}} \cdot \frac{F_0}{m}. \qquad \text{See 7–20}$$

This continues so long as the driving oscillation $F_0 \cos pt$ is maintained. By contrast, the complementary function (y) is a function similar in form to the graph of FIG. 7.4: it is a function which dies away; its role is to adjust the solution to the initial condition. Since this part of the solution dies away, it is often called the 'transient' part. That part of the particular integral which does not die away exponentially, is called the 'forced oscillation'.

7.32 Resonance

For given k and p, the amplitude of the forced oscillation, namely

$$\frac{1}{\sqrt{[(\omega^2 - p^2)^2 + k^2 p^2]}} \cdot \frac{F_0}{m}$$

is greatest when $\omega^2 = p^2$. If kp is small, this amplitude is large, corresponding to possibly violent oscillations: this is the phenomenon known as resonance. We see that it takes place when the periodicity of the applied force ($F_0 \cos pt$) is the same as the natural period of the undamped system ($p = \omega$), and is most marked when the damping is small, i.e. kp is small.

> When a system is only lightly damped, a modelling approximation in which the damping term $k\dot{x}$ is omitted may be acceptable provided that no resonant applied force acts on the system. For applied forces of resonant or near-resonant frequency, inclusion of the damping term, no matter how small k may be, is essential for physically meaningful (non-infinite) results. The validity of the ω^2 term for large amplitudes should also be reconsidered.

7.33 Summary of the method of solution of inhomogeneous linear differential equations

(We summarize the method for the general second-order linear differential equation; only minor modification is required to apply it to equations of higher order.)

> The equation
> $$\ddot{x} + a\dot{x} + bx = f(t) \qquad 7\text{-}29$$
> has general solution
> $$x = y + x_1$$
> where y is the general solution of the equation
> $$\ddot{y} + a\dot{y} + by = 0$$
> (and may be found by the $\exp(mt)$ method), and the function x_1 may be any solution of 7-29 (and is usually discoverable by trial substitutions, e.g. those suggested below).

For $f(t)$ of the form	try
$k_0 \exp(pt)$	$x_1 = K_0 \exp(pt)$
$k_0 \sin pt$	$x_1 = K_1 \sin pt + K_2 \cos pt$
$k_1 \sin pt + k_2 \cos pt$	$x_1 = K_1 \sin pt + K_2 \cos pt$
$k_0 t^n$	$x_1 = K_0 t^n + K_1 t^{n-1} + \cdots + K_n$ (if x-term is missing in differential equation, add t^{n+1} term to the trial particular integral)
$k_0 t^n + k_1 t^{n-1} + \cdots + k_n$	$x_1 = K_0 t^n + K_1 t^{n-1} + \cdots + K_n$
y_1 (i.e. part of the complementary function, CF)	$x_1 = K_0 t y_1$ or $t \times \mathrm{CF}$
sum of terms	sum of corresponding trial terms

Exercise 6

1 Find the complementary functions and particular integrals of each of the following differential equations and deduce their general solutions:
 (i) $\ddot{x} + \dot{x} = \sin 4t$ (ii) $\ddot{x} - 2\dot{x} = t^2$
 (iii) $\ddot{x} + 9x = \sin 5t$ (iv) $\ddot{x} - 6\dot{x} + 5x = 2\cos t - \sin t$
 (v) $\ddot{x} + 4\dot{x} - 5x = 3\exp(t)$ (vi) $\ddot{x} + 9x = \sin 3t$

2 A particle of unit mass lies on a smooth horizontal table. Two springs of natural lengths l_1 and l_2 and stiffness k_1 and k_2 respectively are attached to the mass, the other ends of the springs being fixed to points on the table distance $l_1 + l_2$ apart. Starting from rest at the equilibrium position, the mass is subjected to an applied horizontal force $F_0 \sin pt$ along the line of the springs. Find the position and velocity of the mass at any future time t. For what value of p is the modelling inadequate? Say what you can about the initial increase of displacement with time in this case.

3 Solve the differential equation $\ddot{x} + \omega^2 x = F_0 \cos(\omega + \delta)t$, subject to the initial conditions $\dot{x} = x = 0$ when $t = 0$. If $0 < \delta \ll \omega$, show that the oscillations have the form shown by the full line in FIG. 7.8, the dotted lines corresponding to the graph of the

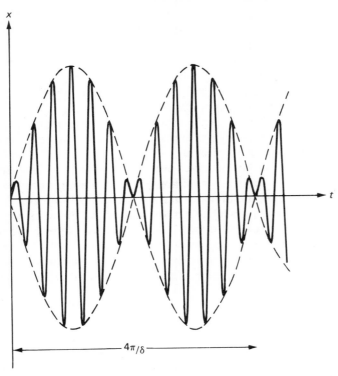

Fig. 7.8

functions
$$\pm \frac{2F_0}{(\omega+\delta)^2-\omega^2}\sin(\tfrac{1}{2}\delta)t$$

period $4\pi/\delta$. A similar effect occurs when two waves of slightly different frequency overlap. With sound waves, a note of fluctuating strength would be heard – the phenomenon of **beats**. The smaller the value of δ, the slower and stronger the beats.

4 A particle of mass m moves along the x-axis under the action of a restoring force $m(k^2+\omega^2)x$, and a resisting force $2mk\dot{x}$, k and ω being constant. Find a general solution for x at any time t, and deduce that at successive positions of instantaneous rest, the values of $|x|$ form a geometric progression of common ratio $\exp(-k\pi/\omega)$.

An additional force $F\cos\omega t$ now acts on the particle in the direction of x increasing. If as $t\to\infty$ the particle oscillates through a constant distance a on each side of the origin, prove that $F=mak(k^2+4\omega^2)^{1/2}$.

5 A microphone may be modelled as a movable vertical plate of mass M mounted so as to be parallel to a fixed vertical plate, but normally separated from it by an air gap. When displaced from its equilibrium position, the moving plate experiences a restoring force equal to $20Mp^2\times$ its deflection x, and a damping force of $4Mp\times$ its velocity $\dot{x}$. Output from the microphone is undistorted if the air gap is never less than $\tfrac{1}{4}a$, where a is the width of the gap in the equilibrium position. Assuming that the microphone is activated by a horizontal periodic force $F_0\sin\omega t$, show that the output is undistorted for all ω provided

$$4p^2 \geqslant F_0/3Ma.$$

(**Hint**: consider only the particular integral since the complementary function rapidly dies away. Maxima and minima are involved in this question.)

7.34 Electric circuits

It would be at variance with the outlook of this book to suggest that the reader could acquire, by reading a few pages, enough understanding of electricity to make him/her an expert modeller in that field. However, electrical characteristics of real circuits are mirrored with peculiar accuracy by simple mathematical expressions, and this greatly reduces the pitfalls of the modelling process. The fact that the equations of harmonic motion which we have already discussed have exact counterparts of practical importance in electric circuit theory provides a further incentive to give a brief description of the successful conventional mathematical model of simple circuits.

An electric circuit may be conceived as a loop of wire (FIG. 7.9a) round which a current flows. The symbol i is commonly used to denote the strength of the current. If there is a simple break in the wire (FIG 7.9b), no current can flow, i.e. $i \equiv 0$. If substantial parallel plates of conducting material are present at opposite sides of the break, as in FIG. 7.9c, current may flow in the wire, at the expense of taking electric charge away from one plate, and building up charge on the other, though of course some driving mechanism will normally be required to do this. Such an arrangement of parallel plates is called a capacitor or condenser.

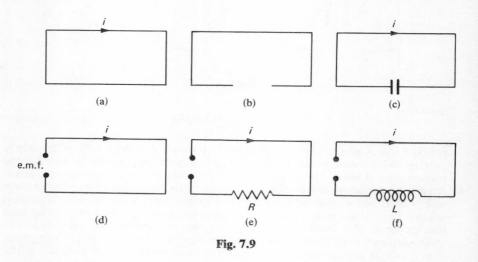

Fig. 7.9

To initiate or maintain current in a circuit there must be some driving force. If i is regarded as the analogue of a displacement x in mechanical motion, then 'electromotive force' (or e.m.f.) is the analogue of force. The corresponding electromotive 'mass' is zero. The current in a closed circuit including an e.m.f. E (FIG. 7.9d) would increase at an uncontrollable rate $(0 \times d^2i/dt^2 = E$, i.e. $d^2i/dt^2 \to \infty)$ if no opposing forces were present: the absence of opposing forces in a circuit, or the presence of only small opposing forces is described by the practical man as a 'short circuit'.

The first kind of opposing e.m.f. is a resistance (of any wire or device through which current flows) **proportional to the magnitude of the current passing**. The

Motion and differential equations

constant of proportionality is called **the resistance** R (of the wire, device or entire circuit, whichever is under consideration); the opposing e.m.f. produced equals Ri. The value of a resistance in a circuit is conventionally indicated against a zig-zag line as in FIG. 7.9e.

The second kind of opposing e.m.f. works against **changes** of current; the magnitude of the opposing e.m.f. is proportional to di/dt, and the constant of proportionality is called **the inductance** L; the counter e.m.f. produced is equal to $L\,di/dt$. Loops in the wire are particularly liable to give rise to this sort of opposition, and the existence of inductance is indicated in a circuit diagram by a symbolic coil, as shown in FIG. 7.9f.

The differential equation for the current in a circuit driven by an imposed e.m.f., $E(t)$, and possessing resistance R and inductance L may now be written down by analogy with its mechanical counterpart:

$$0 \times d^2i/dt^2 = E - Ri - L\,di/dt$$

i.e.

$$L\,di/dt + Ri = E \qquad 7\text{–}30$$

in which L and R are constants, but E may be a function of time. Equation 7–30 is a linear differential equation with constant coefficients and may be solved by the methods of § 7.2 forwards.

Exercise:
Starting from $i = 0$ at time $t = 0$, solve equation 7–30:
(i) for $E = E_0 =$ constant; (ii) $E = E_0 \cos \omega t$.

When a capacitor (condenser) is present, the charge which accumulates on the receiving plate in time t is

$$q = \int_0^t i\,dt,$$

a relationship which may be more conveniently written

$$dq/dt = i.$$

The **condenser** is found to produce an opposition to the flow of current **proportional to** q. For historic reasons the constant of proportionality is written as $1/C$, where C is called the capacity of the condenser, or capacitance.

The current in the L, R, C circuit of FIG. 7.10 is determined by the differential equation

$$0 = E - Ri - L\,di/dt - q/C \qquad 7\text{–}31$$

which, on writing $i = dq/dt$, may be expressed

$$d^2q/dt^2 + (R/L)\,dq/dt + (1/LC)q = (E/L) \qquad 7\text{–}32$$

Since R/L and $1/LC$ are constant coefficients, equation 7–32 is of the form which we have already so extensively discussed in §§ 7.2–7.3.

Exercise 7

Write down the circuit equation corresponding to 7–32 above for the circuit shown in FIG. 7.10. Given that the condenser is initially uncharged and the current initially zero, find the charge q and hence the current i at any subsequent time (i) when $E = E_0 =$ constant, and (ii) when $E = 16(\sin 2t + \cos 2t)$.

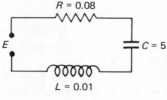

Fig. 7.10

7.4 Models of fall in a resisting medium

7.41 Simplified mathematical models

Most of the mechanical modelling in which we have engaged so far has been based on well-established laws (e.g. Newton's laws of motion or gravitation) of nearly universal validity. In the study of motion in a resisting medium there is no single comparable law. Progress may be made in either of two ways, namely: **Either** by investigating the consequences of mathematically simple hypothetical laws whose correspondence with physical reality is, at the outset, at best unproven and at worst non-existent;
or by carefully examining such experimental evidence as is available, and from it deducing conclusions valid in a strictly limited range.
We shall illustrate both procedures in the pages which follow. We start by investigating postulated laws of resistance of whose applicability to real situations there is as yet no assurance.

7.42 Resistance proportional to the velocity

If a resistance is believed to be velocity-dependent, the simplest form of dependence to investigate is direct proportionality, force per unit mass $= k$ times the velocity say. In terms of a coordinate y measured vertically upwards the equation of motion of free fall through such an atmosphere (liquid or gaseous) would be written

$$m\ddot{y} = -mg - mk\dot{y}$$

(see also Note below) i.e.

$$\ddot{y} + k\dot{y} = -g \qquad 7\text{–}33$$

This may be solved either by the substitution $z = y + gt/k$ or by the complementary function + particular integral method. In either case it may be seen that the general solution of equation 7–33 is

$$y = -gt/k + A + B \exp(-kt) \qquad 7\text{–}34$$

the velocity at any time being given by

$$\dot{y} = -g/k - kB \exp(-kt) \qquad 7\text{-}35$$

The value of B is determined by the initial velocity, and that of A is then determined by the initial position.

Since $\exp(-kt) \to 0$ as $t \to \infty$ (and indeed $\exp(-kt)$ is very small for all values of $kt > 10$), equation 7-35 shows that, according to the model, the downward velocity soon settles to a value differing only slightly from g/k. The value g/k is often called the 'terminal velocity'. Note that at this velocity, the acceleration is zero (equation 7-33 with $\dot{y} = -g/k$).

Note: if the projected body were first thrown upwards, and allowed to rise and fall again under gravity, the same equation 7-33 would hold good throughout the entire upward and downward motion. For, on the upward journey $\dot{y}$ is positive, $\dot{y} = |\dot{y}|$; and the resisting force acts downwards, i.e. $= -mk|\dot{y}| = -mk\dot{y}$. On the downward journey, $\dot{y} < 0$, so $|\dot{y}| = -\dot{y}$; the upward resisting force is equal to $mk|\dot{y}| = -mk\dot{y}$. Equation 7-33 therefore holds in both cases.

7.43 Resistance proportional to the square of the velocity

Under this assumption, the equation of motion for downward motion, measuring coordinate y downwards, has the form

$$m\ddot{y} = mg - mK\dot{y}^2 \qquad 7\text{-}36$$

where mK is written for the constant of proportionality. Since

$$\ddot{y} = \frac{d}{dy}\left(\frac{\dot{y}^2}{2}\right)$$

equation 7-36 may be written in the linear form

$$\frac{d}{dy}(\dot{y}^2) + 2K(\dot{y}^2) = 2g$$

with solution

$$\dot{y}^2 = g/K + A\exp(-2Ky). \qquad 7\text{-}37$$

Although from equation 7-37 it would be possible to express t as an integral of a function of y, it is not possible to derive a simple explicit expression for y as a function of t.

Either equation 7-36 or 7-37 shows that the terminal velocity is equal to $\sqrt{(g/K)}$.

For upward motion, equation 7-36 would have to be replaced. Retaining the original coordinate y (measured downwards) the new equation would be

$$m\ddot{y} = mg + mK\dot{y}^2.$$

A more natural choice of coordinate, z upwards, say, would lead to the equation

$$m\ddot{z} = -mg - mK\dot{z}^2.$$

These two equations are equivalent; but neither is the same as equation 7-36, so, when the resistance is proportional to the square of the velocity, problems involving changes in direction of motion have to be split up into uni-directional sections. The same is true for all resistance laws other than 'resistance proportional to an odd power of the velocity'.

As we shall see later, experiment shows the (velocity)^2 law of resistance to be approximately true over a much wider range of velocities than is the direct proportionality law. Because of its greater convenience, it may nevertheless on occasion be acceptable to use equation 7–33, knowing it to be false, rather than possess no simple way of getting a rough estimate of y at any time.

If models of unproven validity are used – and this applies whatever the problem – the conclusions drawn will be only 'guesstimates'. They may be useful in this role, but the conclusions cannot be accorded the status of predictions. Theories containing unknown constants such as k and K above, should always be treated with great reserve unless and until precise and practical prescriptions for attributing numerical values to those constants are available, e.g. looking up the values in reference tables. If no values are attributable, it is also probable that no range of validity of the law is known to the worker; the mathematical treatment may then assist insight but will not be a source of secure information.

Exercise 8

1   A point moves in a straight line under a retardation kv^2. Show that, if its initial velcoity is u, the distance covered in time t is $(1/k)\ln(1+kut)$.

2   A particle falls vertically under the constant force of gravity, through a low-density medium which exerts a resistance kv per unit mass on the particle. Deduce that the terminal speed is g/k. In a second experiment, the particle is projected vertically upwards with speed $2g/k$; prove that it attains its maximum height after a time $t=(1/k)\ln 3$; and find the time, during the upward journey, at which its speed is one-half the velocity of projection; how far has it then travelled?

3   A particle is projected in a medium whose resistance is proportional to the cube of the velocity, and no other forces act on the particle. While the velocity diminishes from V_1 to V_2, the particle traverses a distance d. Show that the time of travel is $(V_1+V_2)d/2V_1V_2$.

4   A particle of mass m falls under gravity, starting from rest. Its motion is opposed by air resistance which can be represented by a force mkv^2, where v is the speed of the particle and k is a constant. Prove that the speed of the particle can never exceed the value $c=\sqrt{(g/k)}$, and show that the distance fallen (x) and the speed (v) are related by

$$2kx = \ln[c^2/(c^2-v^2)].$$

For a parachute jumper in free-fall, a typical terminal velocity for a 70 kg jumper in a spread-eagle position on a warm summer day is 54 m s^{-1}. At what free-fall distance does his speed reach two-thirds of the terminal speed?

5   The drag force (resistance) on a body moving at speed v through the air may be expressed as $\frac{1}{2}C_D\rho Av^2$, where A is the projected area of the body perpendicular to the direction of motion, ρ is the density of the air, and C_D is a dimensionless number, approximately constant over a considerable range of velocities, called the **drag coefficient**.

A soccer ball is kicked horizontally (but above the ground) at goal from a distance of 20 m at an initial speed of 35 m s^{-1}. Neglecting any vertical motion of the ball, calculate the speed at which the ball reaches the goalkeeper and the time of flight (i) at sea level where $\rho = 1.22$ kg m^{-3}, (ii) at Mexico City where the air density is reduced by a factor of 1.27. Take the ball to have mass 0.43 kg and diameter 0.22 m, with $C_D = 0.2$.

6 An object is projected vertically upwards with initial velocity u in a medium with resistance gv^2/c^2 per unit mass, where v is the velocity, g the acceleration due to gravity and c is a constant. Show that the greatest height of the centre of mass of the object above its point of projection is $(c^2/2g) \ln(1+u^2/c^2)$, and find its speed when it returns to the point of projection.

7.44 Simple laws of resistance and two-dimensional motion

Of all the velocity-dependence laws, the 'resistance proportional to velocity' law is mathematically the most convenient. This is because the components of $-k\mathbf{v}$ in x-, y-, and z-directions are, respectively, $-k\dot{x}$, $-k\dot{y}$, $-k\dot{z}$. For projectile motion in two dimensions, the corresponding equations of motion (taking x and y as horizontal and vertical axes in the plane of motion) reduce to

$$\ddot{x} = -k\dot{x}$$
$$\ddot{y} = -k\dot{y} - g$$

i.e. two ordinary differential equations of the kind whose solution has already been fully discussed.

By contrast, a resistance proportional to the square of the velocity corresponds to force per unit mass $-Kv^2$ in the direction of the unit vector $\mathbf{v}/v$. Since $-Kv^2(\mathbf{v}/v) = Kv\mathbf{v}$, the cartesian equations involve the quantities $-Kv\dot{x}$, $-Kv\dot{y}$ and $-Kv\dot{z}$ respectively, in which $v = \sqrt{(\dot{x}^2 + \dot{y}^2 + \dot{z}^2)}$, i.e. each equation involves derivatives of all the varying coordinates. Such equations are difficult or impossible to solve explicitly.

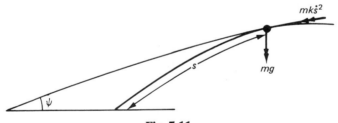

Fig. 7.11

The square root is avoided, but linked equations are still involved, if s, ψ coordinates are used. Thus for a projectile, (see FIG. 7.11 and equation 2–15)

$$\ddot{s} = -K\dot{s}^2 - g\sin\psi$$
$$\dot{s}\dot{\psi} = -g\cos\psi$$

If, by using the second equation, $\dot{s}$ and $\ddot{s}$ were eliminated from the first, an ordinary differential equation would result, but its form would be unpleasant.

The conclusion is drawn that, if the linear law is not regarded as acceptable for the purposes in hand, simultaneous differential equations will arise for which the only practical method of solution is likely to be numerical solution using a computer program.

7.45 More general resistance laws in one dimension

For one-dimensional motion, velocity-dependence laws of some generality can be handled. If the force per unit mass on a body is of the form $f(t) - vg(t)$, the corresponding differential equation is

$$\mathrm{d}v/\mathrm{d}t + g(t)v = f(t) \qquad 7\text{-}38$$

If the force per unit mass is of the form $f(x) - v^2 g(x)$, in which x is the linear coordinate, the corresponding differential equation is

$$\mathrm{d}(v^2)/\mathrm{d}x + 2g(x)v^2 = 2f(x) \qquad 7\text{-}39$$

Equations 7-38 and 7-39 are both ordinary, first-order, linear differential equations with variable coefficients, and are soluble by the methods of the section which follows.

7.5 Linear equations with variable coefficients

The principle of superposition and the uniqueness theorem may be applied to linear differential equations whether the coefficients of the derivatives are constants or functions of the independent variable. Methods of finding simple explicit solutions are not, however, generally available when the coefficients vary. Only in the case of first-order equations is it possible to state a procedure which will lead to the general solution.

7.51 Linear first-order equations with variable coefficients

The linear homogeneous first-order differential equation has the general form

$$\mathrm{d}v/\mathrm{d}t + f(t)v = 0. \qquad 7\text{-}40$$

We have written v for the dependent variable because $\mathrm{d}v/\mathrm{d}t$ arises in equations of motion. Equation 7-40 can be integrated by separating the variables v and t as follows

$$\frac{1}{v}\frac{\mathrm{d}v}{\mathrm{d}t} = -f(t)$$

whence

$$\int \frac{\mathrm{d}v}{v} = \int -f(t)\,\mathrm{d}t;$$

i.e.

$$\log_e v = -\int f(t)\,\mathrm{d}t + C$$

$$v = \exp\left[-\int f(t)\,\mathrm{d}t\right] \cdot \exp C$$

so

$$v = A\,\exp\left[-\int f(t)\,\mathrm{d}t\right], \text{ say.} \qquad 7\text{-}41$$

This answer contains one arbitrary constant A ($=\exp C$), and is the general solution of the given equation 7–40. The constant A looks more like a familiar constant of integration if equation 7–41 is rewritten

$$v \exp\left[\int f(t)\,dt\right] = A.$$

This last form could have arisen as the result of integrating

$$\frac{d}{dt}\left\{v \exp\left[\int f(t)\,dt\right]\right\} = 0$$

which is another way of writing

$$\exp\left[\int f(t)\,dt\right]\left[\frac{dv}{dt} + f(t)\cdot v\right] = 0. \qquad 7\text{–}42$$

Hindsight thus shows us that equation 7–41 could have been arrived at by first changing the original equation 7–40 into the form of 7–42. The important result to note and remember is that

> the expression $dv/dt + f(t)\cdot v$ can be integrated directly with respect to t if it is first multiplied by
>
> $$\exp\left[\int f(t)\,dt\right].$$

For,

$$\exp\left[\int f(t)\,dt\right]\left[\frac{dv}{dt} + f(t)\cdot v\right] = \frac{d}{dt}\left\{v \exp\left[\int f(t)\,dt\right]\right\} \qquad 7\text{–}43$$

7.52 Solution of inhomogeneous linear first-order differential equations

The general inhomogeneous linear first-order differential equation may be written

$$dv/dt + f(t)\cdot v = g(t) \qquad 7\text{–}44$$

In view of 7–43 above, it becomes possible to integrate directly if it is first multipled by $\exp \int f(t)\,dt$. The steps are as follows:

$$\exp\left[\int f(t)\,dt\right]\left[\frac{dv}{dt} + f(t)\cdot v\right] = \exp\left[\int f(t)\,dt\right]g(t)$$

i.e.

$$\frac{d}{dt}\left\{v \exp\left[\int f(t)\,dt\right]\right\} = \exp\left[\int f(t)\,dt\right]g(t)$$

whence

$$v \exp\left[\int f(t)\,dt\right] = \int \exp\left[\int f(t)\,dt\right]g(t)\,dt + C' \qquad 7\text{–}45$$

216 An introduction to mechanics and modelling

and

$$v = \exp\left[-\int f(t)\,dt\right] \int \exp\left[\int f(t)\,dt\right] g(t)\,dt + C' \exp\left[-\int f(t)\,dt\right] \quad 7\text{--}46$$

which is the general solution of the given equation 7–44.

The factor $\exp[\int f(t)\,dt]$ which is the key to finding the solution is called the **integrating factor** of equation 7–44. Note that **any** integral $\int f(t)\,dt$ will serve – the inclusion of a constant of integration in the exponent is allowable but unnecessary (see also note 2 to the example below).

> In any application of the method, great care must be taken to include the very necessary constant of integration C' at the step corresponding to equation 7–45, and to treat it properly in the following line (equation 7–46) where it will appear multiplied by a function of t.

Example

$$dv/dt + 2v/t = 3t^3 \quad 7\text{--}47$$

To get the integrating factor, we first evaluate

$$\int f(t)\,dt = \int \frac{2}{t}\,dt = 2\ln t = \ln t^2$$

from which, the integrating factor is equal to

$$\exp \ln t^2 = t^2$$

(*NB* $y = \ln x \Leftrightarrow x = \exp y$). On multiplying equation 7–47 by the integrating factor t^2, that equation may be written

$$\frac{d}{dt}(v \cdot t^2) = t^2 \cdot 3t^3$$

$$v \cdot t^2 = \int 3t^5\,dt$$

$$= 3t^6/6 + C \quad 7\text{--}48$$

$$v = \tfrac{1}{2}t^4 + C/t^2 \quad 7\text{--}49$$

Note 1: the essential constant of integration, C, in equation 7–48 gives rise to the term C/t^2 in the solution for v.

Note 2: inclusion of a constant of integration in the exponent

$$\int f(t)\,dt \;(=\ln(t^2) + K)$$

would not affect the final result. For, on multiplying both sides of equation 7–47 by the integrating factor etc. the resulting form is

$$v = \tfrac{1}{2}t^4 + [C' \exp(-K)]/t^2$$

Motion and differential equations 217

and $C' \exp - K$ may be replaced by the single constant C without loss of generality.

Note 3: to eliminate ambiguities of notation, the professional mathematician may perform processes of integration using dummy variables [such as q in § 7.1(i)]. Equation 7–45 takes the instructive (even though initially frightening) form

$$\left[v(\tau) \cdot \exp \int_c^\tau f(q)\,dq\right]_{\tau=t_0}^t = \int_{\tau=t_0}^t \left[\exp \int_c^\tau f(q)\,dq\right] g(\tau)\,d\tau$$

to which reference may be made if you are uncertain how to apply equations 7–45 and 7–46.

Exercise 9

1. Solve the following differential equations (general solution).
   - (i) $dy/dt + y = \exp(-t)$
   - (ii) $dy/dx + y/x = x^2$
   - (iii) $x(dy/dx) + 2y = x^4$
   - (iv) $\dot{z} + z \cot t = \sin 2t$
   - (v) $(1/y^2)(dy/dt) + 1/y = t$
     Hint: write $(1/y) = z$; $-(1/y^2)(dy/dt) = dz/dt$.
   - (vi) $y^2(dy/dx) + y^3 \tan x = \sin x \cos^2 x$
     Hint: substitution parallel to that for (v).

2. Find solutions of the following differential equations subject to the prescribed conditions, and state the range of validity of your answers.
   - (i) $x(dy/dx) + 3y = x + 1$; $y = 1$ when $x = 1$.
   - (ii) $dy/dt + y/t = y^3$; $y = 1$ when $t = 1$.
   - (iii) $dv/dt - v^3 = 4v$; $v = 1$ when $t = 0$.
   - (iv) $x \ln x (dy/dx) + y = 2 \ln x$; $y = 2$ when $x = 4$.

3. A spherical object is allowed to fall through a viscous liquid, starting with a downward speed of V. The liquid is subjected to rapid cooling, so that its resistance changes with time, resistance per unit mass $= \kappa(1 + 0.01t) \times$ speed v, say. Find an expression for its speed after time t (one term being left in the form of an integral).

4. To take into account the decrease of atmospheric density with height, its resistance per unit mass of a moving body may be represented by the expression $[\kappa \exp(-ay)]v^2$ where $v = |\dot{y}|$ is the speed of travel. A rocket is projected with upward velocity v_0 from the level $y = 0$. Write down an equation of motion for the initial part of its flight, where the value of g may acceptably be regarded as constant. Deduce an expression (involving an integral) from which the velocity at any height can be derived.

7.6 Fall of a raindrop – an exercise in practical modelling

Step A Modelling the body

We recognize that the shape of a raindrop may vary during fall, and that its mass may change by evaporation or accretion; nevertheless the author makes the choice – 'we will neglect (in our first investigation) all the ways in which the raindrop differs from a rigid body: we identify the motion of its centre of mass as the only item of interest'.

Step B Modelling the forces

(i) Forces other than contact forces: we recognize that gravity acts, and that electrostatic forces may arise in thunderstorms; that energy sources such as the

218 An introduction to mechanics and modelling

sun need not be considered if evaporation is being excluded. The author makes the choice – 'we will confine our attention (in the first place) to situations where gravity is the only non-contact force to be considered.'

(ii) Forces of contact: the only substance in contact with the raindrop is the air around it. The author makes the choice – 'because it is easy to do, we will first look at a crude model in which the air resistance is neglected.' (Subsequently – see step E – we shall look at experimental data on air resistance, and find ways of assessing air resistance effects.)

(iii) Interval of validity of the analysis of forces: the specified force system, viz. gravity only, is appropriate in the time interval between the moment when the drop is fully formed and the moment it hits the ground. We shall use the approximation $g = $ constant.

Step C Choice of coordinates and equations of motion

Three coordinates suffice to describe the position of the centre of mass; their choice is governed (a) by the nature of the forces acting and (b) by the nature of the initial and boundary conditions.

In the present problem, gravity acts vertically downwards, the initial velocity we have not yet prescribed, the spatial boundary is the earth's surface, which it is appropriate to model as a horizontal plane. (Note the acknowledged falsity of the latter model, which does not in the least disturb our confidence in its adequacy for the present discussion.) The situation is most easily represented in a cartesian frame, with the earth's surface corresponding to $z = $ constant $= 0$, say, the z-axis being vertical. If initial conditions involve a horizontal component of velocity, it will be necessary to introduce x-, y-directions as in FIG. 7.12a, otherwise the single coordinate z will suffice (FIG. 7.12b). A figure, such as FIG. 7.12c, showing only an initial position and no coordinate directions, is of little use.

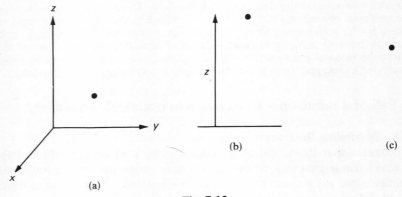

Fig. 7.12

Equations of motion are

$$M\ddot{x} = 0; \quad M\ddot{y} = 0; \quad M\ddot{z} = -Mg; \qquad 7\text{--}50$$

and, with initial conditions $\dot{x} = u_1$, $\dot{y} = u_2$, $\dot{z} = u_3$, $x = y = 0$, $z = h$, we deduce

Motion and differential equations

that (on our model), at any time t in the range of validity of the equations of motion, i.e. before contact is made with the ground,

$$\dot{x} = u_1; \quad \dot{y} = u_2; \quad \dot{z} = u_3 - gt; \quad \bar{x} = u_1 t; \quad \bar{y} = u_2 t; \quad \bar{z} = h + u_3 t - \tfrac{1}{2}gt^2.$$
7–51

Step D Consideration of the predictions of the model

A numerical case usually throws light; suppose the drop starts at time $t = 0$ with a downward velocity of 1 m s^{-1}, consider its velocity (a) 2 seconds (b) $\tfrac{1}{2}$ minute later.

Metres and seconds are SI units, so, provided we measure t in seconds, all units are consistent with $g = 9.8$. We have $u_1 = u_2 = 0$, $u_3 = -1$, so

after t seconds $\dot{z} = -1 - 9.8t$; $\bar{z} = h - t - 4.9t^2$);

i.e. after 2 seconds $\dot{z} = -1 - 19.6 = -20.6 \text{ m s}^{-1}$; $(\bar{z} = h - 21.6 \text{ m})$; 7–52

and after 30 seconds $\dot{z} = -1 - 30 \times 9.8 = -295 \text{ m s}^{-1}$; $(\bar{z} = h - 4440 \text{ m})$; 7–53

Step E Decision about whether to accept the model as an adequate description of the phenomenon we are investigating

In view of the radical oversimplifications adopted in steps A and B, it is highly desirable to estimate the errors involved. The numerical values of step D show that our simple model predicts velocities of 800 km h^{-1} or more for raindrops falling from clouds at 1500 m which seems unlikely, and that even if we consider only fall from small heights, the model fails to explain how drizzle can occur. Since both these failures might well be attributable to air resistance, our next step will be to estimate errors from this source, and, if they are appreciable, to construct an improved model.

Step E′ Experimental information needed to check the validity of assumptions of the first model, and to provide the basis of an improved model

The air resistance of a sphere of unit diameter travelling at unit velocity is found (experimentally) to be almost exactly the same as that of a sphere of diameter say 2, 3 or d times as great travelling with a velocity of $1/2$, $1/3$ or $1/d$ respectively; i.e., the product vd of velocity and diameter rather than either separately, determines the magnitude of air resistance. For this reason the information can be compactly presented on a graph if we write $x = vd$, and plot values of the resistance against x as in FIG. 7.13. [This figure and Table 7.1(a) are derived from the conventional graph of drag coefficient v. Reynold's number for a sphere, see Prandtl *Fluid Dynamics* (3rd edn) p. 191. Values appropriate to 50% relative humidity, 10°C and 1 atmospheric pressure, have been assumed for air density and viscosity. More advanced workers will use tables or graphs of drag coefficients rather than FIG. 7.13.]

If we are interested in drop sizes in the range $0 < d < 1$ cm ($= 10^{-2}$ m), and velocities in the range $0 < v < 300 \text{ m s}^{-1}$ (for we must bear in mind that the velocity of sound is $\approx 330 \text{ m s}^{-1}$ and must expect exceptional behaviour at speeds approaching this value), then we may concentrate on x-values in the range $0 < x < 3 \text{ m}^2 \text{ s}^{-1}$. Bearing in mind that small drops and the lower end of the

220 An introduction to mechanics and modelling

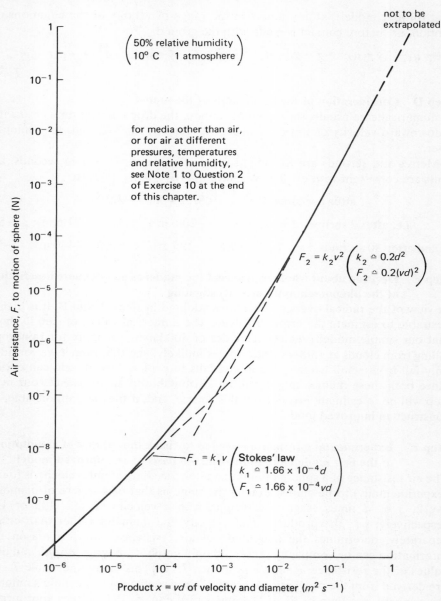

Fig. 7.13

velocity range may be of as much interest as larger drops and higher velocities, we shall present the information in such a way that each range of values of x rising by, say, a factor 10 has similar page space allocated to it; thus equal intervals along the horizontal axis in FIG. 7.13 correspond not to equal intervals of x but of $\ln x$. The range of values of the resistance is also very wide so a logarithmic scale is also used on this axis. When reading the graph at values other than those shown on the scales, this scaling must be taken into account.

Motion and differential equations 221

TABLE 7.1 Weight and air resistance on spherical drop of water
(a) *Air resistance (in newtons) on a non-rotating spherical body, diameter d, velocity v (10°C, 50% relative humidity, 1 atmospheric pressure)*

d (m) $\quad$ v (m s^{-1})	10^{-5} (10 microns)	3.16×10^{-5}	10^{-4} (100 microns)	3.16×10^{-4}	10^{-3} (1 mm)	3.16×10^{-3}	10^{-2} (1 cm)
10^{-3}	1.7×10^{-12}	5.2×10^{-12}†	1.7×10^{-11}	5.2×10^{-11}	1.7×10^{-10}	5.2×10^{-10}	1.8×10^{-9}
3.16×10^{-3}	5.2×10^{-12}*	1.7×10^{-11}	5.2×10^{-11}	1.7×10^{-10}	5.2×10^{-10}	1.8×10^{-9}	6.6×10^{-9}
10^{-2}	1.7×10^{-11}	5.2×10^{-11}	1.7×10^{-10}†	5.2×10^{-10}	1.8×10^{-9}	6.6×10^{-9}	3.2×10^{-8}
3.16×10^{-2}	5.2×10^{-11}	1.7×10^{-10}*	5.2×10^{-10}	1.8×10^{-9}	6.6×10^{-9}	3.2×10^{-8}	1.6×10^{-7}
10^{-1}	1.7×10^{-10}	5.2×10^{-10}	1.8×10^{-9}	6.6×10^{-9}†	3.2×10^{-8}	1.6×10^{-7}	8×10^{-7}
3.16×10^{-1}	5.2×10^{-10}	1.8×10^{-9}	6.6×10^{-9}*	3.2×10^{-8}	1.6×10^{-7}†	8×10^{-7}	4×10^{-6}
1	1.0×10^{-9}	6.6×10^{-9}	3.2×10^{-8}	1.6×10^{-7}*	8×10^{-7}	4×10^{-6}†	3×10^{-5}†
3.16	6.6×10^{-9}	3.2×10^{-8}	1.6×10^{-7}	8×10^{-7}	4×10^{-6}*	3×10^{-5}	2×10^{-4}
10	3.2×10^{-8}	1.6×10^{-7}	8×10^{-7}	4×10^{-6}	3×10^{-5}	2×10^{-4}*	2×10^{-3}*
31.6	1.6×10^{-7}	8×10^{-7}	4×10^{-6}	3×10^{-5}	2×10^{-4}	2×10^{-3}	2×10^{-2}
100	8×10^{-7}	4×10^{-6}	3×10^{-5}	2×10^{-4}	2×10^{-3}	2×10^{-2}	(2×10^{-1})

(b) *Weight (in newtons) of a spherical drop of water, diameter d*

d (m)	10^{-5}	3.16×10^{-5}	10^{-4}	3.16×10^{-4}	10^{-3}	3.16×10^{-3}	10^{-2}
Weight (N)	5.13×10^{-12}	1.63×10^{-10}	5.13×10^{-9}	1.63×10^{-7}	5.13×10^{-6}	1.63×10^{-4}	5.13×10^{-3}

NOTES: * Tabular value for which air resistance $\simeq$ weight.
$\quad\quad\quad\quad$ † Tabular value corresponding to greatest velocity for which the approximation 'resistance $\simeq 0$' is acceptable for free fall.

In any region of the graph where the slope $= m$,

$$\ln F = m \ln(vd) + \text{constant}$$

$$F \propto (vd)^m$$

At small velocities, it can be seen from FIG. 7.13 that $m = 1$, so

$$F_1 \propto v$$

while for the larger velocities shown, $m = 2$,

$$F_2 \propto v^2$$

Regions of relevance of the resistance laws discussed in §§ 7.42 and 7.43 have therefore been ascertained.

A presentation of the data which is directly relevant to our raindrop investigation, is shown in Table 7.1(a). For a sphere of given diameter d, the variation of air resistance with velocity may be read directly from an appropriate column of the table. Tabular values 3.16, 31.6, etc. were chosen in the scales of both v and d because $\sqrt{10} = 3.16$, i.e. 3.16 is midway between 1 and 10 on a logarithmic scale.

Table 7.1(a) may be used for any spherical body moving through air. Table 7.1(b) refers specifically to a sphere of water. The daggers (†) in Table 7.1(a) show where the weight $\simeq 100 \times$ air resistance. When a drop of diameter d has a velocity greater than that at the position of the dagger, we cannot safely neglect the air resistance.

The asterisked values of air resistance in the table are roughly equal to the corresponding drop weight, i.e. the resistance just balances the weight during vertical fall at these velocities, and there is consequently no further acceleration.

Values of velocity beyond the asterisked values, i.e. below them in the table are not relevant to free fall starting from rest (because they will never be achieved), but could be of use in problems relating to spheres projected with high velocity.

Revised step D conclusions

Provided that the drop diameter d and the final speed v correspond to a tabular position above the daggers, the analysis of Step C will closely represent the main features of the fall of a raindrop of constant mass. Thus the fall of a hailstone of diameter 1 cm falling from rest at height h would be adequately described by the equations, $v = -gt$; $z = h - \frac{1}{2}gt^2$; $v^2 = 2g(h-z)$ until it had acquired a speed of about 1 m s^{-1}; thereafter, until a velocity of over 10 m s^{-1} was achieved, a more complicated mathematical model of its motion would be needed; finally, a situation of effective equilibrium would be set up, in which the drop maintained the constant velocity (of the order of 16 m s^{-1}, according to the graph, for a spherical model drop) for which the air resistance equals the weight of the drop.

The fall of a droplet of diameter 0.01 mm, on the other hand, would be controlled almost from the outset by air resistance; such droplets would have speeds of no more than about 0.003 m s^{-1} relative to the air, i.e. the weather man would describe an assembly of such drops as mist rather than rain.

Motion and differential equations

Step E"
The simple mathematical model set up in steps A and B suffices only for severely restricted ranges of drop size and velocity. **Outside** this range, the effect of air resistance is **not** negligible. If we are prepared to model the drop as a sphere, and wish merely to estimate the final (so-called 'terminal') velocity of fall, then a simple balancing method – equating air resistance to weight, and reading the graph of FIG. 7.13, to deduce the corresponding velocity – is an available procedure, giving an order of magnitude result. Corrections and/or error estimates must be made for large drops, for which a shape distortion, or streamlining is to be expected. Yet larger drops may be unstable and break up: a discussion of experimental evidence is to be found in E. G. Richardson's *Dynamics of Real Fluids*. Accretion or evaporation will also affect the rate of fall in humid or dry climatic conditions respectively.

7.61 Lessons drawn from the raindrop exercise
The process of modelling compels the mathematician to think. His model rests on assumptions, and the explicit recognition of these assumptions is one of the rewards accruing from even the crudest original modelling.

Usually problem-solving proceeds as a sequence of attempts, each model rectifying the more serious defects of previous attempts. Sometimes, as in the raindrop example, the initial plan of campaign – to write down and solve equations of motion – may be abandoned or delayed because more profitable procedures open up (e.g. the limit velocity versus raindrop size investigation).

The standard repertoire of applied mathematics consists of studies of simplified situations. Problems from real life are less immediately solved, but familiarity with the standard repertoire acts as a spring board from which to take off – after checking the assumptions which support it!

Exercise 10
1. A spherical body of mass 0.01 g and diameter 1 cm (0.01 m) falls through the air, starting from rest.
   (i) Up to what velocity v is the $F = k_1 v$ approximation (see FIG. 7.13) acceptable?
   (ii) At this value of the velocity, how do the magnitudes of the air resistance and the gravitational force on the body compare?
   (iii) At what velocity, and hence at what (approximate) time after release, does the $F = k_2 v^2$ law become a better approximation than the $F = k_1 v$ law? At this time, approximately how far has the sphere fallen?
   (iv) Given that the sphere falls a total height of 20 m starting from rest, what modelling(s) of the air resistance do you consider appropriate?
   (v) Write down the equation of vertical motion of the sphere assuming the $k_2 v^2$ law of resistance, with $k_2 = 0.2 \times 10^{-4} = 2 \times 10^{-5}$.
   (vi) Solve this equation by the method of § 7.43 to find a relationship between the velocity and the position of the sphere during its fall, and deduce the speed with which the ball hits the ground.
2. Conversion of FIG. 7.13 to refer to other media (or air at temperature, relative humidity or pressure other than those stated). Trace FIG. 7.13 and (using Note 1 overleaf) relabel the scales so that the graph is suitable to
   (i) water (at $\simeq 20°C$), density $\rho = 1000$ kg m^{-3}, viscosity $\mu = 10^{-3}$ SI units,
   (ii) glycerine (at$\simeq 20°C$), density $\rho = 1.25$ gm cm^{-3}, viscosity $\mu = 1.5$ SI units.

Note 1: if spheres move in a medium other than air at the stated temperature, humidity, etc. the law of resistance shown in FIG. 7.13 must be re-scaled as follows: the scale points of x (marked 10^{-6}, 10^{-5}, ... 1) must be remarked with values multiplied by $7 \times 10^4 \, \mu/\rho$, where μ, ρ are respectively the viscosity and density, in SI units of the new medium; the scale points of F must similarly be re-marked with the values shown in FIG. 7.13 multiplied by $4 \times 10^9 \mu^2/\rho$.

The reader will appreciate that it is not possible to read FIG. 7.13, or other graphs derived from it, with any great accuracy. Our purpose in providing them is to assist the reader to make crude estimates, or to judge whether a $k_1 v$ or a $k_2 v^2$ modelling of resistance is likely to be of use in a given situation.

Note 2: after it has been decided whether or not a $k_1 v$ or a $k_2 v^2$ modelling is appropriate, analytical expressions for the drag force will usually be more convenient than the graphical representation: the reader should not assume that their accuracy is necessarily greater than the accuracy of reading values from FIG. 7.13.

The algebraic forms appropriate to a sphere are:

(in the $k_1 v$ range) $F \simeq (3\pi\mu d)v$;

(in the $k_2 v^2$ range) $F \simeq 0.2 \, (\tfrac{1}{4}\pi d^2)\rho v^2$, in which the factor $\tfrac{1}{4}\pi d^2$ indicates that the force when the kv^2 law operates is proportional to the area of the cross-section presented to the medium through which the sphere is moving. Note that although d means the diameter of the *sphere*, ρ stands for the density of the *fluid* through which it moves.

Note 3: algebraic expressions similar to those given for a sphere in Note 2 are also available for the resistance to a circular cylinder whose direction of motion is perpendicular to its axis. The symbol d now denotes the diameter of the cylinder, and l denotes its length. The resistance graph is similar in general shape to that of the sphere, as shown in fig. 7.13, having a curved section which, for motion through air, occurs for vd in the range $10^{-5} < vd < 10^{-2}$. For smaller and larger values of vd, simple algebraic laws are acceptable. The algebraic forms for a cylinder of length l are:

(in the $k_1 v$ range of vd)

$$F \simeq 4\pi\mu lv/(2 - \ln \rho vd/\mu) \quad \text{(Lamb's formula)}$$

(in the $k_2 v^2$ range of vd)

$$F \simeq 0.6(ld)\rho v^2.$$

As for spheres, so also for cylinders, no hard and fast rules can be given about the upper limit of vd for which the $k_2 v^2$ rule can be applied. More detailed calculations, e.g. 'boundary layer' calculations of fluid mechanics, have to be used to discuss the likely behaviour of particular bodies at high velocities.

3 What are the merits and demerits of (a) an experimentally determined stress–strain relationship, such as that sketched in FIG. 7.14, referring to a copper wire sample, and (b) Hooke's Law, viz. $F = k(l - l_0)$, for the same wire?

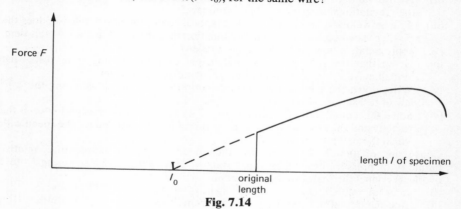

Fig. 7.14

8

Equivalent systems of forces; centre of gravity; forces between rigid bodies

The applied mathematician likes to replace the complexity of an actual system of forces acting on an object by the simplicity of an 'equivalent' single force, if that is possible. If no single force can satisfy his requirements of 'equivalence', he will successively consider pairs or even triads of forces. How and why such equivalents can be found for rigid bodies is made clear in the following paragraphs.

8.1 'Equivalent' systems of force in two dimensions

We first consider the simple case of a plane body acted on by forces in its own plane, which we take as the x-y plane. The general equations of motion of the rigid body are [see Chapter 6, equations 6–8 and 6–27(b)]:

$$M\ddot{\mathbf{r}} = \mathbf{F}, \quad \text{i.e.} \quad \left.\begin{array}{l} M\ddot{x} = X \\ \text{and} \quad M\ddot{y} = Y \end{array}\right\} \qquad 8\text{--}1$$

and $$I_G \ddot{\theta} = M_G \qquad 8\text{--}2$$

From 8–1 we can gain information about the position of the central point G at any time; from 8–2 we can find out how the body has rotated; and these two pieces of information together tell us all we need to know about the body's motion.

Two quite different systems of forces may produce identical motion of the body. For, in order to produce the same accelerations $\ddot{\mathbf{r}}$ and $\ddot{\theta}$ it is only necessary to ensure that $\mathbf{F}$ and M_G are the same for the two sets of forces. Two such systems of force are called 'equivalent', meaning equivalent in their effect on the motion of the single rigid body.

8.11 Example

Show that the single force of magnitude $\sqrt{7}$ shown in FIG. 8.1b and the system shown in FIG. 8.1a are equivalent in their effect on the motion of the rigid regular hexagon with mass centre G.

For the system of FIG. 8.1a,

$$X = -1 - 2\cos 60° - 5\cos 60° + 4 = -\tfrac{1}{2}$$

226 An introduction to mechanics and modelling

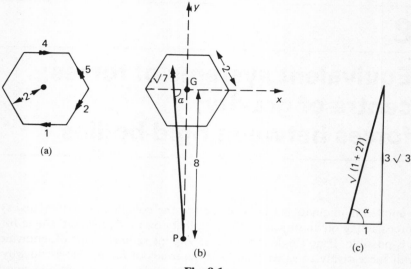

Fig. 8.1

$$Y = -2 \sin 60° + 5 \sin 60° = 3\sqrt{3}/2$$
$$M_G = -2 \times 1 - 2 \times 2 + 2 \times 5 - 2 \times 4 = -4$$

For the system of FIG. 8.1b, namely force $\sqrt{7}$ applied at P,

$'X' = -\sqrt{7} \cos \alpha$, where from FIG. 8.1c $\cos \alpha = 1/2\sqrt{7}$

$$= -\sqrt{7} \times \frac{1}{2\sqrt{7}} = -\tfrac{1}{2}$$

$$'Y' = \sqrt{7} \sin \alpha = \sqrt{7} \times \frac{3\sqrt{3}}{2\sqrt{7}} = \frac{3\sqrt{3}}{2}$$

$'M_G' = x_P 'Y' - y_P 'X'$ (cf. equation 6–31) where $x_P = 0$, $y_P = -8$

$$= 8'X' = -4.$$

Since $X = 'X'$, $Y = 'Y'$ and $M_G = 'M_G'$, the two systems are equivalent. We shall see below, equation 8–8(c), that if M_0 is the moment of a system of forces about an arbitrary origin of coordinates, and G has coordinates $\bar{x}$, $\bar{y}$ with respect to this origin,

$$M_G = M_0 - \bar{x} Y + \bar{y} X \qquad \qquad 8\text{–}3$$

It follows that, since $\bar{x}$, $\bar{y}$ are geometrical items, currently the same for dashed and undashed force-systems,

> two 2-dimensional systems of force with the same X, Y and M_0 values will also have equal M_G values, and so are equivalent systems.

8.12 Equivalent single force or pair of forces

By a slight modification, the method of § 8.11 can be used to find a single force equivalent to almost any two-dimensional system of forces.

For, if the given system of forces has $\Sigma X_i = X$; $\Sigma Y_i = Y$; $\Sigma M_{0i} = M_0$; then the single force $\mathbf{F} = X\mathbf{i} + Y\mathbf{j}$ acting through the point (x, y) has the same X, Y and M_0 as the original system, provided only that the coordinates x, y can be chosen to satisfy

$$xY - yX = M_0 \qquad 8\text{–}4$$

If $X = Y = 0$ and $M_0 \neq 0$, equation 8–4 can be satisfied by no choice of (x, y). But in any other case, equation 8–4 merely prescribes that (x, y) lies on the straight line defined by equation 8–4. The slope of the line is Y/X, i.e. the line runs in the direction of the force $\mathbf{F}$.

> The straight line
> $$xY - yX = M_0$$
> is the line of the body along which the resultant $\mathbf{F}$ must act if it is to constitute a single force equivalent to the given system of forces. The line is called the 'line of action' of the resultant.

Thus, in the example of § 8.11, the single force equivalent of the system of forces shown in FIG. 8.1a is the force $X = -\tfrac{1}{2}$; $Y = 3\sqrt{3}/2$ acting along the straight line

$$\tfrac{1}{2}(3\sqrt{3})x - \tfrac{1}{2}y = -4$$

i.e. $\qquad 3\sqrt{3}x - y = -8.$

In the special case $\mathbf{F} = 0$ (i.e. $X = Y = 0$), $M_0 \neq 0$, there is no single equivalent force, but a simple equivalent system may be provided by any force $\mathbf{P}$ in the plane of forces, arbitrarily placed, together with a suitably placed equal and opposite force $-\mathbf{P}$.

For, suppose that the force $\mathbf{P}$, of magnitude P, acts along a line at distance p from the origin. Then its anticlockwise moment about the origin is either pP (FIG. 8.2a) or $-pP$ (FIG. 8.2b). To produce an arbitrary positive, i.e. anticlockwise moment M about O we need only place the equal and opposite force along the dotted line indicated in each case with an appropriate value of d. In case (a) we require

$$pP - (p - d)P = M;$$

i.e.

$$Pd = M$$

and for case (b),

$$(p + d)P - pP = M;$$

i.e.

$$Pd = M$$

228 An introduction to mechanics and modelling

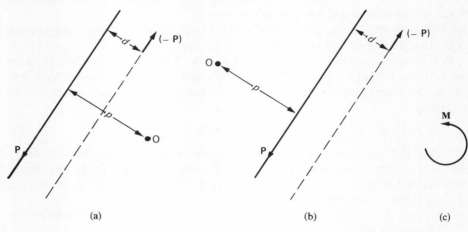

Fig. 8.2

We conclude that an arbitrary positive moment may be produced by a pair of equal and opposite forces whose magnitude and direction may be arbitrarily chosen, but whose distance apart must be related to the magnitude by the relation $Pd = M$.

By interchanging the position of **P** and $-$**P** in either FIG. 8.2a or b, the arbitrary negative moment $-M$ may be obtained.

A pair of forces **P** and $-$**P** acting along lines distance d apart is called a 'couple', and $\pm Pd$ is called the moment of the couple. Since the particular choice of **P** is immaterial, a couple is often represented as in FIG. 8.2c.

> Any two-dimensional system which cannot be represented by a single equivalent force (including zero force) is equivalent to a couple, of which only the moment, not the individual constituents, need be specified.

The result stated earlier, that any two-dimensional system is equivalent either to a single, suitably placed force, or else to a couple (whose placing is immaterial so long as its moment is right) has thus been proved. It is often unnecessary to reduce the system quite so far as this: when **F** and M_0 have been evaluated for a chosen origin O, it suffices to represent the system by the equivalent system consisting of force **F** applied at O, together with a couple M_0 (whose placing is immaterial, but which it is natural to think of as a torque or twist about O).

Exercise 1

1. The two-dimensional system of forces shown in FIG. 8.3 acts on the triangle OAB in which OA = OB = 2 units. The magnitudes of the forces $\mathbf{F}_1$, $\mathbf{F}_2$ and $\mathbf{F}_3$ are 2, 4 and 3 respectively. Show that a single equivalent force may be found whose point of application may be taken on the line AO. Identify the position of this point on AO (extended if necessary).

2. Show that the force system given in question 1 is equivalent to a single force applied at a point Q of the side OB (extended if necessary). Identify the position of Q.

3. Show that the force system given in question 1 is equivalent to a single force applied at point (x, y) provided that x, y lies on a certain straight line. Verify that the slope

Equivalent systems of forces 229

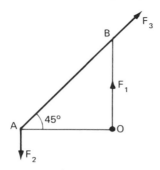

Fig. 8.3

of this line is the same as that of the single force, i.e. that the line is appropriately described as the line of action of the force.

4 Gravitational forces of magnitude $3g$, $4g$ and $5g$ act vertically downwards in the same vertical plane. Taking cartesian axes Ox, Oy in that plane (with the y-axis pointing vertically upwards) the points of application of the given forces are $(2, 0)$, $(-1, 3)$ and $(1, 2)$ respectively. Find the magnitude and line of action of the single equivalent force.
5 Repeat question 1 for a general system of gravitational forces in a vertical plane, viz. $m_1g, m_2g, m_3g, \ldots$ acting at points (x_1, y_1), (x_2, y_2) and $(x_3, y_3) \ldots$ respectively.
6 The vertices A, B, C of an equilateral triangle are at the points $(0, 0)$, $(2, 0)$ and $(1, \sqrt{3})$ respectively of the x-y plane. Forces of magnitudes 4, 5 and 6 act in the lines CA, AB and BC respectively. Find the magnitude of the resultant force, and the x-coordinate of the point at which the line of action of the resultant cuts the x-axis.
7 Prove that the algebraic sum of the moments of two concurrent forces about any point in their plane is equal to the moment of their resultant about that point if the line of action of the resultant passes through the point of concurrence of the given forces.

  ABC is an equilateral triangle of side a. The perpendicular from A to BC meets BC at D. Forces 1, 2, 3 and $\sqrt{3}$ act along AB, BC, CA and AD respectively. Show that the system reduces to a couple, and find its magnitude.

  If the force along AD is altered to $2\sqrt{3}$, find where the resultant of the new system of forces meets BC and its magnitude.
8 ABCD is a square of side 2 units which is placed with AB along the positive x-axis and AD along the positive y-axis. Forces of magnitudes 3, 4, 2 and 1 act in the lines AB, BC, CD, DA respectively and a force of magnitude $7\sqrt{2}$ is directed from the mid-point of BC to the mid-point of AB. Find the x-coordinate of the point at which the line of action of the resultant force meets the x-axis. Indicate the line of action and the direction of the resultant force on a sketch.

8.2 Equivalent systems of force in three dimensions

For motion in three dimensions the equation

$$M\ddot{\mathbf{r}} = \mathbf{F} \qquad \qquad 8\text{--}5$$

(where $\mathbf{F}$ is known) continues to supply all the information required to determine the way in which the point G, starting from a known position and velocity, will subsequently move.

  The three-dimensional equation of angular motion, derived in Chapter 6 on the basis of the particle model of a body, was stated in equation 6–22(b) in the

230   An introduction to mechanics and modelling

form

$$\frac{d}{dt}\sum \mathbf{s}_i \wedge m_i \dot{\mathbf{s}}_i = \sum \mathbf{s}_i \wedge \mathbf{F}_i \qquad 8\text{--}6$$

$$= \text{vector sum of moments of external forces about G}$$

$$= \mathbf{M}_G \quad \text{say}$$

This may be neatened by writing

$$\mathbf{H}_G = \sum \mathbf{s}_i \wedge m_i \dot{\mathbf{s}}_i \qquad 8\text{--}7$$

whereupon 8–6 takes the form

$$\frac{d}{dt}(\mathbf{H}_G) = \mathbf{M}_G \qquad 8\text{--}8$$

The quantity $\mathbf{H}_G$ is known as the angular momentum vector and, from equation 8–7, it is determined by the geometry and motion of the body.

Equation 8–6 was deduced from a model: it strongly suggests that the total moment of forces, $\mathbf{M}_G$, is very important for the motion of a rigid body. If the geometrical consequences were further pursued, we should find that the model indicates that the angular motion of a rigid body, starting from a known situation, is completely determined by $\mathbf{M}_G$. The ultimate test, however, is experimental; and for the purposes of the present chapter all we need is the postulate, tested directly and indirectly by a multitude of experiments and emerging unscathed, that

> the angular motion of a given rigid body, starting from given conditions, is determined by $\mathbf{M}_G$, the vector sum of the moments about G of all external forces. 8–9

Hence, two systems of force are equivalent (for any motion of a rigid body, whether uni-, two- or three dimensional) if they have the same $\mathbf{F}$ and $\mathbf{M}_G$.

If the mass distribution on which the two systems of force acted had constituted a body capable of deformation, then knowledge of $\mathbf{F}$ and $\mathbf{M}_G$ alone would not fully determine the consequent motion. Two systems of force which are equivalent for a rigid body are not equivalent for a non-rigid mass unless additional conditions are also satisfied. Equivalence is dependent on the nature of the mass system as well as the force system. It is unprofitable to enumerate the full equivalence conditions relevant to jointed systems, elastic materials and the like. In elementary mechanics the procedure is to subdivide every complex system into rigid body components (i.e. parts, not components in the technical sense!), and consider each component part separately.

8.21 Moments about different points

Theorem 1
If $\mathbf{F}$ and $\mathbf{M}_G$ are known for a system of forces, the current position of G being known, the moment $\mathbf{M}_P$ of the system about any other specified point P in space

can be calculated:

$$M_P = M_G + (\bar{r} - R) \wedge F \text{ (FIG. 8.4)}.$$

Conversely, if F and M_P are known, the value of M_G is determined.

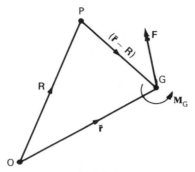

Fig. 8.4

Proof

Let G have position vector $\bar{r}$; let P have position vector R; let the forces of the system be $F_1, F_2, \ldots$, acting at points $r_1, r_2, \ldots$, of the body. The displacement of the point of application of force F_i from point P is

$$r_i - R$$

and, similarly, the displacement of the point of application from the centre of mass G, is

$$r_i - \bar{r}$$

$$M_G = \sum_i (r_i - \bar{r}) \wedge F_i$$
$$= \sum r_i \wedge F_i - \bar{r} \wedge \sum F_i \qquad 8\text{--}10$$
$$M_P = \sum_i (r_i - R) \wedge F_i$$
$$= \sum_i r_i \wedge F_i - R \wedge \sum F_i$$

whence

$$M_P - M_G = (\bar{r} - R) \wedge \sum F_i$$
$$= (\bar{r} - R) \wedge F \qquad 8\text{--}11$$

i.e.

$$M_P = M_G + (\bar{r} - R) \wedge F$$

where all the symbols on the right-hand side represent known quantities; and this proves the theorem.

If the point P is taken as origin, $R = 0$, equation 8–11 may be written

$$M_O = M_G + \bar{r} \wedge F \qquad 8\text{--}12\text{(a)}$$

232 An introduction to mechanics and modelling

or
$$\mathbf{M}_G = \mathbf{M}_0 - \bar{\mathbf{r}} \wedge \mathbf{F} \qquad \text{8–12(b)}$$

If the third component of the vector $\mathbf{M}_0$ is written N_0, and that of $\mathbf{M}_G$ written N_G, then the third (i.e. **k**- or z-component) of equation 8–12(b) may be written
$$N_G = N_0 - \bar{x}Y + \bar{y}X \qquad \text{8–12(c)}$$
which is, in a slightly different notation, equation 8–3 used in the two-dimensional discussion of § 8.1.

Theorem 2
Two systems of force with the same total **F** and same $\mathbf{M}_0$ (where the choice of the particular origin O at any instant is immaterial) are equivalent systems of force.

For, if **F** and $\mathbf{M}_0$ are the same, then by equation 8–12(b), **F** and $\mathbf{M}_G$ are the same: the equations of motion 8–5 and 8–8 are indistinguishable for the two systems. The conditions for **equilibrium** are $\mathbf{F} = \mathbf{0}$ and $\mathbf{M}_G = \mathbf{0}$.

Hence, for the purpose of investigating the equilibrium, or any motion, of a rigid body (whether in one, two or three dimensions) it suffices to represent any set or subset of forces acting on the body by a force **F** applied at any chosen origin O together with a couple of moment $\mathbf{M}_0$.

Example 1
A body whose centre of mass is at the point $(1, 2, -1)$ is acted on by the following forces: $(2, 0, 0)$ at point $(0, 1, 2)$; $(1, 0, 1)$ at point $(-1, -1, -1)$ and $(1, 2, 2)$ at point $(1, 0, 1)$. Find **F** and $\mathbf{M}_G$ for the body.

For **F**, we get the vector sum by adding components, thus
$$X = 2+1+1 = 4; \quad Y = 0+0+2 = 2; \quad Z = 0+1+2 = 3;$$
so **F** is the force with components $(4, 2, 3)$, i.e. $\mathbf{F} = 4\mathbf{i} + 2\mathbf{j} + 3\mathbf{k}$. For $\mathbf{M}_G$, equation 8–12(b) is easier to use than equation 8–9. We first find

$$\mathbf{M}_0 = \begin{vmatrix} \mathbf{i} & \mathbf{j} & \mathbf{k} \\ 0 & 1 & 2 \\ 2 & 0 & 0 \end{vmatrix} + \begin{vmatrix} \mathbf{i} & \mathbf{j} & \mathbf{k} \\ -1 & -1 & -1 \\ 1 & 0 & 1 \end{vmatrix} + \begin{vmatrix} \mathbf{i} & \mathbf{j} & \mathbf{k} \\ 1 & 0 & 1 \\ 1 & 2 & 2 \end{vmatrix} \quad \text{[cf. equation 2–31(b)]}$$

$$= (4\mathbf{j} - 2\mathbf{k}) + (-\mathbf{i} + \mathbf{k}) + (-2\mathbf{i} - \mathbf{j} + 2\mathbf{k})$$

$$= -3\mathbf{i} + 3\mathbf{j} + \mathbf{k}$$

$$\bar{\mathbf{r}} \wedge \mathbf{F} = \begin{vmatrix} \mathbf{i} & \mathbf{j} & \mathbf{k} \\ 1 & 2 & -1 \\ 4 & 2 & 3 \end{vmatrix} = 8\mathbf{i} - 7\mathbf{j} - 6\mathbf{k}.$$

So, using equation 8–12(b)
$$\mathbf{M}_G = -11\mathbf{i} + 10\mathbf{j} + 7\mathbf{k}$$

Equivalent systems of forces

Example 2
Show that (unlike the two-dimensional system of forces shown in FIG. 8.1a), the three-dimensional system of forces described in Example 1 above has no single-force equivalent.

For, if there were a single equivalent force, (X, Y, Z) say, acting at point (x, y, z), we should have to satisfy

(i) $X = 4;\ Y = 2;\ Z = 3;$

(ii) $\begin{vmatrix} \mathbf{i} & \mathbf{j} & \mathbf{k} \\ x & y & z \\ X & Y & Z \end{vmatrix} = \mathbf{M}_0;$

i.e. $\begin{vmatrix} \mathbf{i} & \mathbf{j} & \mathbf{k} \\ x & y & z \\ 4 & 2 & 3 \end{vmatrix} = -3\mathbf{i} - 5\mathbf{j} + \mathbf{k}$

i.e. $3y - 2z = -3$

$4z - 3x = 3$

$2x - 4y = 1$

These three relations are incompatible: for the first two (eliminating z) require $2y - x = -1$; while the third relation requires $2y - x = -\frac{1}{2}$. We conclude that there is no single-force equivalent of the given three-dimensional system, nor is the system equivalent to a couple.

That any system of forces is equivalent to a single force together with a couple is almost self-evident. For, the single force $\mathbf{F}$ applied at the centre of mass G of the body concerned, together with any couple of moment $\mathbf{M}_G$, give the right values to ensure that we have an equivalent system for the rigid body in question.

A simple vectorial expression for the moment of a given couple is next derived.

8.22 Moment of a couple in three dimensions

The moment about the origin of force $-\mathbf{F}$ acting at $\mathbf{r}$ and $\mathbf{F}$ acting at $\mathbf{r} + \mathbf{R}$, as in FIG. 8.5, is

$$\mathbf{M}_0 = \mathbf{r} \wedge (-\mathbf{F}) + (\mathbf{r} + \mathbf{R}) \wedge \mathbf{F} = \mathbf{R} \wedge \mathbf{F} \qquad 8\text{--}13$$

Similarly the moment of the same forces about any point P with position vector $\mathbf{r}_P$ is

$$\mathbf{M}_P = (\mathbf{r} - \mathbf{r}_P) \wedge (-\mathbf{F}) + (\mathbf{r} + \mathbf{R} - \mathbf{r}_P) \wedge \mathbf{F} = \mathbf{R} \wedge \mathbf{F} \qquad 8\text{--}14$$

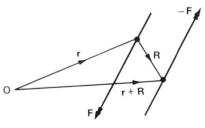

Fig. 8.5

234 An introduction to mechanics and modelling

A couple therefore has the same moment $\mathbf{R} \wedge \mathbf{F}$ about every point in space. It is an unlocalized quantity whose magnitude and direction are determined by $\mathbf{F}$ and any vector $\mathbf{R}$ joining a point on the force $\mathbf{F}$ to a point on the force $-\mathbf{F}$.

8.23 Simplest equivalent system in three dimensions

The vector sum $\mathbf{F}$ of any three-dimensional system of forces is a well-defined quantity. Consider a single force $\mathbf{F}$ placed in turn at a number of trial positions. The value of its moment $\mathbf{r} \wedge \mathbf{F}$ about the origin depends on its position; but, whatever the value of $\mathbf{r}$, $\mathbf{r} \wedge \mathbf{F}$ is perpendicular to $\mathbf{F}$. It follows that a moment $\mathbf{M}_0$ which has a component **parallel** to $\mathbf{F}$ in FIG. 8.6a can be produced by no placing of $\mathbf{F}$, i.e. that a three-dimensional system of forces $(\mathbf{F}, \mathbf{M}_0)$ for which $\mathbf{M}_0$ has a non-zero component in the direction of $\mathbf{F}$ is equivalent to no single-force placement.

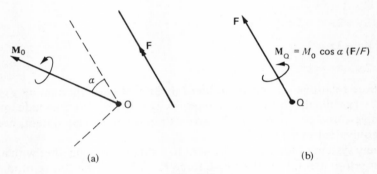

Fig. 8.6

> In general, the simplest system 'equivalent' to a three-dimensional system of forces consists of both a force $\mathbf{F}$, **and** a couple whose moment is a vector parallel to $\mathbf{F}$. Such a force and couple (or torque) combination is called a 'wrench' (see FIG. 8.6b). Only in special cases does a three-dimensional system 'reduce' to a single force or to a couple.

Example
(Central axis of a wrench).

With a chosen origin O, a system of forces is characterized by $\mathbf{F}$, $\mathbf{M}_0$. Find the equation of the line of points P for which $\mathbf{M}_P$ is parallel to $\mathbf{F}$. (This line is called the axis of the wrench $\mathbf{F}$, $\mathbf{M}_P$.)

(i) If $\mathbf{M}_0$ is parallel to $\mathbf{F}$, consider a point P for which $\mathbf{r}_P = \lambda \mathbf{F}$. Relative to P, the system of forces is characterized by $\mathbf{F}$ and $\mathbf{M}_P = \mathbf{M}_0 - \mathbf{r}_P \wedge \mathbf{F} = \mathbf{M}_0$ (which is parallel to $\mathbf{F}$). Hence $\mathbf{r} = \lambda \mathbf{F}$ is the equation of the axis of the wrench, $(-\infty < \lambda < \infty)$.

(ii) If $\mathbf{M}_0$ is not parallel to $\mathbf{F}$, then $\mathbf{F}$, $\mathbf{M}_0$ and $\mathbf{F} \wedge \mathbf{M}_0$ are three non-coplanar vectors. Write

$$\mathbf{r}_P = \alpha\mathbf{F} + \beta\mathbf{M}_0 + \gamma\mathbf{F}\wedge\mathbf{M}_0$$
$$\mathbf{M}_P = \mathbf{M}_0 - (\alpha\mathbf{F} + \beta\mathbf{M}_0 + \gamma\mathbf{F}\wedge\mathbf{M}_0)\wedge\mathbf{F}$$
$$= \mathbf{M}_0 - \beta\mathbf{M}_0\wedge\mathbf{F} - \gamma[(\mathbf{F}\wedge\mathbf{M}_0)\wedge\mathbf{F}]$$

Now

$$(\mathbf{F}\wedge\mathbf{M}_0)\wedge\mathbf{F} = F^2\mathbf{M}_0 - (\mathbf{F}\overset{\circ}{\wedge}\mathbf{M}_0)\mathbf{F}$$

so

$$\mathbf{M}_P = (1 - \gamma F^2)\mathbf{M}_0 - \beta\mathbf{M}_0\wedge\mathbf{F} + \gamma(\mathbf{F}\cdot\mathbf{M}_0)\mathbf{F}$$

This vector is parallel to $\mathbf{F}$ only if $\beta = 0$ and $1 - \gamma F^2 = 0$, so $\gamma = 1/F^2$. Then

$$\mathbf{M}_P = (\mathbf{F}\cdot\mathbf{M}_0)\mathbf{F}/F^2$$

Since $\mathbf{M}_P$ is a multiple of $\mathbf{F}$, the coefficient $(\mathbf{F}\cdot\mathbf{M}_0)/F^2$ is of some interest. It is called the **pitch** of the wrench. The equation of the axis of the wrench is obtained by inserting the values $\beta = 0$, $\gamma = 1/F^2$ in the expression for $\mathbf{r}_P$

$$\mathbf{r} = \alpha\mathbf{F} + (\mathbf{F}\wedge\mathbf{M}_0)/F^2 \quad -\infty < \alpha < \infty$$

We notice that this equation is valid for case (i), where $\mathbf{M}_0$ is parallel to $\mathbf{F}$, as well as for case (ii).

The equation of the axis is sometimes quoted in the equivalent form $(\mathbf{r} - \mathbf{F}\wedge\mathbf{M}_0/F^2)\wedge\mathbf{F} = \mathbf{0}$. The student may establish this as an exercise.

8.3 Centre of gravity

The reader's attention is drawn to questions 4 and 5 of Exercise 1 where simple two-dimensional calculations reveal a lot about the nature of the 'centre of gravity' concept. Corresponding work will now be pursued for the gravitational forces acting on a three-dimensional body, which we model as a system of particles.

We assume that the overall dimensions of the body are small compared with the radius of the earth. The body is modelled as a system of point masses, and the gravitational forces on those particles are represented, as in FIG. 8.7, by forces acting in parallel lines (vertically downwards).

Let masses $m_1, m_2, m_3, \ldots$ be situated at points with position vectors $\mathbf{r}_1, \mathbf{r}_2, \mathbf{r}_3, \ldots$, etc. Then, using $\mathbf{k}$ to denote unit vector in the direction of the upward vertical, the gravitational forces on the body consist of

$$-m_1 g\mathbf{k} \text{ at } \mathbf{r}_1; \quad -m_2 g\mathbf{k} \text{ at } \mathbf{r}_2; \text{ and so on}$$

The system is therefore characterized by

$$\mathbf{F} = \sum_i -m_i g\mathbf{k} = -(m_1 + m_2 + m_3\cdots)g\mathbf{k}$$
$$= -Mg\mathbf{k} \text{ say,}$$

236 An introduction to mechanics and modelling

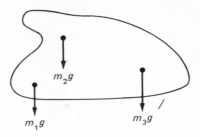

Fig. 8.7

and
$$\mathbf{M}_0 = \sum_i (\mathbf{r}_i \wedge -m_i g \mathbf{k}) = -\sum m_i \mathbf{r}_i \wedge g \mathbf{k}$$

(because g is a constant and the m's are numerical magnitudes; so we can say
$$\mathbf{r}_i \wedge m_i g \mathbf{k} = m_i \mathbf{r}_i \wedge g \mathbf{k}$$
just as we might write $\mathbf{r}_i \wedge 6\mathbf{k} = 3\mathbf{r}_i \wedge 2\mathbf{k}$).

The combination $\sum m_i \mathbf{r}_i$ is familiar from its occurrence in Chapter 6 in connection with equations of motion and the centre of mass property. It was there shown that a point G of a body exists which is fixed in the body if the body is itself rigid, and whose position vector satisfies

$$M\bar{\mathbf{r}} = \sum m_i \mathbf{r}_i \quad \text{where } M = \sum m_i \qquad 8\text{-}15$$

Using the simple expression $M\bar{\mathbf{r}}$ instead of the summation $\sum m_i \mathbf{r}_i$, the quantities $\mathbf{F}$ and $\mathbf{M}_0$ characterizing the system of gravitational forces may be written

$$\mathbf{F} = -Mg\mathbf{k}$$

and
$$\mathbf{M}_0 = -M\bar{\mathbf{r}} \wedge g\mathbf{k} \equiv \bar{\mathbf{r}} \wedge -Mg\mathbf{k} = \bar{\mathbf{r}} \wedge \mathbf{F}$$

> The set of gravitational forces on a rigid body in the proximity of the earth, of dimensions small compared to those of the earth, is equivalent to a single force $-Mg\mathbf{k}$ (Mg acting downwards) through the centre of mass. This is the standard way of modelling the gravitational forces on a rigid body of ordinary size. 8-16

Because of this property, the centre of mass is also referred to as the **centre of gravity**. The position and the significance of the centre of mass are independent of the existence of any gravitational forces, uniform or otherwise: but it is only when parallel gravitational forces act on the constituent particles that the concept of a centre of gravity is of use.

By similar arguments, any system of parallel forces of non-zero sum can be shown to have the same effect on the motion of a rigid body as a single parallel force, suitably sited.

8.31 Modelling of a pendulum

In steps $A, B, \ldots$ below the author gives an account of the modelling steps involved in a discussion of the conditions under which a pendulum beats seconds. The mathematician-in-a-hurry would usually think through the steps without necessarily committing them to paper.

Step A

A clock pendulum is usually a rigid body swinging about a fixed horizontal axis. The term 'pendulum' is also used for a heavy bob swinging at one end of a string. Provided the string remains taut and the motion is confined to a vertical plane, the bob pendulum is mechanically equivalent to the rigid pendulum, for (i) the distance between any two points of the string and/or bob remains constant throughout the motion, which means the bob apparatus is technically 'rigid'; and (ii) since the motion lies in a plane, and one end is fixed, the motion of the bob is correctly described as rotation about a (horizontal) axis, through its point of suspension.

The bob at the end of a string is usually called a 'simple pendulum', the strictly rigid object a 'compound pendulum': the same mathematics applies to both. Our model for either is a rigid body moving in a vertical plane with a fixed, horizontal, axis of rotation, the rigid body being acted on only by gravity and reaction forces. We choose to consider the simplified case in which frictional forces at the axis as well as air-resistance effects are negligibly small.

Step B

The only non-contact force is gravity, and the effect of gravity is represented by the force Mg acting vertically downward through the centre of mass G.

The only materials in contact with the body are the air and the axis. The chosen model attributes the value zero to the force of the air. The model also puts the frictional moment about the horizontal axis of rotation equal to zero, an assumption more securely based for a simple pendulum than for a compound pendulum. Thus the forces at the axis will be represented by the equivalent pair of unknown components X', Y' acting at O, see FIG. 8.8a.

Step C

We have available the two rigid body equations,

$$I_0 \ddot{\theta} = N_0 \qquad \text{8-17(a)}$$

$$M\ddot{\mathbf{r}} = \mathbf{F} \qquad \text{8-17(b)}$$

where, to specify the inclination θ of the pendulum, we shall use the angle between the downward vertical through O and the line OG (see FIG. 8.8a). Note that the former, the vertical, is a direction fixed in space, and that the latter, OG, is a direction fixed in the body: the value of θ is measured **from** the former direction **into** the latter and expressed in radians; the positive sense is anticlockwise, as shown in our diagram.

238 An introduction to mechanics and modelling

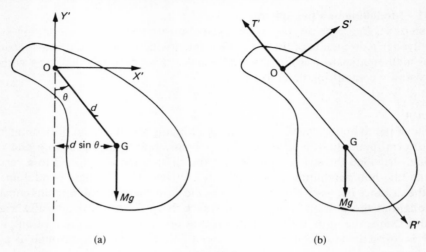

Fig. 8.8

Writing $OG = d$,

$$N_0 = -Mg \cdot d \sin \theta \qquad 8\text{-}18$$

F involves the unknowns X', Y'.

When the value from equation 8–18 is inserted, equation 8–17(a) takes the form,

$$I_0 \ddot{\theta} = -Mgd \sin \theta$$

i.e.

$$\ddot{\theta} = -(Mgd/I_0) \sin \theta \qquad 8\text{-}19$$

Integrating equation 8–19 with respect to θ gives a $\dot{\theta} - \theta$ relation but does not provide direct information about the time taken; nor is a second integration possible in terms of elementary functions.

An approximation gets us out of this difficulty: for a pendulum, we may restrict the angle of swing to, say, 5°; this ensures that throughout the motion $\sin \theta \simeq \theta$. (5° = 0.0873, sin 5° = 0.0872). Then equation 8–19 may be approximated as

$$\ddot{\theta} = -(Mgd/I_0)\theta \qquad 8\text{-}20$$

which is the equation of simple harmonic motion (corresponding to equation 7–9) with $\omega^2 = Mgd/I_0$ and solution

$$\theta = A \cos(\omega t + \varepsilon) \qquad 8\text{-}21$$

A change of $\omega t + \varepsilon$ by 2π requires a change in t of $2\pi/\omega$, so the motion repeats itself at intervals of $2\pi/\omega$. The time between extreme positions – i.e. for the half journey – usually regarded as constituting the beat, is π/ω.

For a 'beat' of 1 second, we require

$$\pi/\omega = 1$$

$$Mgd/I_0 = \omega^2 = \pi^2 \qquad 8\text{-}22$$

$$I_0 = Mgd/\pi^2$$

For a bob pendulum of mass M and string length l,

$$d = l \quad \text{and} \quad I_0 = Ml^2$$

so 8–22 requires,

$$l^2 = gl/\pi^2$$

i.e.

$$l = g/\pi^2 = 9.8/\pi^2 = 0.993$$

$$\simeq 1 \text{ (metre)}$$

Step D
To beat seconds, a rigid pendulum must have a moment of inertia about its axis of rotation,

$$I_0 = Mgd/\pi^2$$

where d is the distance from the axis to its centre of mass. To beat seconds, a simple pendulum must have a length $l \simeq 1$ metre.

In the first case (rigid pendulum), fine corrections should be applied to take account of actual friction at the axis (assumed zero in the model) and of air resistance. In the second case (simple pendulum), fine corrections should be applied to take account of air resistance, the mass of the string, and the finite size of the bob (assumed to be a point mass in the model, with $I_0 \equiv Ml^2$). In both cases it should be noted that the approximation $\sin \theta = \theta$ has been used, and fine corrections may be necessary on this count also.

Step E
No calculations of greater refinement will be entered on here: the fine corrections, if needed, are probably better achieved by practical means than by further theory.

8.32 Calculation of reaction at the axis
If the mass M of a pendulum such as that just considered is large, the variations in X', Y' during the motion may be of importance, particularly if the point of suspension is not very rigidly anchored.

Since the angle θ is directly given by the solution 8–21, the position of G is most conveniently specified by polar coordinates, namely (d, θ); and the associated components of reaction, T' and S' (as shown in FIG. 8.8b), are rather easier to evaluate than X', Y'.

In polar coordinates (see § 2.33) equation 8–17(b) reduces to the pair of equations

(radial equation)

$$Md\dot{\theta}^2 = T' - Mg \cos \theta$$

(transverse equation)

$$Md\ddot{\theta} = S' - Mg \sin \theta$$

> Had we not been aware from everyday experience that pendulum motions do confine themselves to small angles, it would have been necessary to verify that, as time went on, the predicted value of θ would at no subsequent time violate the assumption $\sin \theta \simeq \theta$. In our case, provided A (see equation 8-21) is less than about 0.1, we can be assured of the approximate validity of the solution at all times until the cumulative effects of air resistance can be no longer ignored.

The values of θ, $\dot{\theta}$, $\ddot{\theta}$ are obtainable from equation 8-21, which we may simplify by choosing a time origin which makes $\varepsilon = 0$; we may also write the amplitude A as θ_0, as a reminder of its significance, and use the approximations $\sin \theta \simeq \theta$, $\cos \theta \simeq 1$, whereupon the reaction components have the form

$$T' \simeq Mg + M\theta_0^2 \omega^2 d \sin^2 \omega t$$

$$S' \simeq Mg\theta_0 \cos \omega t - M\theta_0 \omega^2 d \cos \omega t$$

The order of magnitude and the periodicity of the variable forces on the support are apparent from the form of these equations. Also see qn. 18 of Ex. 2.

8.4 Dynamically equivalent bodies in two dimensions

Just as the right-hand sides in the basic equations $M\ddot{\mathbf{r}} = \mathbf{F}$ and $I_G \ddot{\theta} = N_G$ involve only the force combinations denoted by $\mathbf{F}$ and N_G, so the left-hand sides are affected by the shape, material and dimensions of the body only in so far as these attributes affect the position of G in the body and the values of the constants M and I_G. Two bodies with equal M and I_G values may be called dynamically equivalent because, under the action of equal $\mathbf{F}$ and N_G, the two bodies will have equal accelerations $\ddot{\mathbf{r}}$ and $\ddot{\theta}$.

If two dynamically equivalent bodies are subjected to (unknown) reaction forces at identical contact points as well as equal applied forces ($\equiv \mathbf{F}'$, N_G' say), the identical form of the equations of motion and the geometrical conditions in the two cases together ensure that the reaction forces as well as the applied forces are equal in the two cases.

For example, a body capable only of pure translation in the x-direction, and subjected to applied force X' in addition to a normal reaction $X'' = R$ at a fixed inelastic plane surface perpendicular to the x-axis (FIG. 8.9) has equation of motion

$$M\ddot{x} = X' + R$$

By § 5.32, $\ddot{x} = 0$ provided that the value of R required for this is non-negative, i.e. $R = -X'$ provided $-X' \geq 0$; otherwise $R = 0$ and contact will be lost.

Conclusions of this kind depend only on equations of motion and geometry and can be applied equally to all dynamically equivalent bodies capable of being acted on by the given forces.

The use of dynamical equivalence is thus not confined to situations where all the forces are of prescribed value, but extends also to situations where reaction forces arise.

Equivalent systems of forces 241

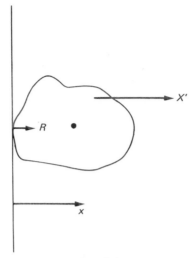

Fig. 8.9

The infelicity of showing the weight of a gramophone record acting at the middle of its central hole can be avoided by working in terms of a dynamically equivalent body without a hole. The details of construction of a ladder – uprights and rungs – are best ignored when considering its overall motion or statics: the ladder is suitably represented by a dynamically equivalent rectangle or even, for two-dimensional problems, by a straight line, to which are attributed not only the correct mass and moment of inertia, but also the same length and initial position as the ladder, so that there is a proper correspondence in the reaction forces which will be evoked.

When rotation about a fixed axis is in question, the correspondence of the contact point O as well as the position of G for the two equivalent bodies ensures that $OG = d$ is the same for both bodies. Because of the parallel axis theorem $I_0 = I_G + Md^2$, two such bodies having the same I_G and M values also have identical I_0 values.

Exercise 2

1. Given that force **F** lying in the x-y plane has components X, Y, and that (x_1, y_1) is a point on the line of action of **F**, show that

   > the pair of forces $X\mathbf{i}$ and $Y\mathbf{j}$, both acting at (x_1, y_1) is **equivalent** to the single force **F**.

2. Consider the factors affecting the safety of a ladder – what should you bear in mind when choosing a position for it, and when you climb it. If you can't think how to tackle this assignment, work through questions 3–10.
   Those who have given a substantial answer to question 2, supported by sound mathematical treatment, may omit questions 3–10.

3. A ladder is modelled as a rigid, linear body with centre of mass at the mid-point of its length. Consider a ladder placed in the usual way with one end resting on horizontal ground, and the other end resting against a vertical wall. A weight $2W$ equal to twice the ladder's own weight is applied at a point one-third of the way up the

242   An introduction to mechanics and modelling

ladder. Draw a diagram, mark in all the forces acting **on the ladder.** (**Note:** do not show forces which act on the wall or on the ground; remember the three main categories of force to look for, namely forces at a distance, especially gravity; known applied forces; and contact forces not already accounted for, which may be of unknown magnitude and direction. When the magnitude and direction of a force are unknown, remember that replacement by its components, as in question 1 above, is often advantageous.)

4   Write down equations of equilibrium for the situation of question 3. (**Note** that the $X = 0$ and $Y = 0$ equations are unaffected by choice of origin, but that some choices of origin give rise to simpler $N = 0$ equations than others – the simplest choice is the best choice.)

5   If the ground is perfectly smooth, what values are required (by the $X = 0$, $Y = 0$ condition) for (a) the normal reaction at the wall, (b) the frictional force at the wall (remembering that $F \leq \mu R$)? If the wall is perfectly smooth, but the ground is rough, with coefficient of friction equal to μ_1, what values of (c) the normal reaction at the ground and (d) the friction at the ground are required in order to satisfy $X = 0$, $Y = 0$?

6   In the case of smooth ground (question 5), can the equations $X = Y = N = 0$ be simultaneously satisfied? In the case of a smooth wall, is equilibrium possible? If the answer in either case is 'yes', what restrictions must be imposed on the value of μ, the angle of inclination of the ladder, etc.? If the load ($2W$) is applied at a higher point of the ladder, how will the restrictions alter?

7   A uniform ladder (centre of mass at mid-point) rests with one end on the ground and the other against a vertical wall in a plane perpendicular to the line of intersection of wall and ground. The coefficient of friction at both the wall and the ground is μ. Show that the inclination of the ladder to the ground cannot be less than the positive root of $\mu^2 + 2\mu \tan \alpha - 1 = 0$. If $\mu = \tan 15°$, show that the angle $\geq 60°$.

8   A light ladder (i.e. one for which the approximation $W = 0$ is acceptable in the present context) rests at one end against a rough vertical wall ($\mu = \frac{1}{4}$) and at the other on rough horizontal ground ($\mu = 4/9$). If the ladder makes an angle of $45°$ with the ground, show that a man cannot climb more than half-way up the ladder without the ladder slipping.

9   In the light of questions 6, 7 and 8, make hypotheses about the effect on ladder stability of (a) roughness of ground and wall, (b) angle of the ladder, (c) ratio of ladder weight to load, (d) position of load on ladder. Verify (or disprove) your hypotheses for situations of as great a generality as the degree of mathematical complication warrants.

10   If you previously assumed the man's centre of mass to be vertically above the rung on which he stood, consider what modification of your conclusions would be required for a more natural posture.

11   Two equal uniform ladders AB and BC are smoothly joined at B, and stand with the ends A and C on a rough horizontal plane. Each is of length a and makes an angle α with the vertical. A man whose weight is equal to that of one of the ladders ascends one of them. Prove that the other will slip first. If slipping begins when the man is a slant distance x up the ladder, prove that the coefficient of friction is $(a + x) \tan \alpha / (2a + x)$.

12   Envisage a ladder situation not covered by the foregoing questions, e.g. sloping ground, centre of mass not at mid-point, different kind of upper support, etc. Formulate a problem and solve it.

13   A uniform plank AB of length $2L$ is firmly fixed at one end A to an immovable support S, the plank projecting horizontally from the support. The system of forces exerted by the support is represented by a single force (X_1, Y_1) at A together with a couple of moment N_1. Given that the weight of the plank is W, find the values of X_1, Y_1 and N_1 required for equilibrium.

14   A man of weight $4W$ walks slowly from end A to end B of the plank in question 13. Describe the forces provided by support S during his progress. (**Note**: you may interpret 'walks slowly' as 'without acceleration' – this is an appropriate reinterpretation, not a true equivalent.)

15 The firm fixture of end A to support S in question 13 is replaced by a smooth joint. Rotation of the plank is prevented by providing a trestle at end B for the plank to rest on. If the system of forces exerted by the joint on the plank is represented by X_2, Y_2 and N_2, to which of the three symbols can a value be assigned, and what value, because the joint is smooth? Find the vertical reaction force at the trestle.

16 Discuss the options open for evaluating the horizontal component of the reaction force exerted by the trestle in question 15.

17 A uniform horizontal plank AB of weight W and length $2L$ is supported on smooth trestles at its ends. C is a point distance x from end A. (a) Find the support forces at the trestles. (b) Find the values X_3, Y_3 and N_3 which represent the system of forces which section AC of the plank must exert on section CB in order to sustain equilibrium: it assists thought if AC is mentally isolated by an imaginary cut at C. (**Notes:** (i) 'Uniform' implies that the centre of mass of any section is at the mid-point of that section. (ii) N_3 is called the 'bending moment' at point C of the plank. The ability of every cross-section to maintain the necessary bending moment is crucial in the design of bridges, cranes, etc.)

18 Following § 8.32, verify that for a simple pendulum, $S' = 0$.

8.5 Representation of the forces of contact between rigid bodies (in two dimensions)

When two rigid bodies, A and B say, are in contact, the forces between them are essentially unknown. The forces may be concentrated near a point, or distributed evenly or unevenly near a curve or surface of contact (cf. FIG. 8.10a).

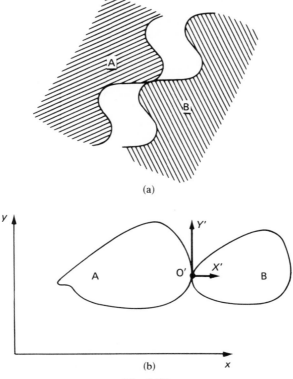

Fig. 8.10

244 An introduction to mechanics and modelling

The complications of the actual forces are avoided by discussing the contact forces (a subset of all the forces on the body) in terms of the equivalent system consisting of X', Y' located at any conveniently chosen point O', together with a couple N_0' (cf. § 8.1 above). The point O' could in theory be chosen anywhere in the plane, but in practice it assists clarity of thought to relate O' to the actual boundary; special choices often assist the incorporation of additional information about the nature of the contact, with corresponding simplifications, as below.

8.51 Rigid bodies in contact at a single point
Take O' at the point of contact (FIG. 8.10b). No force of contact has a moment about a point on its own line of action, so $N'=0$, and the forces exerted by body A on body B are equivalent to a single force (X', Y') located at O'. If the motion of body A is under discussion, we require also a representation of the forces exerted by B on A. Taking the same position of O', the required equivalent system is, by Newton's third law, $(-X', -Y')$ at O'.

8.52 Rigid bodies connected by a hinge
In FIG. 8.11 the forces exerted by A on B are equivalent to X', Y' at the centre of a hinge O', together with couple N'. The advantage of this choice of O' is that the value of N' must equal the moment of all the hinge forces about the axis, i.e. N' is zero for a freely turning joint, and when the joint or hinge is not perfectly free (or 'frictionless', or 'smooth') such information is most likely to be expressed in terms of N', the frictional moment about O'.

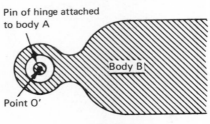

Fig. 8.11

(The forces X', Y' themselves are usually only deducible from considering simultaneously

(i) the equations of motion (both translation and rotation) of body B;
(ii) the equations of translation and rotation of body A;
(iii) the conditions of constraint, ensuring that the point O' of body A has zero velocity relative to point O' of body B.

In all there are six scalar equations of motion from (i) and (ii), and two geometrical relations from (iii). Starting from a known situation, these eight relations suffice to determine the five unknown quantities $\bar{x}$, $\bar{y}$, θ, X' and Y' for body B and the further three unknowns $\bar{x}$, $\bar{y}$ and θ for body A. Part of the art of problem-solving lies in the avoidance of the necessity for dealing with so many unknowns simultaneously: a good start is made by incorporating geometrical relations in the chosen coordinates wherever possible.)

8.53 Multiple or continuous contact with a plane

The plane is a constraint which obstructs rotation. Hence the equivalent single force exerted is so located as to ensure zero total moment of all forces on the body about the centre of mass G. If the only other forces on B are those of gravity (resultant Mg through G), then O' is conveniently taken at G, and the plane's contact is modelled by X', Y' through G, $N'_G = 0$. Otherwise it is best to choose an O' for the contact forces which simplifies not so much the mechanics as the geometry of the problem.

In FIG. 8.12a, b we might well represent the whole net effect of all contact forces exerted by the plane on body B by couple N'_2 and force X', Y' located at the point O'_2 shown, largely because O'_2 is a point which it is easy to identify

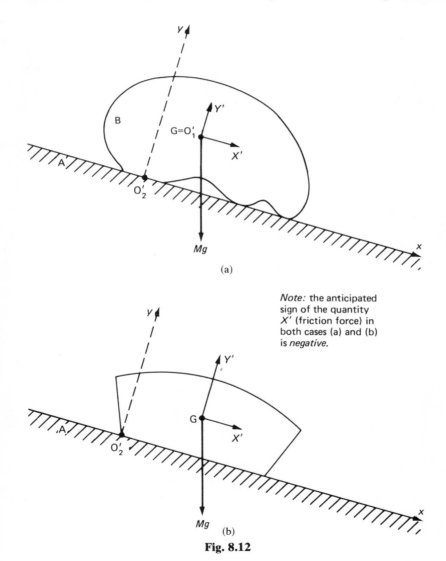

Note: the anticipated sign of the quantity X' (friction force) in both cases (a) and (b) is *negative*.

Fig. 8.12

246 An introduction to mechanics and modelling

on the diagram. The directions chosen for the axes of x and y, on the other hand, are chosen so that the normal and tangential components of the reaction correspond to Y' and X'; the useful approximation to reality incorporated in the laws of friction is then simply expressible as

$$|X'| \leq \mu Y' \qquad 8\text{-}23$$

(Note: Y' is necessarily ≥ 0, § 5.32) and it is an intrinsic part of the laws of friction that this inequality holds for the equivalent force components X', Y' **irrespective of how the constituents X_i' and Y_i' are distributed over the points of contact or interface** between the two bodies A and B.

If the points of contact or interface are 'smooth'

$$X' = 0$$

In this case the contact forces may either be represented in the way we have just described, by force $(0, Y')$ at O_2', together with unknown couple N'; or equivalently, but more graphically, by a single force $(0, Y')$ at an unknown x-value, as in FIG. 8.13a and b. In this representation, the unknown x' $(= N'/Y')$ replaces the former unknown N'.

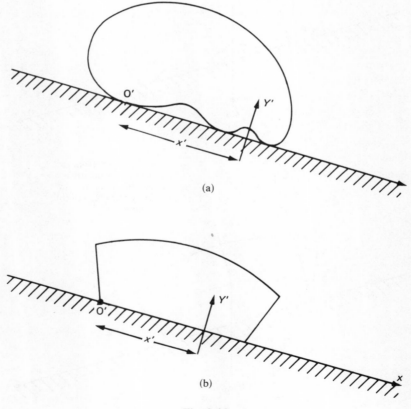

Fig. 8.13

Equivalent systems of forces   247

8.54   General situations

The cases of §§ 8.41–8.43 are common in problems. In unfamiliar circumstances, all three unknowns X', Y' and N' must be shown in one way or another in the force diagram and allowed for in the equations of motion. The reader may wonder where he/she can find enough equations to evaluate or eliminate all the unknowns.

> It is usually the case that an unknown force arises because of a geometrical constraint. That geometrical condition commonly provides the additional equation whereby (in conjunction with equations of motion) the unknown force may be evaluated or eliminated.

The relation $s = a\theta$ performs such a service in problems where rolling occurs, see questions 3, 4 and 5 of Exercise 3 below. The relation $y = $ constant (or $\ddot{y} = 0$) is the equation of constraint which determines the normal reaction exerted by the plane $y = $ constant.

It seems appropriate to remind the reader at this point of the earlier discussions of point contacts, strings, etc. in §§ 4.4, 5.32 and 5.33, so that he/she feels confident about dealing (mathematically speaking) with all common kinds of constraint.

In dealing with complicated systems, the procedure should be split up into systematic stages as follows.

1   Draw a diagram to show the general configuration of the system. It is sometimes possible to choose coordinates at this stage and show them in the same diagram.
2   Identify the rigid bodies concerned; make a deliberate act of mental recognition of the boundaries of each; identify also any points where strings are joined together, or other crucial moving points not part of the identified rigid bodies.
3   Consider the rigid bodies one at a time, and the points one at a time. For each body or point capable of movement, mark on the diagram the forces acting on that body or point. (Remember the three categories: invisible forces, known applied forces, other contact forces.) Where appropriate, use Newton's third law, or the results for equal tensions at the two ends of a string, or $F = \mu R$ when there is actual slipping, to keep the number of unknowns from proliferating unnecessarily.

The importance of stage 2 can hardly be overemphasized.

The student who becomes proficient in the art of modelling forces in two dimensions is unlikely to have much difficulty in modelling forces in three dimensions when he becomes familiar with three-dimensional rigid dynamics. With the knowledge of two-dimensional mechanics which has been built up,

> the student should now be equipped to tackle a wide range of real-life problems.

The modelling example of the next chapter will show how simple principles may be put to use.

Exercise 3

1. A rectangular gate of mass M, height h and width b is supported by two hinges symmetrically placed on the end of the gate, distance d apart. The gate is painted brown and has nine equally-spaced horizontal bars. Given that the upper hinge provides no vertical support, find the magnitude of the force exerted by each hinge on the gate when the gate is at rest. Which items of information have you chosen to neglect, and what unauthorized assumptions have you made?

2. In FIG. 8.14, (X_1, Y_1, N_1) and (X_2, Y_2, N_2) represent the forces exerted on the body of a goods wagon by the two axles which support it. If the force P (acting along a line joining the mid-points of the two axles) **causes** the wagon to move, state (i) the signs which you would expect the quantities $X_1, Y_1, N_1, X_2, Y_2, N_2$ to have, and state (ii) what simplification could be made if it were known that the bearings were in good condition and well-oiled. Given that the axles are distance d apart, that the mass of the wagon and contents is equal to M, and that the centre of mass is at height h above the axles, estimate the values of Y_1 and Y_2 when the acceleration of the wagon (forwards) is a.

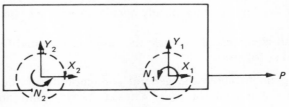

Fig. 8.14

3. A circular hoop runs downhill starting from rest. Investigate the factors determining its motion. If you do not know where to start, work through questions 4–6 which follow.

4. (i) Draw a diagram of a hoop of radius a rolling (without leaning sideways) down a slope of angle α.

   (ii) Choose x-, y-directions which will assist your discussion of motion, and mark in the diagram the coordinates x, y of the centre of the hoop, noting the sense of measurement of each, viz. **from** a fixed axis, e.g. through the starting point **to** the general position shown on your diagram. If one of the coordinates is constant, note this.

   (iii) Assuming no slipping (i.e. assuming arc length = distance of new contact point on the plane from the initial contact point), relate the angle θ turned through by the hoop to the variable coordinate which characterizes the position of the centre of the hoop.

   (iv) Note that the normal to the plane is a direction fixed in space, and that the radius of the hoop which was originally in contact with the plane is a fixed line in the body which originally lay along this direction fixed in space. Mark the sense of the angle θ which corresponds to the direction of rotation $\dot{\theta}$ for the body.

   (v) Mark in all the forces acting on the hoop (noting that the contact with the plane is reasonably modelled as a point contact).

   (vi) Write down the three equations of motion for two-dimensional motion of the hoop.

   (vii) Using the three equations of motion and the geometrical constraint identified in (iii) above, evaluate the linear acceleration of the hoop.

Equivalent systems of forces

5 In question 4, what is the least coefficient of friction required to ensure that there is no slipping between hoop and plane?

6 If the actual coefficient of friction μ is less than the value required in question 5 what equation governs the variation of θ? [**Note:** the relation (iii) is no longer applicable.]

7 A circular hoop made of thin wire is of radius a and mass m. It is smoothly pivoted so as to rotate in its own vertical plane about a point A on its circumference, and is loaded with a particle of mass m placed on the circumference at the point B diametrically opposite to A. Show that the period of small oscillations of the body is $2\pi\sqrt{(2a/g)}$.

Show that if the oscillation is just large enough for AB to reach the horizontal through A, then in this extreme position the pivot exerts a force on the wire of magnitude $\tfrac{1}{2}mg$.

8 A thin uniform rod of length $2a$, attached to a smooth fixed hinge at one end O, is held in a horizontal position and then allowed to fall from rest. Show that the horizontal component of force on the hinge is greatest when the rod is inclined at an angle $\pi/4$ to the vertical, and that the vertical component of force is then $11/8$ times the weight of the rod.

9 Two particles of mass m and $3m$ are connected by a light inextensible string which passes over a pulley of radius a. The pulley is free to turn about a fixed horizontal axis through its centre, and there is no friction at the bearings. The moment of inertia of the pulley about its centre is I. Given that the string does not slip on the pulley, and that the system starts from rest, show that the time taken for the mass $3m$ to descend a distance x is

$$[(4ma^2 + I)x/mga^2]^{1/2}$$

(**Note:** the pulley is neither 'small' nor 'light'.)

10 A uniform thin cylindrical shell of diameter $2a$ is set in skidding motion on a rough horizontal plane, so that the axis moves at right angles to its length with an initial velocity V_0; the cylinder also has an initial angular velocity V_0/a in the sense opposite to that required for rolling. Show that the cylinder will skid for a certain time, and then stop dead. Find the distance moved by the axis in terms of V_0, g and the coefficient of friction.

11 A thin circular hoop of radius a is projected up a rough inclined plane which makes an angle α with the horizontal, the coefficient of friction being $\tfrac{1}{4}\tan\alpha$. The initial velocity of the centre of the hoop is V and the initial angular velocity is Ω $(< V/5a)$ in such a sense as to oppose pure rolling.

Show that the point of contact on the hoop first has zero velocity (a condition for rolling) after time $2(V+a\Omega)/3g\sin\alpha$. Write down equations to determine the subsequent motion on the assumption that it is one of pure rolling, and by solving for the frictional force, show that in fact there is insufficient friction to maintain pure rolling. Deduce that the hoop stops moving up the plane after time

$$4(2V - a\Omega)/9g\sin\alpha$$

12 A rectangular block ABCD, of mass M, has two small studs fixed at A and B. The block is placed with A and B resting on a horizontal table. The coefficients of friction at A and B are μ and μ' respectively. A string attached at C runs in the direction DC to a small smooth pulley whose axis is fixed above the end of the table, and over that pulley, after which it is attached to (and supports) a pan which can contain weights.

Draw a diagram showing the forces (a) on the block, (b) on the pan. If weights are added to the pan until motion takes place, what will the initial motion be? (If you have difficulty with the sliding/tipping alternative, refer to the method of thinking used in § 5.332. Some examples involving slipping/tipping alternatives are also discussed in the next chapter.)

Assignment

The swing door of a hotel consists of four uniform rectangles of height 3 m and width 1.2 m, set to form four vanes rotating about a common axis which abuts a long edge of each. A girl pushing on one of the rectangular panels at a point 1 m from the axis of rotation, with a force of 5 kgf, can just maintain the rotation of the door at a constant rate of $\frac{1}{2}$ revolution per second. When she leaves the door to come to rest by itself, it is observed that the door turns through a total of 3 right angles in $2\frac{1}{2}$ seconds before coming to rest.

Investigate simple hypotheses about the law governing the resisting moment, restricting yourself in the first place to a single constant (e.g. of proportionality) in each case. Write down the relations which must be satisfied between this constant and the mass of the door with a view to estimating the mass. Comment on your findings. (If necessary extend your investigations to include resistances proportional to $1/\dot{\theta}$ or $1/\dot{\theta}^2$, and perhaps to linear combinations of resistances already considered singly.) Your summary should include a discussion of the effect of different resistance laws on the shape of the $\dot{\theta} - t$ graph; suggestions about how a resisting moment of the preferred kind might arise; and an estimate of the range of values within which you think the door's mass must lie.

9
Mechanics applied to the real world: frames of reference in the real universe

9.1 A 'real' problem (loaves of bread down a chute), and the application of mechanics to it

A factory wishes to instal a chute which will convey wrapped loaves of bread from one level to another 2 m below. What design advice can be given?

This is not a problem with a single 'right' solution. The desired chute may be a cheap inclined plane, or an expensively engineered job incorporating a curve. On-site consultation is necessary. But to gain insight into the way in which design affects the motion of the loaves, it is very useful to look at calculations for a simple model – e.g. the straightforward inclined plane.

We leave the first stage of inquiry to the reader to pursue – we suggest that he uses a simple inclined plane model to discover what degree of roughness is required to control the velocity of arrival within appropriate limits. The author's discussion can be found in § 9.5. The break in continuity which is caused by this relegation of material to an Appendix bears witness to the importance which the author attaches to the reader making his/her own attempt on the problem at this juncture.

The student who successfully completes the first stage of inquiry should also try his hand at a second stage. One type of follow-up calculation is presented in § 9.2 below; it concerns the collision of the loaf at the bottom of the plane. The student's choice of assumptions and situations to consider need not be those of the author at either the first or second stage, but his assumptions and conclusions should be stated with equal clarity.

Both in § 9.2 and in § 9.5 the steps are laid out with some formality: step A specifies the model used for the objects under consideration, step B considers forces and geometry. The reader should identify and express in words the nature of steps C, D and E and check against § 7.6 where the same notation was used. These steps are common to most problem-solving in mechanics, though rarely made quite so explicit. The virtue of clearly stating the assumptions is most apparent when a solution fails, i.e. when the conclusions are incompatible with experimental findings (cf. the discussion of the direct impact of billiard balls in § 9.41), for they provide a focus for critical reappraisal; but when no experimental tests are available, the clear statement of assumptions is even more necessary. Non-mathematicians are only too likely to repose excessive faith in any page of symbols whose algebra has been checked by a competent mathematician; but the mathematician can assert only that certain conclusions follow from such and such assumptions. Of the validity of the assumptions, the mathematician does not have to make himself sole arbiter – he should lay the assumptions bare for all to see.

9.2 The loaf of bread problem (stage 2) and the idea of 'impulse'

The mathematical content of this section, dealing with impulses, is embodied in the transition from equations 9–2 to equations 9–13 below. We first lay out the modelling steps in detail.

9.21 Step A: Modelling the objects

FIG. 9.1 illustrates our model – a uniform rigid rectangular block striking a horizontal surface with velocity v from angle α. The resulting motion of the block will depend on the roughness of the horizontal surface, so we introduce coefficient of friction μ' to model this roughness.

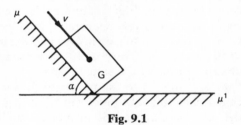

Fig. 9.1

A real loaf of bread – particularly if it is a wrapped sliced loaf – may suffer damage if it arrives with too great a speed at the bottom of the slope.

In the preliminary discussion set out in § 9.5, the author points out that the velocity v with which the loaf arrives at the bottom of the slope is not a good measure of the amount of damage likely to be sustained by the loaf at impact. We must appeal to engineering experience to provide us with a better criterion, preferably a numerical criterion, or statistic. One statistic commonly used is the magnitude of the 'impulse' sustained, i.e. of the total change of momentum suffered by the body in the very small interval of time occupied by the collision. Another statistic, which is often more appropriate for deformable bodies, is the change in a quantity known as the 'kinetic energy' of the body, defined in the present case as

$$\text{KE} = \tfrac{1}{2}M\bar{v}^2 + \tfrac{1}{2}I_G\dot{\theta}^2 \qquad 9\text{--}1$$

where $\bar{v}$ is the velocity of the centre of mass G. This quantity is discussed in more detail in the next chapter.

The two statistics are not equivalent. It would seem prudent to formulate two criteria, one in terms of impulse and one in terms of kinetic energy, and aim at satisfying both.

Mechanics applied to the real world

The calculation of either statistic requires a knowledge of the velocity of all the points of the loaf just after as well as just before the impact at the bottom of the inclined plane. The same calculations will help to inform us about the likelihood of toppling.

9.22 Step B: Modelling the forces and geometry

The only 'invisible forces' (see item 3 of the summary at the end of § 8.4), are those of gravitation, equivalent to Mg acting downwards through G.

There are no deliberately applied, known forces at the instant of collision.

The appropriate representation of the unknown impact forces depends on the geometrical configuration just after the loaf collides with the plane. There are four logical possibilities for our rigid body model – represented by its rectangular mid-section ABCD (FIG. 9.2a, b) – and separate mathematical treatments relevant to each. The four possibilities are the combinations of

A remains in contact with plane OX;		D remains in contact with inclined plane OP;
or, A does not remain in contact;	**and**	**or,** D does not remain in contact with plane OP.

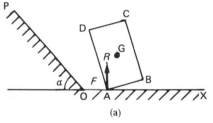 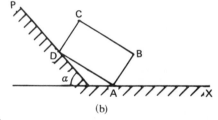

(a) (b)

Fig. 9.2

It would be time-consuming and probably unrewarding to pursue the theoretical analysis of all four possibilities. For the purposes of this book the author will make the assumption that the contingency that point A jumps off the plane can be disregarded, i.e. that the amount of rebound is insignificant. This leaves the two logical possibilities represented in FIG. 9.2a, b, namely, either the edge through D lifts, or the loaf continues supported on the two edges through A and D. We shall here discuss the former possibility, and leave the reader to follow through a similar discussion of the latter. Only after the analyses have been completed can we determine the conditions necessary for the occurrence of each type.

Case (a): the edge through A remains in contact with plane OX, but the edge through D loses contact with plane OP. The contact forces at A are equivalent to R, F at A as shown. $R \geqslant 0$; and **either** A at rest **or** $F = \mu R$. (Note how the either/or statement gives us one piece of information either way.)

9.23 Step C: Condition that point A shall not be halted

Take O as origin, take axes of x, y parallel and perpendicular to OX, write $AG = c$, $\theta = \angle OAG$ in FIG. 9.3. Note that θ increasing corresponds to **clockwise**

254 An introduction to mechanics and modelling

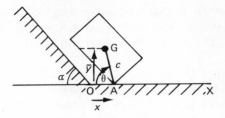

Fig. 9.3

rotation, i.e. clockwise moments will be taken as positive in the θ equation. The equations of motion of our model rigid body are:

$$M\ddot{x} = -F \qquad 9\text{--}2(a)$$

$$M\ddot{y} = R - Mg \qquad 9\text{--}2(b)$$

$$I_G\ddot{\theta} = Fc \sin \theta - Rc \cos \theta \qquad 9\text{--}2(c)$$

With two unknown constraint forces R and F, we look for two corresponding geometrical relations imposed by the constraints. The available constraint information enables us to say:

1. R ensures that A has no y-component of velocity, and therefore:

$\dot{y}$ = y-component of velocity of A + y-component of the velocity of G relative to A (see § 1.9)

$$= 0 + c\dot{\theta} \sin \theta \qquad 9\text{--}3(a)$$

2. F ensures that A has no x-component of velocity, unless the F required for this is greater than μR, i.e.

$$\dot{x} = c\dot{\theta} \cos \theta \text{ unless this requires } F > \mu R \qquad 9\text{--}3(b)$$

To utilize the constraint conditions 9–3(a) and 9–3(b), we integrate equations 9–2 with respect to time. From the resulting five equations, $\dot{x}$, $\dot{y}$, $\int R \, dt$ and $\int F \, dt$ can be eliminated to give equation 9–6, to which many readers may wish to proceed immediately. The detailed line of thought runs as follows:

Let Δt represent a short interval of time covering the period of collision; then writing the values of $\dot{\theta}$, θ at time Δt as $\dot{\theta}_1$, θ_1,

$$[M\dot{x}] = M(c\dot{\theta}_1 \cos \theta_1 - v \cos \alpha) = -\int_0^{\Delta t} F \, dt \qquad 9\text{--}4(a)$$

unless this requires $F > \mu R$;

$$[M\dot{y}] = M(0 + v \sin \alpha) = \int_0^{\Delta t} (R - Mg) \, dt \qquad 9\text{--}4(b)$$

$$[I_G\dot{\theta}] = c \int_0^{\Delta t} (F \sin \theta - R \cos \theta) \, dt \qquad 9\text{--}4(c)$$

Mechanics applied to the real world 255

The quantities F and R are time-dependent. From equation 9–4(b) we observe that

$$\int_0^{\Delta t} R \, dt = Mv \sin \alpha + Mg \, \Delta t.$$

Since Δt is small, $Mg \, \Delta t$ is small in comparison with $Mv \sin \alpha$, and it is convenient to introduce the modelling fiction $\Delta t \to 0$. This necessitates a parallel fiction about R so that its integral does remain equal to $Mv \sin \alpha$. In fact we accept the unknown constant $\int_0^{\Delta t} R \, dt$ rather than the time-dependent unknown R as our unknown (see also § 9.25 below).

Note: in the remainder of this section we shall use the abbreviation

$$\int R \, dt \quad \text{for} \quad \int_0^{\Delta t} R \, dt$$

From the finiteness of $\int R \, dt$ we may deduce the finiteness of $\int F \, dt \ (\leqslant \mu \int R \, dt)$ and hence of $\dot{x}$ and $\dot{\theta}$ during the collision; from this it also follows that throughout the collision interval the value of θ satisfies

$$\theta - \theta_0 = \int \dot{\theta} \, dt \leqslant \dot{\theta}_{\max} \Delta t \to 0$$

i.e. the value of θ in equations 9–4 may everywhere be written as equal to θ_0, its initial value. Thus equation 9–4(c) becomes

$$I_G \dot{\theta}_1 = c \sin \theta_0 \int F \, dt - c \cos \theta_0 \int R \, dt \qquad 9\text{--}5$$

The two quantities $\int F \, dt \ (= F' \text{ say})$ and $\int R \, dt \ (= R')$ may be eliminated from equations 9–4(a), (b) and 9–5 so that, again putting $Mg \, \Delta t = 0$, $\theta = \theta_0$,

$$(I_G + Mc^2 \sin \theta_0 \cos \theta_0) \dot{\theta}_1 = Mcv(\sin \theta_0 \cos \alpha - \cos \theta_0 \sin \alpha) \qquad 9\text{--}6$$

which gives $\dot{\theta}_1$ and hence $\dot{x} \ (= c\dot{\theta}_1 \cos \theta_0)$ and $\dot{y} \ (= c\dot{\theta}_1 \sin \theta_0)$ in terms of geometrical, and therefore notionally 'known' quantities. This solution is valid only if it is consistent with $0 \leqslant F \leqslant \mu R$ (note this requires $R > 0$), i.e. with

$$0 \leqslant \int F \, dt \leqslant \mu \int R \, dt \qquad 9\text{--}7$$

i.e. $0 \leqslant F' \leqslant \mu R'$.

It is possible to test whether condition 9–7 is satisfied in the circumstances that led to the value of $\dot{\theta}_1$ given in equation 9–6 for, using equations 9–4 and 9–7, the condition takes the form,

$$0 \leqslant M(v \cos \alpha - c\dot{\theta}_1 \cos \theta_0) \leqslant \mu Mv \sin \alpha \qquad 9\text{--}8$$

If the value of $\dot{\theta}_1$ given by equation 9–6 satisfies condition 9–8, then we conclude that point A is indeed initially halted at O, and that the loaf begins to rotate

256 An introduction to mechanics and modelling

about the edge through A. This is clearly undesirable in practice, so in our design recommendations

> we shall require the coefficient of friction at the plane OX to be such that
> $$\mu < (v \cos \alpha - c\dot{\theta}_1 \cos \theta_0)/v \sin \alpha \qquad 9\text{-}9$$
> with $\dot{\theta}_1$ given the value obtained from equation 9–6.

9.24 Step C (continued): Sliding and lifting?

Having ensured that point A of the loaf does not stop dead at O, it still remains for us to discuss the motion which we are actually going to allow, and to consider the question of damage. When inequality 9–9 is satisfied, we no longer have 'the x-component of A's velocity is zero', so the geometrical deduction $\dot{x} = c\dot{\theta} \cos \theta$ is no longer valid. Instead we have: 'A is in actual motion along plane OX, so $F = \mu'R$ where μ' is the coefficient of sliding friction at the plane OX'.

Equations 9–4(b) and (c) remain unaffected, but equation 9–4(a) must be discarded in favour of the alternative

$$\int F \, dt = \mu' \int R \, dt \qquad 9\text{-}10$$

Using 9–9 and 9–4(b) in 9–4(c), we find

$$I_G \dot{\theta}_1 = Mcv \sin \alpha (\cos \theta_0 + \mu' \sin \theta_0) \qquad 9\text{-}11(a)$$

The value of $\dot{\theta}_1$ is thus determined in terms of known geometrical quantities and the velocity of approach v which is 'evaluated' in step C(i) of § 9.53, equation 9–28.

Knowing $\dot{\theta}_1$, the value of $\dot{y}$ follows from $\dot{y} = c\dot{\theta}_1 \sin \theta \qquad 9\text{-}11(b)$

and $\dot{x}$ is derived from the equation which replaces 9–4(a),

$$M(\dot{x} - v \cos \alpha) = -\int F \, dt$$

$$= -\mu' \int R \, dt$$
$$= -\mu' Mv \sin \alpha \qquad 9\text{-}12$$

(We have used equation 9–4(b) and subsequently written $\Delta t = 0$).

The new velocity of any point of the body is expressible in terms of $\dot{x}$, $\dot{y}$, and $\dot{\theta}$, so the mathematical part of this discussion has been successfully completed. It remains only to discuss the relevance of our calculations to the assessment of probable damage.

Mechanics applied to the real world 257

9.25 Step D: Assessing the likelihood of damage

> If we write $\mathbf{P}=$ the reaction force $[=(-F, R)$ in our case, or in general $=(X, Y)$ say], then the vector $\mathbf{J} = \int \mathbf{P}\,dt$, whose components are $\int -F\,dt$ and $\int R\,dt$ in our case, or in general $\int X\,dt$ and $\int Y\,dt$, is the impulse of the total reaction force $\mathbf{P}$.

The magnitude J of the impulse $\mathbf{J}$ is the first statistic which we may use as a tentative indicator of likely damage.

We note that use of symbols J_x, J_y to denote components of impulse would allow the algebra of § 9.23 to be written down more neatly. Thus, in place of equations 9–4 we should have

$$M(c\dot{\theta}_1 \cos \theta_1 - v \cos \alpha) = J_x \qquad 9.13(a)$$

$$M(0 + v \sin \alpha) = J_y \qquad 9\text{–}13(b)$$

$$I_G \dot{\theta}_1 = -c(J_x \sin \theta_0 + J_y \cos \theta_0) \qquad 9\text{–}13(c)$$

with the proviso $|J_x| \not> \mu J_y$.

A second simple indicator in common use is the amount of energy absorbed by the body. The concept of energy will be briefly discussed in the next chapter. The 'energy-absorbed' statistic takes into account more features of the body than have been mentioned in the work we have attempted so far; in particular, the elastic properties, i.e. the **non**-rigid characteristics of the body, are involved. Reluctantly, in spite of its relevance to our inquiry, we can find no place for such a discussion here. (The examples of § 10.243 show in a modest way how absorbed energies can be estimated). In these circumstances we revise our step D statement thus.

9.26 Step D (revised)

The motion of the body just after the collision at the bottom of the slope is given by the values of $\dot{x}$, $\dot{y}$, $\dot{\theta}_1$ obtained from equations 9–12, 9–11(a) and 9–11(b). The magnitude of the impulse,

$$J = \sqrt{(J_x^2 + J_y^2)}$$

where J_x and J_y have the values obtained by inserting the calculated value of $\dot{\theta}_1$ in equations 9–13(a) and (b). These results have been derived on the model of the loaf as a rigid body for which point A does not leave plane OX during the collision, and for which the friction coefficient with plane OP satisfies condition 9–9. An experimental value for the highest acceptable value of J ($=\max J$, say) must be adduced from observations with any available apparatus, e.g. by simply dropping loaves from measured heights. The design recommendations for the projected new chute must specify geometry and materials, i.e. values of μ, α, μ' which ensure that $J < \max J$ for all the masses M and semi-diagonal dimensions c of loaves to be sent down the chute.

9.27 Step E

Let us assume that tentative values of μ, α have been chosen, and the corresponding $\dot{\theta}$, $\dot{x}$ values immediately after impact have been calculated. With these values, the question of whether the loaf will rotate clockwise about the moving point A until it topples, or whether point D retains or regains contact with the supporting surface, can be examined. The reader may attempt this inquiry himself (cf. question 4 of Exercise 4 at the end of this chapter).

This seems a convenient place to end the discussion of the loaf problem. Not every possibility has been explored, not every assumption verified. This is typical of real problems as distinct from exercises.

From the production-line view of mathematics, we have merely performed a string of stereotyped exercises. It is the formulation of the problem – the selection of the beads which are to go on the string, their fitting together, the balance of the whole design – which call for creative initiative, while technical proficiency ensures the proper functioning of the individual parts. The final decision to be made in any mathematical enterprise must not be evaded: it is the decision to end this piece of work here.

Exercise 1

(This may be made into anything from a 'quickie' to a full-scale project). Investigate one or more aspects of the game of tennis by use of mathematical models, some, at least, of which should be mechanical models. **Hint:** first identify different parts of the motion of the ball. What questions might be asked and answered? A look at § 9.4 might give you ideas. As a last resort, look at the questions which follow.

(i) Find the relation between height, speed and angle of projection in order that the service ball may clear the net and land in the correct court. (Simplify as much as possible in your first attempt, to make it easy to handle; later consider refinements. It would be a good thing first to check tennis court dimensions, net height, etc. so that meaningful, numerical values may result from your calculations.)

(ii) If the service ball has spin, what spin will the ball have after bouncing? What data do you need in order to attempt such a calculation? Why not estimate the coefficient of restitution by dropping a tennis ball vertically on to the court and measuring rebound height? An equally crude estimate of coefficient of friction might be useful. Can you calculate the path of the ball after bouncing?

(iii) (This is not mechanics.) Given that player B is twice as good a player as A, what are A's chances of winning a game? (First model what you mean by 'twice as good a player'. After completing a simple calculation, state all the ways in which you think the question is unsatisfactory, and the reservations you wish to make about your solution.)

(iv) Discuss quantitatively the differences in time and amount of movement required by a player at net and at the back of the court respectively.

9.3 Laboratory, earth-centred and sun-centred frames of reference

9.31 The problem

In the problems discussed so far, we have tacitly assumed that the frame of reference provided by the room in which we work, or the patch of ground on which we stand, may be regarded as 'stationary' for the purposes of mechanics. Owing to the rotation of the earth about its axis, FIG. 9.4, different points on the earth's surface are not stationary relative to one another: each has a distinctive

Mechanics applied to the real world 259

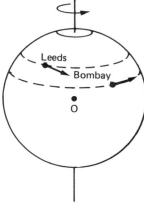

Fig. 9.4

velocity relative to the earth's centre O. Choice of O as the origin of an overall terrestrial frame of reference makes it easier to compare the descriptions of mechanical events given by observers at different points. In the next section we shall find out whether equations of motion formulated on the assumption that, say, the city of Leeds is stationary are consistent with equations of motion formulated on the assumption that the centre of the earth is stationary.

We will discuss in § 9.32 the effect, if any, which the rotation of the earth about its axis has on equations of motion, and then widen our view to include the whole solar system (FIG. 9.5). The various motions of the planets are most easily described, and their systems of mechanics therefore most easily related, by choosing the sun's position, S, as origin in an overall system of coordinates. The consistency of mechanics formulated in the earth-centred frame with mechanics formulated in the sun-centred frame is discussed in § 9.33. Finally we may extend our view outwards into the depths of the universe (§ 9.34).

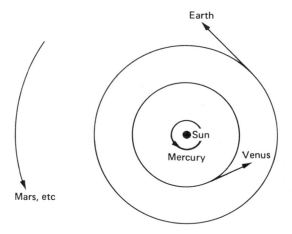

Fig. 9.5

9.32 Mechanics in the earth-centred and in laboratory-centred frames

9.321 Correction to motion of a particle

Step A(1). We shall model the earth as a rotating sphere (FIGS. 9.4 and 9.6) and we shall consider the motion of a point particle P near O', a fixed point on

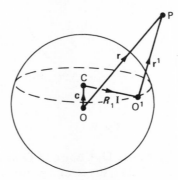

Fig. 9.6

the sphere. We shall use the symbol **r** to denote the position of P relative to the earth's centre, O, and the symbol **r'** to denote its position relative to O'. On the assumption that the equation of motion

$$m\ddot{\mathbf{r}} = \mathbf{F} \qquad 9\text{--}14$$

is 'correct', we shall make an estimate of the error introduced by substituting for equation 9–14 the incorrect equation of motion

$$m\ddot{\mathbf{r}}' = \mathbf{F} \qquad 9\text{--}15$$

Only the order of magnitude of the estimate will be regarded as significant.

Step B(1). The forces on the particle are modelled by the single equivalent force **F**, whose magnitude and direction are otherwise unspecified except that, since gravity acts on every particle, the force **F** includes mg and is typically, though not necessarily, of a magnitude comparable to mg.

Step C(1). In FIG. 9.6, O' moves in a circle of radius CO', taking 24 hours for one complete circuit. Except at the equator, CO' = R_1 < the earth's radius. The vector $\overrightarrow{CO'}$ may be written

$$\overrightarrow{CO'} = R_1\mathbf{I}$$

where **I** is a unit vector rotating in a plane with constant angular velocity Ω equal to 1 revolution (2π radians) in each 24 hours, i.e. 7.27×10^{-5} radians/sec. We note that both Ω and R_1, as well as the vector $\overrightarrow{OC} = \mathbf{c}$, are independent of time t on our model.

The relation

$$\mathbf{r} = \mathbf{c} + R_1\mathbf{I} + \mathbf{r}' \qquad 9\text{--}16$$

Mechanics applied to the real world   261

is true at all times (see FIG. 9.6) and can therefore be differentiated twice with respect to t, giving

$$\ddot{\mathbf{r}} = R_1\ddot{\mathbf{I}} + \ddot{\mathbf{r}}' \qquad 9\text{-}17$$

owing to the constancy of $\mathbf{c}$ and R_1.

The discussion of rotating unit vectors in § 2.29 shows that

$$\ddot{\mathbf{I}} = -\Omega^2 \mathbf{I}$$

($\dot{\theta} = \Omega$ in our case).

Thus, from equation 9-17 we find that $\ddot{\mathbf{r}}$ differs from $\ddot{\mathbf{r}}'$ by a quantity of magnitude $\Omega^2 R_1$ where $\Omega \simeq 7.27 \times 10^{-5}$ and $R_1 \leqslant$ earth's radius, 6.38×10^6 m. $\ddot{\mathbf{r}}'$ therefore differs from $\ddot{\mathbf{r}}$ by a quantity of magnitude $\leqslant 3.37 \times 10^{-2}$ m s^{-2}. This may be compared with the value 9.8 m s^{-2} used for acceleration due to gravity near the earth's surface.

The discrepancy caused by using the equation $m\ddot{\mathbf{r}}' = \mathbf{F}$, instead of $m\ddot{\mathbf{r}} = \mathbf{F}$ is thus comparable to that caused by an error of $\leqslant 0.3\%$ in the value of g. This is insignificant in a large majority of practical applications.

Step D(1)

> We conclude that the use of local systems of coordinates in the basic equations of motion of a particle gives acceleration values differing only slightly from those which would be predicted using a terrestrial coordinate system with origin at the centre of the earth. [But see step E(1).]

Step E(1). We have established only that discrepancies in **acceleration** values predicted by using (a) an uncorrected local frame and (b) an earth-centred frame of reference will be less than about 0.03 m s^{-2}. When large intervals of space or time are involved, as, for example, in weather calculations or space flight, the process of integration with respect to space or time may lead to large discrepancies in the predicted changes of velocity

$$[\mathbf{v}] = \int \mathbf{a}\, dt \quad \text{or position} \quad [\mathbf{r}] = \int \mathbf{v}\, dt$$

Under these circumstances it may well be imperative to state a clear preference for one system rather than the other, and to make appropriate 'corrections' if the less fundamental system is adhered to.

Long before space flight was possible, one large-time-interval phenomenon at variance with the predictions of Newtonian mechanics in a stationary laboratory frame was known. This was the Foucault pendulum phenomenon described in Feather, *Matter and Motion* (see Further Reading in the Introduction). An observed long-term rotation of a pendulum relative to the laboratory frame apparently contradicted Newtonian mechanics. The contradiction disappeared when Newtonian mechanics was applied in an earth-centred frame wherein the rotation of the earth was recognized. The original apparent contradiction of Newtonian mechanics was thereby transformed into a corroboration!

262 An introduction to mechanics and modelling

The earth's rotation leads not only to an acceleration of any **origin** on the earth's surface relative to the centre, but also to a rotation of laboratory **axes** relative to the fixed space determined by the background of the stars. The discrepancies in mechanics attributable to such rotations (if not taken into account) are of the order of magnitude indicated by the calculations which we have already made.

Another possible source of error in the step C(1) calculations may be dismissed quite quickly. The earth was assumed to be a perfect sphere, which it is not. However, the general conclusions are unaffected whether we take the least or the greatest earth radius. The minor asymmetries of the earth are immaterial to the overall picture.

We conclude that provided times of the order of seconds or minutes rather than days are involved, and distances of the order of metres rather than hundreds of kilometres, then the step D(1) conclusion justifies us in regarding 'mechanics of a particle' modelled in laboratory and earth-centred frames respectively as nearly equivalent. After a glance at the implications for rigid body mechanics, the next step will be to consider sun-centred frames.

9.322 Correction to the motion of a rigid body

The results of step C(1) refer to a particle. Acceptance of the particle model of rigid bodies leads to the conclusion that

> for rigid bodies, as for particles, we may apply Newtonian mechanics in a local or laboratory frame to small-scale events, without appreciable correction for the rotation of the earth being necessary.

9.33 Comparison of earth-centred and sun-centred frames

Step A(2). We model the earth's centre as a point E moving in a circular orbit round the sun, FIG. 9.7, and consider the description of the motion of a point particle (i) in a solar frame of reference, origin S, and (ii) in the terrestrial frame centred at E.

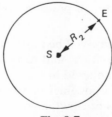

Fig. 9.7

Step B(2). Let **F** be the resultant applied force on the particle.

Step C(2). The analysis proceeds in just the same way as step C(1), but now

$$R_2 = 93 \text{ million miles} = 149 \times 10^9 \text{ metres}$$

$$\Omega_2 = 1 \text{ revolution in 365 days} = 2 \times 10^{-7} \text{ radians/sec}$$

Mechanics applied to the real world

Hence
$$\Omega_2^2 R_2 = 5.9 \times 10^{-3} \text{ m s}^{-2}$$
and $\ddot{\mathbf{r}}'$ differs from $\ddot{\mathbf{r}}$ by a quantity of magnitude
$$\simeq 6 \times 10^{-3} \simeq 0.06\% \text{ of } g$$

Step D(2). This is the parallel of step D(1) above, and an extension to rigid bodies parallel to § 9.322 is equally valid.

Step E(2). As far as the local motion of objects near the surface of the earth is concerned, correction for the motion of the earth round the sun is (even) less necessary than is correction for the rotation of the earth about its axis.

The motion of the solar system relative to the Galaxy and, beyond that, to the other galaxies of the universe, is less easy to model as there is no recognizable 'centre' to the universe. As measured from the solar system, however, there are recognizable 'fixed' directions: the angles between the directions of the most distant stars and nebulae, as viewed from earth, do not detectably vary with time. The distant stars provide a natural 'fixed' reference system for direction. The sun-centred frame in which directions are specified in relation to the directions of distant stars, is known as the **sidereal** frame.

> It is customary to accept Newtonian mechanics as valid in the sidereal frame of reference, and to make the corresponding, usually minor, corrections to mechanics as necessary, when pursued in terms of local coordinates.

9.34 Cosmological frame
If, notwithstanding the practical adequacy of the above decision for all purposes except perhaps theoretical astronomy and cosmology, we want to pursue the question of motion in a yet larger frame, we must fall back on very crude estimates. One such estimate of relative acceleration can be deduced from 'Hubble's law', the observational law which states that the more distant bodies of the universe – the external galaxies – are apparently receding from the solar system with velocity V proportional to their distance R from earth.

$$V (= \dot{R}) \simeq kR \qquad 9\text{--}18$$

where $k \simeq 3 \times 10^{-12} \text{ s}^{-1}$. This, if true for the same value of k at all epochs of time, would imply

$$\ddot{R} \simeq k\dot{R} \simeq k^2 R \qquad 9\text{--}19$$

Most of the visible nebulae have $R < 10^{18}$ m, so that $k^2 R \leqslant 10^{-5} \text{ m s}^{-2}$, an acceleration which we can happily ignore for earthly doings. We need not even restrict ourselves to that part of the universe which is actually observed through telescopes: if equation 9–18 is accepted, there is a finite value of R, namely 10^{20} m at which the velocity of recession would be equal to the speed of light $(3 \times 10^8 \text{ m s}^{-1})$, and beyond which heavenly bodies would be, in principle as well as in practice, invisible. Even at this hypothetical limiting value, the acceleration

264 An introduction to mechanics and modelling

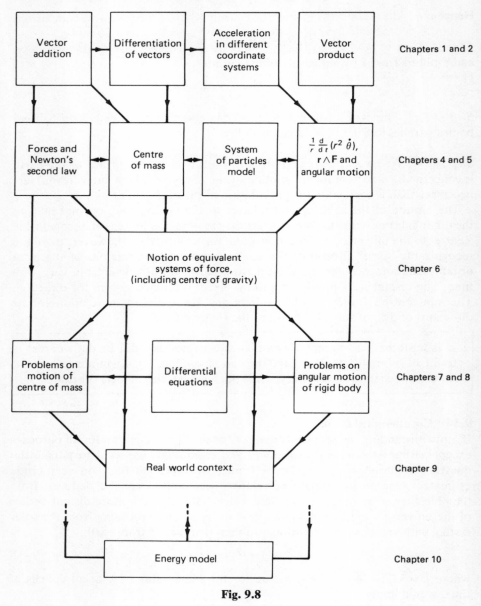

Fig. 9.8

correction $\ddot{R} \simeq 10^{-3}$ m s^{-2}, is not a large number; moreover, no observational evidence suggests that the solar system is nearer to the 'edge' than to the 'centre' of the universe. The reader may be made uneasy by the conceivable existence of immense realms beyond the visible limit; but problems far more urgent than this arise in the vast space–time of astronomy.

Mechanics applied to the real world   265

Summary: the reader may choose for himself which of the following conclusions it is more appropriate for him to draw from the work we have done.
Either the use of local or laboratory frames for the solution of everyday problems has been vindicated.
Or the success of mechanics in local frames may be regarded as evidence that Newtonian mechanics will also have some validity in wider fields, e.g. in the context of planetary or stellar motion considered in a sun-centred frame; and vice versa.

9.35 Structure of the modernized Newtonian model of mechanics
The logical line of mechanical theory, as it has been developed in Chapters 1–9 can now be appreciated. It is shown in the flow-diagram, FIG. 9.8.

9.4 'Pot Black': a discussion of the motion of billiards/snooker balls

An expert snooker or billiards player would find his skills of little avail if the standard balls were replaced by, say, lead spheres, the table by a polished dining-room table, the chalk on his cue by butter. The physical characteristics of balls, table and cue have an important effect on the game.

In older textbooks, instead of looking at such realities, the motion of colliding billiard balls is discussed without preamble, as in § 9.41 below.

9.41 Trial model of 'direct collision' of two equal billiard/snooker balls
Let the first ball (the cue ball) have initial and final velocities u_1 and u_2 along the line of centres AB (FIG. 9.9a); and let the second ball, which is struck by the first, have initial and final velocities v_1 and v_2.

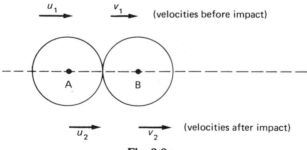

Fig. 9.9a

If the balls have equal mass M, the principle of conservation of linear momentum (§ 5.22) tells us that

$$Mu_2 + Mv_2 = Mu_1 + Mv_1$$

i.e.

$$u_2 + v_2 = u_1 + v_1 \qquad \qquad 9\text{–}20$$

266 An introduction to mechanics and modelling

To get a second equation relating the unknown final velocities u_2 and v_2 to the presumed known initial velocities u_1 and v_1, we invoke the coefficient of restitution concept (§ 5.4) whence

$$v_2 - u_2 = e(u_1 - v_1) \qquad 9\text{--}21$$

For known e, equations 9–20 and 9–21 give the final velocities in terms of the initial velocities,

$$v_2 = \tfrac{1}{2}(1+e)u_1 + \tfrac{1}{2}(1-e)v_1$$
$$u_2 = \tfrac{1}{2}(1-e)u_1 + \tfrac{1}{2}(1+e)v_1$$

so that, in the common case $v_1 = 0$,

$$v_2 = \tfrac{1}{2}(1+e)u_1; \qquad u_2 = \tfrac{1}{2}(1-e)u_1; \qquad 9\text{--}22$$

and

$$v_2/u_2 = (1+e)/(1-e) = \text{constant} \qquad 9\text{--}23$$

(constant for different collisions between the same pair of balls).

The prospective billiard player would be ill-advised to give uncritical credence to this result. The modelling assumptions have not been clearly stated, let alone checked. Both the player and the mathematician would be wise to check against observation, when they will see that a skilled player can vary the ratio of the final velocities of the two balls almost at will by imparting more or less spin to the cue ball, A.

At this point the scientist with easy access to high-speed photographic equipment might turn to it to find out what is going wrong. The mathematician, however, like the intelligent novice who is puzzled by the 'solution' given above, will first identify the assumptions implicit in the quick-fire textbook-type solution given above, i.e. he will belatedly perform the step A, step B processes without which the status of equations 9–20 to 9–23 is very questionable.

9.411 Assumptions affecting the validity of the work of § 9.41
The step B assumptions about forces, on which the validity of equations 9–20 and 9–21 depend might have been stated as follows.

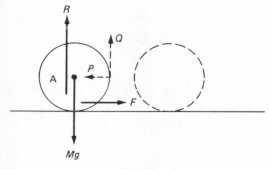

Fig. 9.9b

Mechanics applied to the real world   267

The forces on ball A (FIG. 9.9b) consist of:

Mg vertically downwards (due to gravity);
R = normal reaction between table and ball;
F = frictional force at point of contact with the table, $F \leq \mu R$;
P, Q = normal and tangential components of the contact force with ball B;
Air resistance neglected;

and similarly for ball B.

Of these we assume $Q = 0$, i.e. we model the contact between the balls as 'smooth'. With this assumption, $R = Mg$. We accept that the contact between the cloth of the billiard table is unlikely to be 'smooth', but we note that

$$\int_0^{\Delta t} F \, dt \leq \int_0^{\Delta t} \mu Mg \, dt = \mu Mg \, \Delta t$$

and assume that the time Δt of contact is sufficiently small for this frictional impulse to be negligible.

The coefficient of friction at the table is undoubtedly much greater than that between the two balls, so the validity of the modelling assumption

$$\int F \, dt = 0$$

is the first to be scrutinized.

Velocities of the order of 1 m s^{-1} are of interest, so the validity of equation 9–20 requires

$$\mu Mg \, \Delta t \leq M$$

$$\Delta t \leq 1/g \approx 1/10 \text{ sec}$$

The human eye is unlikely to detect what is happening in a period so small as 1/10 second, but this is ample time to accommodate the impact of two bodies with large elastic modulus. (**Note:** when $\ddot{x} = -\lambda x$, period $= 2\pi/\sqrt{\lambda}$ i.e. the half-period corresponding to the collision time, $= \pi/\sqrt{\lambda}$, is small when λ is large.)

We are left with three questions:

(i)   Is the restitution law (equation 9–21) valid in our chosen circumstances?
(ii) Does the spin of the cue ball A (or B) invalidate equations 9–20 and 9–21?
(iii) Do the speeds assessed visually by the observer correspond to the time $t = 0$ (before impact) and $t = \Delta t$ (after impact) which are relevant to equations 9–20 and 9–21?

We take these in turn.

(i)   The restitution law is an experimental law based on collision experiments of which billiard ball collisions uncomplicated by friction at a table are wholly typical; there are no grounds for suspecting its validity here.
(ii) No. The basic equations of Newtonian mechanics tell us that translation and rotation may be separately considered.

268 An introduction to mechanics and modelling

(iii) The velocities assessed by the observer are averages over several tenths of a second, i.e. over a period $\tau \gg \Delta t$ during which

$$\int_0^\tau F\, dt$$

may cause momentum changes corresponding to several metres per second.

We conclude that the appropriate next step in our investigation is to study velocity changes caused by the friction of the cloth. Such friction arises when there is relative motion between the ball and the table at their contact point: the spin of the ball is a key factor determining this relative motion.

9.42 Spin and acceleration/deceleration of a billiard ball

For pure rolling with no tendency to slip (FIG. 9.10), $F = 0$, and the equations of linear and angular motion give

$$M\, dv/dt = 0$$

i.e. $v = $ constant

$$I\, d\omega/dt = 0$$

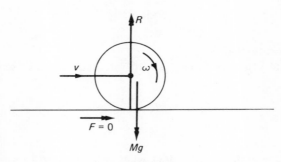

Fig. 9.10

i.e. $\omega = $ constant (ω denotes the angular velocity, $\omega = \dot\theta$, of the ball). In time t, the centre advances distance vt, so the point of contact is distance vt from the initial contact point. Meanwhile the ball has rotated through angle $\theta = \omega t$, so the arc length on the ball between initial and final contact points $= a\theta = a\omega t$ (a is the ball's radius). Pure rolling requires that the arc length $a\omega t$ is laid down, with no contraction or expansion, on the line on the table of length vt between initial and final contacts, i.e.

rolling without slipping requires $a\omega = v$.

In any motion with $\omega \ne v/a$, slipping will take place at the point of contact, and a friction force $F = \mu R$ will arise, the direction of F being such as to oppose the relative motion. (These assertions should be regarded, not as ex-cathedra pronouncements about the physical situation, but as fallible modelling decisions.) The acceleration due to excessive spin ($\omega > v/a$) is given by $M\ddot{x} = \mu Mg$, i.e.

Mechanics applied to the real world 269

$\ddot{x} = \mu g$; deficient spin ($\omega < v/a$), which includes negative or 'bottom' spin, is accompanied by a friction force in the opposite direction and corresponding deceleration μg.

Simultaneously with the acceleration which occurs with excessive spin (when $\omega > v/a$ and consequently $F > 0$) there will be an angular deceleration, as predicted from the angular equation of motion

$$I \, d\omega/dt = -aF = -a\mu Mg.$$

The ball accelerates and its angular velocity concurrently diminishes until $\omega = v/a$, upon which the force F changes to zero, and pure rolling with constant velocity and angular velocity is thereafter predicted by the model.

Thus, according to the model now proposed,

> a cue ball which has top spin after its centre has been instantaneously reduced to rest by a collision, will subsequently follow the ball which it has just hit; but a cue ball with bottom spin in the same circumstances will return towards the player.

More detailed calculations can be carried out using a simple model in which,

(i) The balls are equal spheres of uniform density, so we may write $I = 2Ma^2/5$ (cf. § 6.95).
(ii) Contacts between balls are smooth.
(iii) Collisions between balls are perfectly elastic, i.e. $e = 1$.

It has been found by Daish, in *The Physics of Ball Games*, p. 153, that such a model gives very good agreement with observation. Using the modelling postulates (i), (ii) and (iii) above, establish the results stated in Exercise 2 which follows.

Exercise 2

1 Show that, under the action of a horizontal blow whose line of action passes through the centre, a billiard ball acquires an initial velocity without rotation, followed by a period of deceleration of duration

$$t = 2V/7\mu g$$

(where V is the ball's velocity immediately after the cue stroke) and that its velocity of pure rolling after this time is $5V/7$.

2 Show that the distance travelled by the ball of question 1, before slipping at the contact ceases, is $12V^2/49\mu g$.

3 Using the value $\mu = 0.2$ (suggested by Daish as typical of billiard ball–table contacts), find the time and distance before pure rolling occurs for (a) a ball of initial velocity 3 m s^{-1}, (b) a ball of initial velocity 1 m s^{-1}.

4 Find the height h above the table at which a horizontal blow must be applied to the ball if pure rolling is to result with no intervening period of slipping. (The jargon phrase for the point at height h on the vertical diameter is 'the centre of percussion'.)

5 A cue ball rolling with velocity V strikes a stationary billiard ball. Show that an observer, basing his calculations on the observed final value of the velocity of the struck billiard ball, in conjunction with the momentum and restitution equations, might be led to the false conclusion that the coefficient of restitution for collision, $e = 3/7$.

6 (i) Given that the cue ball at the instant before impact has angular velocity $\omega = V/a$ (pure rolling), find its eventual velocity (see method used in question 1). (ii) With what value of the spin before impact would the cue ball stop dead and remain stationary?

7 If, in the situation of questions 5 and 6 it is assumed that the contact between balls is not smooth, but has coefficient of friction 0.1, find the final velocities of the two balls.

9.43 How to make a billiard ball swerve

In order to achieve a path such as that shown dotted in FIG. 9.11, the ball must acquire, through friction, a velocity component at right angles to the initial line of travel. This it can do, provided that the cue imparts an angular velocity component (ω_1 say) about the line of the initial velocity, as shown in FIG. 9.11: the method of doing this is by cueing (downwards) at an off-centre point on the upper hemisphere of the ball.

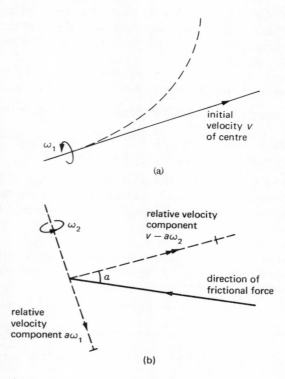

Fig. 9.11

The angular velocity component ω_1 is not, of course, the entire angular velocity of the ball. A component ω_2 assisting or opposing the initial forward velocity v arises during any actual motion; the axis of ω_2 is horizontal and at right angles to the velocity v. Finally, since a spin component ω_1 can be produced at the same time as a forward velocity v only by cueing at an off-centre point of the near side of the ball, a spin about the vertical axis, ω_3 say, is also simultaneously present in a swerving billiard ball.

Mechanics applied to the real world   271

The reader should recognize that this situation lies outside the realm of the two-dimensional mechanics whose theory we built up in Chapter 6. Even the question of whether we may talk of components of angular velocity – thereby implying that angular velocities are vectorial in character, i.e. obey the vector law of addition – even this question has not yet been discussed. We begin by making a simple calculation of the initial curvature of the path.

Let the relative velocity of the ball at the point of contact with the table be represented by components $a\omega_1$, $v - a\omega_2$, 0 as shown in FIG. 9.11. (Spin ω_3 about a vertical axis produces no motion at the point of contact.) This velocity of the point of contact relative to the table makes an angle $\alpha = \tan^{-1}[a\omega_1/(v - a\omega_2)]$ with the direction of motion of the centre of the ball (see FIG. 9.11b). The friction force is thus $\mu Mg \cos \alpha$ opposing the velocity and $\mu Mg \sin \alpha$ at right angles to the current velocity v, so that (with our choice of the direction of rotation of ω_1) the ball will swerve to the left.

The centre of mass of the ball moves in a plane, so to it we may apply the equations of motion

$$M \, dv/dt = -\mu Mg \cos \alpha$$

and

$$Mv^2/\rho = \mu Mg \sin \alpha$$

(cf. equations 2–16). The initial radius of curvature of the path is thus

$$\rho = v^2/\mu g \sin \alpha$$

where v is the velocity of the centre, and α is the value of $\tan^{-1}[a\omega_1/(v - a\omega_2)]$. The initial value of the curvature $(1/\rho)$ is $\mu g \sin \alpha/v^2$, and perhaps the structure of this expression alone gives sufficient guidance to the beginner at billiards who wishes to use his knowledge of mechanics to improve his game.

General equations determining the variation of ω_1, ω_2 and ω_3 for any rigid body cannot be written down until three-dimensional motion has been studied. When frictional contacts with other bodies are involved, as in rolling, the solution is often a very formidable problem [see sections on non-holonomic systems in Chapter 8 of E. T. Whittaker, *Analytic Dynamics of Particles and Rigid Bodies* (CUP, 1937; Dover, 1944).]

Happily, the spherical symmetry of the billiard ball makes possible a quite simple solution for the swerving ball problem. The reader must take on trust that, because of the special symmetry of the billiard ball, its three-dimensional equations of angular motion reduce to the especially simple form,

$$I_x \, d\omega_x/dt = M_x$$
$$I_y \, d\omega_y/dt = M_y$$
$$I_z \, d\omega_z/dt = M_z$$

where x, y, z refer to fixed orthogonal cartesian axes, ω_x, ω_y, ω_z are components of the ball's angular velocity parallel to these axes, and I_x, I_y, I_z; M_x, M_y, M_z

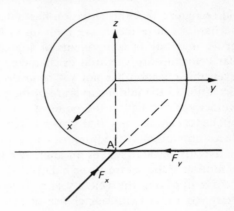

Fig. 9.12

are respectively the moments of inertia and moments of forces about axes through the centre of mass, parallel to x, y, z. We take the z-axis vertically upwards; we write the initial values of $(\dot{x}, \dot{y}, \dot{z})$ and of $(\omega_x, \omega_y, \omega_z)$ as $(u, v, 0)$ and $(\Omega_x, \Omega_y, \Omega_z)$ respectively; and the contact force at the plane is represented by F_x, F_y, R as shown in FIG. 9.12.

The equations of motion are

$$\left. \begin{aligned} M\ddot{x} &= -F_x \\ M\ddot{y} &= -F_y \\ 0 &= R - Mg \end{aligned} \right\} \qquad 9\text{--}24$$

and

$$\left. \begin{aligned} \frac{2Ma^2}{5}\frac{d\omega_x}{dt} &= -aF_y \\ \frac{2Ma^2}{5}\frac{d\omega_y}{dt} &= aF_x \\ \frac{2Ma^2}{5}\frac{d\omega_z}{dt} &= 0 \end{aligned} \right\} \qquad 9\text{--}25$$

Since friction opposes any motion of the contact point A of the ball relative to the plane, so long as slipping occurs,

$$\frac{F_y}{F_x} = \frac{\dot{y} + a\omega_x}{\dot{x} - a\omega_y} \qquad 9\text{--}26$$

where

$$F_x^2 + F_y^2 = (\mu Mg)^2 \qquad 9\text{--}27$$

Mechanics applied to the real world 273

Equations 9–24 and 9–25 give

$$\frac{F_y}{F_x} = \frac{\ddot{y}}{\ddot{x}} = -\frac{\dot{\omega}_x}{\dot{\omega}_y}$$

$$= \frac{\ddot{y} + a\dot{\omega}_x}{\ddot{x} - a\dot{\omega}_y} \qquad 9\text{–}28$$

(by the laws of manipulation of fractions).
Using equation 9–26 with 9–28,

$$\frac{\ddot{y} + a\dot{\omega}_x}{\dot{y} + a\omega_x} = \frac{\ddot{x} - a\dot{\omega}_y}{\dot{x} - a\omega_y}$$

whence

$$\ln(\dot{y} + a\omega_x) = \ln(\dot{x} - a\omega_y) + \text{constant},$$

i.e.

$$\frac{\dot{x} - a\omega_y}{\dot{y} + a\omega_x} = \text{constant}$$

$$= \frac{u - a\Omega_y}{v + a\Omega_x} \qquad 9\text{–}29$$

Equation 9–29 springs to life, and gives illuminating information if we choose the direction of the y-axis parallel to the direction of the initial relative motion at A, i.e. so that there is no initial slipping in the x-direction, $u - a\Omega_y = 0$. Equation 9–29 now tells us that at all subsequent times also

$$\dot{x} - a\omega_y = 0 \qquad 9\text{–}30$$

i.e. that there is no slipping in the x-direction at any time. Moreover, from equation 9–26, $F_x = 0$ at all times.

The ball therefore moves under the action of the constant force $(0, -\mu Mg, Mg)$. At time t after starting from the origin,

$$x = ut$$
$$y = vt - \tfrac{1}{2}\mu g t^2 \qquad 9\text{–}31$$

Elimination of t shows that y is expressible as a quadratic function of x, i.e. the path is a parabola as long as slipping occurs at the point of contact. Thereafter $\dot{x} = a\omega_y = \text{constant}$; $\dot{y} = -a\omega_x = \text{constant}$; $F_x = F_y = 0$; and motion is in a straight line (with constant velocity).

9.431 Conclusion (step E)

The author would expect the path of a suitably cued, real, billiard ball to be approximately parabolic. This expectation derives from a model in which friction effects have been treated fairly crudely. Only experiment can establish how much credit should be given to the predictions when a real billiard ball, table, etc. are involved.

9.44 Direct and oblique impact of two billiard balls

The reader should be able to write out his own discussion of this subject, proceeding (unless he prefers to be independent) along the lines suggested by the following questions.

Exercise 3

1. Assuming (an assumption which is justified by the equations of three-dimensional motion) that during rolling and/or slipping on a horizontal plane, the angular velocity component ω_3 about the vertical remains unaltered, show that in any collision between smooth balls the magnitude of ω_3 is immaterial, i.e. the subsequent motion is not affected by it.
2. Just before striking a stationary billiard ball directly, a cue ball has angular velocity components ω_1, ω_2 and ω_3, the reference directions being, respectively, the direction of travel, the horizontal direction perpendicular to the direction of travel, and the vertical. Assuming a small coefficient of friction, $\mu_1 = 0.02$ say, between the two balls, describe in general terms how the motion of the object ball immediately after impact is affected by the values of ω_1, ω_2 and ω_3.
3. Reverting to the smooth contact model, consider the oblique impact problem illustrated just before impact in FIG. 9.13.

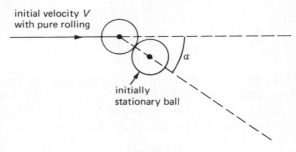

Fig. 9.13

Readers who have access to a billiard table, or who have some previous experience of the game, should now consider how collisions at the cushion (the wall at the edge of the table) might appropriately be modelled. If possible, check your predictions for the ball's path and velocity after hitting the cushion by observation in collaboration with a player.

9.5 Appendix: loaves of bread down a chute (first model)

9.51 Step A(1)

We model the chute as an inclined plane down which the loaf slides under the force of gravity. The chute must be such that (i) loaves do not get stuck, motionless, on the incline, and (ii) loaves must not arrive at the bottom with a speed so great that damage results. These criteria involve the velocity at specific positions on the slope, so we shall seek a velocity–position relationship.

We shall model the loaf as a rigid body which starts from rest at the top, and whose base remains wholly in contact with the plane during the descent. The motion may be affected by the mass of the loaves, the inclination of the plane,

Mechanics applied to the real world

and the coefficient of friction between loaf and plane; and it is appropriate to represent these quantities by symbols M, α, μ to which values can be attributed later. The specification of suitable values for α and μ is part of the design advice which can subsequently be offered.

Upon impact at the bottom of the slope, conditions change radically. The direction of motion alters and tipping may occur; this will require a separate discussion, starting at step A § 9.21.

9.52 Step B(1)

The only forces on the body in motion, other than forces of reaction, are the forces of gravity. The gravity forces are equivalent to Mg acting vertically downwards through the centre of mass.

The reaction forces may be represented by a single equivalent force, components R, F (cf. FIG. 9.14a). We write R, F rather than Y', X' for the unknown components to remind ourselves that, for contact forces, a relation exists between the two components, viz. $F \leq \mu R$, where μ is the relevant coefficient of friction.

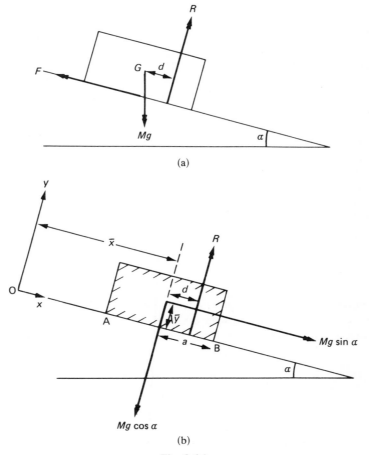

Fig. 9.14

276 An introduction to mechanics and modelling

(That F is represented by an X'-component alone, with zero Z'-component, is a consequence of (a) our postulate that the initial motion takes place down a line of greatest slope, and (b) the law of friction which states that the direction of F is antiparallel to the direction of motion.)

So long as the body is in motion $F = \mu R$. When there is no relative motion $F \leq \mu_0 R$, where the coefficient of static friction $\mu_0 \geq \mu$. As we are concerned with the motion of the centre of mass in a straight line down the plane, it might be thought that only one of the component equations of motion is of relevance, viz. the equation of motion parallel to the plane. However, this equation involves the force F; when motion occurs, the only information about the magnitude of F, viz. $F = \mu R$, involves R; and the only way of determining R is to consider the form taken by the equation of motion perpendicular to the plane in view of the requirement that the loaf stays on the plane without penetrating it. Thus the normal as well as the tangential component equation of $M\ddot{\mathbf{r}} = \mathbf{F}$ must be used. We shall also look at the significance for our problem of the equation of angular motion, $I_G \ddot{\theta} = N_G$.

9.53 Step C(1)
Specification of coordinates (see FIG. 9.14b) are as follows:

Origin: starting point of centre of mass.
x-, y-axes: parallel and perpendicular to the plane as shown.
d = auxiliary unknown characterizing the line of action of the resultant normal reaction R. We note that the resultant of parallel 'push' forces located on the surface of contact represented by AB can have only a clockwise moment about B, so $d < a$, where $a = x_B - \bar{x}$.
θ = notional coordinate specifying the (constant) orientation of the loaf in the plane shown, $\theta = \alpha$ say.
Forces: Mg, F, R as discussed above.
Equations of motion:

$$M\ddot{x} = -F + Mg \sin \alpha \qquad 9\text{-}32$$

$$M\ddot{y} = R - Mg \cos \alpha \qquad 9\text{-}33$$

$$I_G \ddot{\theta} = F\bar{y} - dR \qquad 9\text{-}34$$

and we bear in mind that

$\ddot{y} = 0$ provided this does not imply a negative value of R,
$\ddot{\theta} = 0$ provided this is consistent with $d < a$,
$\ddot{x} = 0$ provided this is consistent with $F \leq \mu_0 R$, but if we require $\dot{x} > 0$, then $F = \mu R$.

Using equation 9-33, $\ddot{y} = 0$ implies $R = Mg \cos \alpha$ which is non-negative, so we conclude that contact is maintained and

$$R = Mg \cos \alpha \qquad 9\text{-}35$$

Using equation 9-34, $\ddot{\theta} = 0$ implies $F\bar{y} = dR$, i.e. $d = \bar{y}F/R$. The condition $d \leq a$ is satisfied provided $a \geq \mu \bar{y}$ and this is satisfied for all normal surface

Mechanics applied to the real world 277

materials ($\mu < 1$) provided the loaf is placed with its least dimension perpendicular to the slope, so we take no further notice of equation 9–34.

Using equation 9–32, $\ddot{x} = 0$ requires $F = Mg \sin \alpha$, so

---

the loaf will not slide down the plane if

$$Mg \sin \alpha < \mu_0 Mg \cos \alpha$$

i.e. if $\mu_0 > \tan \alpha$.

If $\mu_0 < \tan \alpha$, (so that also $\mu < \tan \alpha$), $F = \mu R$ and $M\ddot{x} = -\mu Mg \cos \alpha + Mg \sin \alpha$, whence

$$\ddot{x} = g(\sin \alpha - \mu \cos \alpha) \qquad 9\text{–}36$$

Before integrating this equation we check the right-hand side to make sure there are no hidden variables. Neither g, μ nor α varies during the transit of the loaf down the plane, so the right-hand side of equation 9–36 is constant in the range of the integration we wish to perform. We wish to find the velocity with which the loaf arrives at the bottom of the slope, i.e. from equation 9–36 we seek a velocity–distance relation. The appropriate integration (which is manageable because of the constancy of the right-hand side) is with respect to x. Since the body starts from rest, this gives

$$\tfrac{1}{2}(\dot{x})^2 = g(\sin \alpha - \mu \cos \alpha)\bar{x} \qquad 9\text{–}37$$

At the bottom of the slope, having descended a vertical height of 2 metres,

$$2/\bar{x} = \sin \alpha, \quad \text{i.e. } \bar{x} = 2/\sin \alpha$$

and so equation 9–37 becomes

$$v^2 = (\dot{x})^2_{\text{final}} = 4g(1 - \mu \cot \alpha) \qquad 9\text{–}38$$

---

where $g = 9.8 \text{ m s}^{-2}$.

9.54 Step D(1)

Criterion (i) of step A is satisfied for our model if $\tan \alpha > \mu_0$. The second criterion to be satisfied was that loaves must not arrive at the bottom with a speed so great that damage results. The magnitude of the velocity of arrival has been calculated [$y = 2\sqrt{g(1 - \mu \cot \alpha)}$], but information must be sought on whether this will cause damage.

If we were prepared to make a broad statement 'whatever the value of α, the loaves will be undamaged if their velocity at the bottom of the slope is less than, say, 3 m s^{-1},' then we could terminate our inquiry at this point with specific design recommendations. Thus equation 9–38 shows that we should require

$$4g(1 - \mu \cot \alpha) < 9, \quad \text{whence } \tan \alpha \leq 1.3\mu$$

so both criteria are satisfied provided

$$\mu_0 \leq \tan \alpha \leq 1.3\mu \qquad 9\text{–}39$$

278 An introduction to mechanics and modelling

> The model indicates that if μ, μ_0 are too variable, so that $\mu_0 < 1.3\mu$ cannot be relied on, the chute will be satisfactory for no value of the angle α.

The model indicates (still using the 3 m s^{-1} terminal velocity assumption) that if steel were used for the chute, with a hypothetical value $\mu = \mu_0 = 0.2$ say, for the wrapper–steel coefficient of friction, an appropriate angle of inclination might be obtained by putting $\tan \alpha = 1.15\mu$, $\alpha = 13°$. Alternatively, if a well-finished wood or plastic surface with $\mu_0 = \mu = 0.5$ were used, $\alpha = 30°$ could be an appropriate inclination.

9.55 Step E(1)

As in so many practical problems, we may rest at the end of the first stage of inquiry, or we may engage on a second stage to remove one or more of the less satisfactory features of the first attempt.

One particularly unsatisfactory assumption was incorporated – one might almost say was smuggled into step D(1) above, in the suggestion that the magnitude v alone determined the amount of damage. The inadequacy of such a supposition will be obvious to any motorist who has at different times encountered both a slight change of gradient (say 10°), and a sudden change of gradient (say 90°, at a brick wall) at similar speeds.

A second unsatisfactory feature, which we have already pointed out and promised to remedy, was our failure to indicate whether or not the loaf would topple on reaching the change of gradient, or whether it would proceed without tumbling. Both these matters are discussed in § 9.2.

Exercise 4 (Miscellaneous exercises on Chapters 8 and 9)

1. In FIG. 9.15a, a force P acts in a line whose distance from O is p as shown. A second force of equal magnitude acts in the opposite direction, its line of action being distance d from that of the first force.

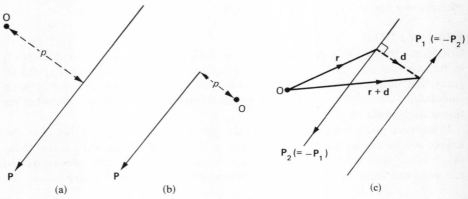

Fig. 9.15

(i) Given that the two forces are together equivalent to the system $\mathbf{F} = 0$, $N_0 = 20$, find d, and show the position of the two forces in a diagram.

Mechanics applied to the real world   279

(ii) Given that the two forces are together equivalent to $\mathbf{F}=0$, $N_0=-10$, find the position of the second force, and show it on a diagram.
(iii) With the two forces as in (i) above, state the general relationship between N_0, P and the algebraic value of d; note that the relation does not involve p.
(iv) With the original force P on the other side of O, as in FIG. 9.15b, verify that the relationship between N_0, P and d is unchanged provided the sense of d remain unaltered.
(v) Using the two-dimensional FIG. 9.15c, we define

$$\mathbf{N}_0 = (\mathbf{r}+\mathbf{d}) \wedge \mathbf{P}_1 + \mathbf{r} \wedge \mathbf{P}_2$$

Given that $\mathbf{P}_2 = -\mathbf{P}_1$, verify that $\mathbf{N}_0$ is independent of $\mathbf{r}$; also show that the sense of the moment $\mathbf{N}_0$ is correctly given by the right-hand screw convention (§ 2.53) whether $\mathbf{N}_0$ is expressed in terms of $\mathbf{P}_1$ or in terms of $\mathbf{P}_2$.

2 The gravitational force on a mass m_i may be represented by the vector $-m_i g\mathbf{k}$. Write down the contribution of this force to the moment $\mathbf{N}_0$ about the origin O (use $\mathbf{r}_i$ to denote the position of the mass). Hence find the resultant $\mathbf{F}$, and $\mathbf{N}_0$, for the system of gravitational forces acting on a rigid body which is modelled as a three-dimensional array of particle masses m_i. Verify that this force system is equivalent to a single force of magnitude Mg ($M = \sum m_i$) acting downwards through the centre of mass G.

3 Ladders: (a consolidation of Exercise 2, Chapter 8).
(i) A uniform ladder OA of length $2a$ and weight W leans against a smooth vertical wall at A and rests on rough horizontal ground at O (coefficient of friction μ). The ladder is inclined at angle θ to the horizontal. Prove that it will not slip provided $\cot\theta < 2\mu$. Does this mean that a steeply inclined ladder position is safer than a flatter position, or vice versa?
(ii) A uniform ladder leans against a **rough** vertical wall at one end, and rests on **smooth** horizontal ground at the other. The ladder is inclined at angle θ to the horizontal. For what angles θ is equilibrium possible?
(iii) A uniform ladder rests with one end against a rough vertical wall and the other on a rough horizontal floor. The coefficients of friction at the wall and floor are respectively μ_1 and μ_2. Prove that the ladder will slip if the angle which it makes with the horizontal is less than $\tan^{-1}[(1-\mu_1\mu_2)/2\mu_2]$. [**Hint:** when there are friction forces at both ends, the general equations of equilibrium, supplemented by the two inequalities $F \leq \mu_1 R$, $F_2 \leq \mu_2 R_2$, are awkward to manipulate. It is therefore easiest to concentrate attention on the critical position – when equilibrium is about to be broken. The ladder will not, from a stationary start, spring away from the wall or the ground (for $R_1 = 0$ implies $F_1 = 0$, whence N_0 would be in the sense which would restore contact); and if contact is maintained at both ends, **both ends must slide if one slides**. Hence in the critical position, limiting friction is in operation at both ends simultaneously. Using the two equalities $F_1 = \mu_1 R$, $F_2 = \mu_2 R_2$ in conjunction with the three equations of equilibrium enables us to evaluate the critical angle θ without manipulating inequalities.

The more ambitious reader wishing to obtain the complete solution (involving the inequalities) may first derive a relation of the form $F_1 + 2F_2 \tan\theta = R_2$, and from it deduce the inequality $\mu_1 + 2\tan\theta \geq 1/\mu_2$, from which it follows that θ_0 as derived for the critical case is the smallest angle with the horizontal consistent with the ladder's equilibrium, but that all positions with $\theta > \theta_0$ are possible equilibrium positions.]
(iv) A ladder inclined at 60° to the horizontal, rests with one end on horizontal ground and the other against a smooth vertical wall. The centre of mass of the ladder is one-third of the way up its length. A man whose weight is twice that of the ladder stands on a rung half-way up the ladder. Find the friction force and the normal reaction at the foot of the ladder, given that $\mu = 5\sqrt{3}/27$. Show that he cannot climb beyond the two-thirds point of the ladder without causing the ladder to slip. What practical moral can you draw about the pitching of a ladder?

280   An introduction to mechanics and modelling

4   The following exercises (i–iv) may be regarded either as further investigations relevant to the loaf of bread problem, or as exercises in the use of the equations $M\ddot{\mathbf{r}} = \mathbf{F}$; $I_G \ddot{\theta} = N_G$ for the motion of a rigid body. We consider a body one point, A, of which slides along a rough (μ') horizontal plane, starting from O as indicated in FIG. 9.3. Both the initial velocity of A and the initial angular velocity are non-zero.

(i)   Writing $OA = x$, $AG = c$ and $\angle OAG = \theta$, write down expressions for $\bar{x}$, $\bar{y}$ and hence for $\ddot{x}$, $\ddot{y}$ in terms of x, θ and derivatives of x and θ. (Alternatively, derive the $\ddot{x}$, $\ddot{y}$ expressions by using 'acceleration of G = acceleration of A + acceleration of G relative to A'.)

(ii)   Bearing in mind that M, I_G and c are measurable constants, show that the three equations of motion corresponding to equations 9-2 of § 9.23 may be expressed in terms of three variables only, and their time derivatives. (**Note:** equations 9-2 involve **five** variables x, y, θ, F and R as well as derivatives of these.)

(iii)   Deduce an equation involving only θ and its derivatives (and the measurable constants).

(iv)   Since the equation derived in (iii) is difficult to solve, consider the particular case $\mu' = 0$, i.e. a **smooth** plane. Verify that the equation obtained by writing $\mu' = 0$ in (iii) is the same as the equation obtained when the relation

$$\dot{\theta}^2(I_G + Mc^2 \cos^2 \theta) + 2Mgc \sin \theta = \text{constant}$$

is differentiated with respect to θ. (The given equation is the form taken by the equation of energy of the body – see Chapter 8 – when conservation of linear momentum parallel to the plane is also taken into account.) The given energy equation is thus the integrated form of the differential equation (in θ) demanded in (iii).

(v)   From (iii) with $\mu' = 0$, verify that the body rights itself provided $\dot{\theta} = 0$ before $\theta = \pi/2$ (i.e. verify that $\ddot{\theta}$ is negative when $\dot{\theta} = 0$, so no greater value of θ is achieved).

(vi)   In the integrated equation of motion (equation of energy) of (iv), substitute the following values: $I_G = \tfrac{1}{3}Mc^2$; initial values of θ and $\dot{\theta}$ are 30° and $\sqrt{(2g/11c)}$ respectively. Prove that under these conditions the greatest value of θ which occurs in the ensuing motion is $\sin^{-1}(2/3)$.

5   A uniform circular disc of radius a rolls without slipping, and with its plane vertical, down a line of greatest slope of a plane inclined at angle α to the horizontal. Show that the coefficient of friction must be at least $\tfrac{1}{3} \tan \alpha$. [**Hint:** choose coordinates x, θ for the linear and angular positions respectively of the disc in such a way that $x = a\theta$ follows from the rolling postulate – because the arc length on the disc matches the distance moved down the plane by the point of contact. This gives one relation between $\ddot{x}$ and $\ddot{\theta}$. Three other relations are available between the four quantities $\ddot{x}$, $\ddot{\theta}$, F and R (where F and R are components of reaction at the point of contact), namely the two relevant components of equations 6-8 and 6-27. From these four relations F/R may be determined.]

6   One end of a thread, which is wound on to a reel, is fixed, and the reel falls in a vertical line, its axis being always horizontal and the unwound part of the string vertical. If the reel is modelled as a solid cylinder of radius a and mass M, show that the acceleration of the centre of the reel is $2g/3$ and that the tension of the thread is $Mg/3$. Verify that the postulated motion is consistent with the thread being 'perfectly flexible' (§ 4.4).

7   A circular cylinder of mass M and radius b has its centre of mass in its axis and its moment of inertia about its axis is k^2M. (**Note** that k has the dimension of a length; the length k, such that Mk^2 is equal to the moment of inertia of a body about a stated axis, is called the **radius of gyration** of the body about the axis.) Prove that when the cylinder rolls down a plane of inclination α, the acceleration of its axis is $b^2 g \sin \alpha / (b^2 + k^2)$.

All the problems below can be solved using the fundamental rigid-body equations of motion in the form $M\ddot{x} = X$; $M\ddot{y} = Y$; $I_G\ddot{\theta} = N_G$. However, when one point (O) of the rigid body is stationary during an interval of time, solution is usually made easier by using $I_O\ddot{\theta} = N_O$ instead of the last of the three fundamental equations; when this is done, it is often also advantageous to express the equation $M\ddot{\mathbf{r}} = \mathbf{F}$ not in terms of cartesian coordinates x, y but in terms of the polar coordinates r, θ of the centre of mass, so that the equations of motion of the point G take the form:

$$M(\ddot{r} - r\dot{\theta}^2) = R \quad \text{and} \quad \frac{M}{r}\frac{d}{dt}(r^2\dot{\theta}) = S \quad \text{(see § 2.33 equation 2–18)}$$

8 A uniform rod of weight W, free to turn about a fixed smooth pivot at one end, is held horizontally and released. Prove that when, in the subsequent motion, the rod makes an angle θ with the downward vertical, the pressure on the pivot is $\tfrac{1}{4}W/(1 + 99\cos^2\theta)$. (**Hints:** the equation of angular motion about the fixed axis gives $\ddot{\theta}$ and, by integration $\dot{\theta}^2/2$; the equations of motion of the centre of mass then give the two components of reaction. Although the weight W is given, you may prefer to work in terms of mass M, putting $Mg = W$ only at the end.)

9 A uniform rod falls under gravity from a position of rest in which it is inclined at an angle of $\pi/3$ to the horizontal with its lower end on a rough horizontal plane. Show that the rod will not initially slip on the plane if the coefficient of friction between the rod and the plane exceeds $3\sqrt{3}/13$ (cf. § 3.22).

10 A uniform rod is held so that one-third of its length rests on a horizontal table at right angles to an edge, the other two-thirds projecting beyond the edge. The rod is then released. Show that it will rotate through an angle $\tan^{-1}\tfrac{1}{2}\mu$, where μ is the coefficient of friction between the rod and the edge, before it begins to slip.

Assignment

A popular executive toy consists of a set of equal steel spheres suspended from a fixed horizontal support by equal parallel strings (FIG. 9.16). When at rest, each sphere just touches its neighbours. The toy is set in motion by pulling aside one of the end spheres, and allowing it to drop back. It is observed that in the ensuing motion, all the spheres except the two end ones remain virtually at rest, while those two end spheres take it in turn to rise and fall, with an amplitude of swing which remains almost constant over several repetitions.

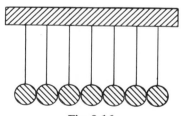

Fig. 9.16

(i) Suggest a simple model which would describe the phenomenon by postulating a small air-gap between each pair of spheres in the equilibrium position.

(ii) Discuss the modelling of the action of the toy without postulating such air-gaps.

(iii) If your discussion under (ii) seems unsatisfactory or incomplete, consider the role of the elasticity of the spheres, and the time of impact, checking your theory by rough quantitative estimates. (All the elastic moduli of steel

282   An introduction to mechanics and modelling

are of the order of $10^{11}\,\text{N}\,\text{m}^{-2}$. There is no recognized 'right' model, so any attempt you make may be illuminating.)

(iv) Are there any other approaches or extensions you would wish to try? To what extent would the concept of energy (discussed in the next chapter) or wave motion be of use?

Part III
Energy modelling in elementary mechanics

10
The energy model

Ideally the student should have some knowledge of line integrals before studying this chapter. However, the concept of energy is of such great importance that its introduction will not be further delayed. The reader will gain some notion of the nature and importance of line integrals as he goes along, and may fill out his knowledge later.

10.1 You pays your money and you takes your choice

Everyone today knows that different forms of energy are largely convertible one into another. From electricity we can get heat or light or sound or motion. From heat we can produce pressure and hence motion. From motion, e.g. that of a running stream or a moving bicycle wheel, we can generate electricity. From the chemical energy of coal or gas, or the nuclear energy of plutonium we can derive heat and hence electricity or locomotion. The energy stored in the water of a dam by reason of its height above a chasm may be released and transformed into the energy of motion by opening the dam gates, or converted less catastrophically, by a controlled process, into electricity. When you hear a noise, you know something has moved. When a cyclist's brakes fail, you expect him to gather speed as he goes downhill; on the other hand, if brakes are used continually on a bicycle or car, keeping the speed constant during the descent, you expect the vicinity of the brake to get hot.

Even though no definition of energy has been given – only an enumeration of types – the concept of something called energy being an essential common component of heat, motion, pressure, etc. provides the beginning of an 'explanation' of the phenomena mentioned above; each phenomenon is seen as a transformation of the essence 'energy' from one form into another. That no form of energy can be created without expenditure of some other resource is an important characteristic of nature which is reflected in the energy model.

> The central idea is given formal expression in the principle of conservation of energy. 'Energy can be neither created nor destroyed; the total energy of any system remains constant except in so far as energy is transferred to it/from it by an external source/receiver.' A receiver of energy is also known as a sink.

The rudimentary idea of energy which we have used so far will become a satisfactory scientific concept only when given quantitative definition. Mechanics can give us a starting point; a study of the region of overlap of the energy model and the $\mathbf{F} = m\mathbf{a}$ model may suggest – though it cannot logically necessitate – quantitative definitions of energy of motion and energy of position which are compatible with both pictures.

10.2 Particle mechanics and energy

10.21 Kinetic energy of a particle
From the equations of motion
$$m\ddot{x} = X; \quad m\ddot{y} = Y; \quad m\ddot{z} = Z$$
we have, by integration (cf. equation 7–1),
$$\tfrac{1}{2}m\dot{x}^2 = \int X\,dx + C_1;$$
$$\tfrac{1}{2}m\dot{y}^2 = \int Y\,dy + C_2;$$
$$\tfrac{1}{2}m\dot{z}^2 = \int Z\,dz + C_3;$$
and so, by addition,
$$\left.\begin{aligned}\tfrac{1}{2}mv^2 \equiv \tfrac{1}{2}m(\dot{x}^2 + \dot{y}^2 + \dot{z}^2) &= \int (X\,dx + Y\,dy + Z\,dz) + C \\ &= \int \mathbf{F}\cdot d\mathbf{r} + C, \quad \text{say;}\end{aligned}\right\} \quad 10\text{–}1$$
(where $C = C_1 + C_2 + C_3$), i.e.
$$\tfrac{1}{2}mv^2 - \int \mathbf{F}\cdot d\mathbf{r} = \text{constant} \qquad 10\text{–}2$$

We note that $\mathbf{F}\cdot d\mathbf{r} \equiv (X\,dx + Y\,dy + Z\,dz)$ is the scalar product of $\mathbf{F}$ and $d\mathbf{r}$. If the magnitudes of $\mathbf{F}$ and $d\mathbf{r}$ are written as F and ds respectively, and if θ denotes the angle between them (FIG. 10.1) we may write $\int \mathbf{F}\cdot d\mathbf{r} = \int F\cos\theta\,ds$, in which there is no reference to a particular choice of x-, y-, z-axes. Both $\tfrac{1}{2}mv^2$ and $\int \mathbf{F}\cdot d\mathbf{r}$ are coordinate-independent quantities.

Equation 10–1 can be written in a form which avoids the use of the arbitrary constant C. Using definite integrals we may write
$$[\tfrac{1}{2}mv^2]_1^2 = \int_{\mathbf{r}_1}^{\mathbf{r}_2} \mathbf{F}\cdot d\mathbf{r} \qquad 10\text{–}3$$

While equations 10–2 and 10–3 are mathematically equivalent, they correspond to slightly different mental pictures. In equation 10–2, the sum of the two quantities $\tfrac{1}{2}mv^2$ and $(-\int \mathbf{F}\cdot d\mathbf{r})$ is constant, i.e. we have a conservation law. In equation 10–3, the change in the quantity $\tfrac{1}{2}mv^2$ is presented as the result of the action of force $\mathbf{F}$. In either case the quantities $\tfrac{1}{2}mv^2$ and $\pm\int \mathbf{F}\cdot d\mathbf{r}$ behave in ways appropriate to energy modelling.

The quantity $\tfrac{1}{2}mv^2$ is given the name 'kinetic energy' of the point mass m moving with velocity v; it represents energy of motion.

The units of $\int \mathbf{F} \cdot \mathbf{dr}$ are force × distance, i.e., in SI units, newton-metres (N m).

The SI unit of energy is called the joule (J), $1\,\text{J} = 1\,\text{N m}$.

We note that this is the energy required to give a mass of 2 kg a velocity of $1\,\text{m s}^{-1}$ (for $\tfrac{1}{2}mv^2 = 1\,\text{m}^2\,\text{kg s}^{-2} = 1\,\text{N m}$).

10.22 Work done by a force, and potential energy

The force $\mathbf{F}$ on a particle may depend on the coordinates of the particle, on the velocity of the particle, on the time of day or a host of other factors. Provided $\mathbf{F}$ is known at every point of the actual path taken from $\mathbf{r}_0$ to $\mathbf{r}_1$, the integral

$$\int_{\mathbf{r}_0}^{\mathbf{r}_1} \mathbf{F} \cdot \mathbf{dr}$$

can, in principle, be evaluated (see § 10.243), but the value of the integral will generally depend on the path chosen between the two end-points $\mathbf{r}_0$ and $\mathbf{r}_1$. The necessity of evaluating **path-dependent** integrals constitutes a major hurdle to the application of equations 10–2 and 10–3.

The value of $\int \mathbf{F} \cdot \mathbf{dr}$ along a path C may be written $\oint \mathbf{F} \cdot \mathbf{dr}$ (to remind us of its dependence on the path chosen), and is known as **the work done** by the force $\mathbf{F}$ on the particle: it represents the amount of energy which has been supplied by means of the force (cf. equation 10–2).

For some types of force of common occurrence in nature, it is possible to express $\mathbf{F} \cdot \mathbf{dr}$ in the form $f(p)\,dp$ where p is a spatial coordinate. In this case $\int \mathbf{F} \cdot \mathbf{dr}$ is an ordinary integral whose value depends only on the initial and final values of p, i.e. is determined by the initial and final positions $\mathbf{r}_0$ and $\mathbf{r}_1$ of the particle, regardless of the path taken in travelling from one to the other. Such a force $\mathbf{F}$ is called a **conservative force**.

When the value of

$$\int_{\mathbf{r}_1}^{\mathbf{r}_2} \mathbf{F} \cdot \mathbf{dr}$$

is expressible as a function of $\mathbf{r}_1$ and $\mathbf{r}_2$ only, i.e. when the value is determined by the end-points only, not the path, then the result of going from $\mathbf{r}_1$ to $\mathbf{r}_2$ and back again, beginning and ending at $\mathbf{r}_1$, is the same as the result of staying at $\mathbf{r}_1$ (FIG. 10.1). So, **for any conservative force, F,**

$$\int_{\mathbf{r}_1}^{\mathbf{r}_2} \mathbf{F} \cdot \mathbf{dr} + \int_{\mathbf{r}_2}^{\mathbf{r}_1} \mathbf{F} \cdot \mathbf{dr} = 0$$

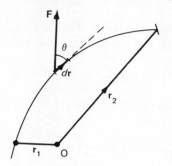

Fig. 10.1

whence

$$-\int_{r_1}^{r_2} \mathbf{F} \cdot d\mathbf{r} = \int_{r_2}^{r_1} \mathbf{F} \cdot d\mathbf{r} \quad \text{(for conservative } \mathbf{F}\text{, whatever the paths).}$$

The integral

$$\int_{r_1}^{r_2} -\mathbf{F} \cdot d\mathbf{r}$$

may therefore be regarded as the work which would be done on the particle (by force $\mathbf{F}$) if it were at any future time to return from its current position $\mathbf{r}_2$ to $\mathbf{r}_1$.

If, for some arbitrarily chosen position vector $\mathbf{c}$, the value of

$$\int_r^c \mathbf{F} \cdot d\mathbf{r} \left(= -\int_c^r \mathbf{F} \cdot d\mathbf{r} \right)$$

is written as $V = V(\mathbf{r})$, then equation 10–2 may be written

$$\tfrac{1}{2}mv^2 + V = \text{constant} \qquad \qquad 10\text{–}4$$

The right-hand side of equation 10–3 may likewise be expressed in terms of V:

$$\int_{r_1}^{r_2} \mathbf{F} \cdot d\mathbf{r} = \int_{r_1}^{c} \mathbf{F} \cdot d\mathbf{r} - \int_{r_2}^{c} \mathbf{F} \cdot d\mathbf{r} = V(\mathbf{r}_1) - V(\mathbf{r}_2) = -[V(x)]_1^2$$

Equation 10–3 therefore takes the form

$$[\tfrac{1}{2}mv^2]_1^2 = -[V(\mathbf{r})]_1^2$$

and this may be rewritten as a zero change, or conservation law,

$$[\tfrac{1}{2}mv^2 + V(\mathbf{r})]_1^2 = 0 \qquad \qquad 10\text{–}5$$

The function $V(\mathbf{r})$ measures the energy which would become available to the particle if it were to return to the standard position $\mathbf{c}$. It is called the potential energy (PE) of the particle.

10.221 Summary

> The potential energy of a particle, possessed by virtue of the operation of a position-dependent force $\mathbf{F} = \mathbf{F}(\mathbf{r})$, is defined only if the work done by the force in any displacement is independent of the path taken between the two end-points, i.e. *only if the force is conservative*. When $\mathbf{F}$ is a conservative force, the corresponding potential energy of the system is defined as
>
> $$V = \int_{\mathbf{r}}^{\mathbf{c}} \mathbf{F} \cdot d\mathbf{r},$$
>
> where $\mathbf{c}$ may be chosen in the way most convenient to the investigator, but, once chosen, must be adhered to throughout the problem in hand.

The point with position vector $\mathbf{c}$ is called the chosen **zero of potential energy**. That the choice of a different zero at the outset makes no difference to the final result may be demonstrated explicitly as follows.

For a conservative force $\mathbf{F}$ and any $\mathbf{c}, \mathbf{c}'$,

$$\int_{\mathbf{r}}^{\mathbf{c}'} \mathbf{F} \cdot d\mathbf{r} = \int_{\mathbf{r}}^{\mathbf{c}} \mathbf{F} \cdot d\mathbf{r} + \int_{\mathbf{c}}^{\mathbf{c}'} \mathbf{F} \cdot d\mathbf{r}$$

The energy equation with $\mathbf{c}'$ as zero of potential energy is

$$\left(\tfrac{1}{2}mv^2 + \int_{\mathbf{r}}^{\mathbf{c}'} \mathbf{F} \cdot d\mathbf{r}\right)_{\text{final}} = \left(\tfrac{1}{2}mv^2 + \int_{\mathbf{r}}^{\mathbf{c}'} \mathbf{F} \cdot d\mathbf{r}\right)_{\text{initial}}$$

i.e.

$$\left(\tfrac{1}{2}mv^2 + \int_{\mathbf{r}}^{\mathbf{c}} \mathbf{F} \cdot d\mathbf{r}\right)_{\text{final}} + \int_{\mathbf{c}}^{\mathbf{c}'} \mathbf{F} \cdot d\mathbf{r} = \left(\tfrac{1}{2}mv^2 + \int_{\mathbf{r}}^{\mathbf{c}} \mathbf{F} \cdot d\mathbf{r}\right)_{\text{initial}} + \int_{\mathbf{c}}^{\mathbf{c}'} \mathbf{F} \cdot d\mathbf{r}$$

since the integral

$$\int_{\mathbf{c}}^{\mathbf{c}'} \mathbf{F} \cdot d\mathbf{r}$$

depends only on $\mathbf{c}, \mathbf{c}'$, not the initial and final values of $\mathbf{r}$ in the motion under consideration. Consequently, the energy equation with $\mathbf{c}'$ as zero of potential energy is equivalent to

$$\left(\tfrac{1}{2}mv^2 + \int_{\mathbf{r}}^{\mathbf{c}} \mathbf{F} \cdot d\mathbf{r}\right)_{\text{final}} = \left(\tfrac{1}{2}mv^2 + \int_{\mathbf{r}}^{\mathbf{c}} \mathbf{F} \cdot d\mathbf{r}\right)_{\text{initial}}$$

> The conclusions drawn from use of the equation of energy are the same whatever choice is made for the zero of potential energy, e.g. whether potential energy is defined as
>
> $$\int_{\mathbf{r}}^{\mathbf{c}} \mathbf{F} \cdot d\mathbf{r} \quad \text{or as} \quad \int_{\mathbf{r}}^{\mathbf{c}'} \mathbf{F} \cdot d\mathbf{r}$$

290 An introduction to mechanics and modelling

The concept of potential energy becomes of practical use, and an unadulterated energy model of mechanics can be used, only when the form of $V(\mathbf{r})$ can be quoted, from memory or reference table, thus avoiding the need to integrate. We proceed to derive the form of the potential energy of a particle due to various types of conservative force which occur in nature.

10.23 Potential functions for standard types of force

10.231 Constant terrestrial gravity
When $\mathbf{F} = -mg\mathbf{k}$, and $d\mathbf{r} = \mathbf{i}\,dx + \mathbf{j}\,dy + \mathbf{k}\,dz$, $-\mathbf{F} \cdot d\mathbf{r} = mg\,dz$, so

$$\int_\mathbf{c}^\mathbf{r} -\mathbf{F} \cdot d\mathbf{r} = [mgz]_{c_1 c_2 c_3}^{x,y,z} = mg(z - c_3)$$

i.e.

$$V = mgh \qquad 10\text{-}6$$

where h is the particle's height above the chosen standard position. (We could write $c_3 = 0$ and $V = mgz$, but the loss of flexibility incurred would be accompanied by no compensating advantage.) The initial and final values of x and y have no effect on the potential energy.

Example 1
A man fires a rifle held at a height of 1.7 m above ground level. The bullet hits a window 30 m above the ground. Given that the bullet left the rifle with a speed of 100 m s^{-1} find the greatest speed with which it can hit the window.

Neglecting air resistance, the only force on the bullet during flight is that of gravity (FIG. 10.2). Writing v for the velocity with which the bullet strikes the window, equation 10–5 takes the form

$$\tfrac{1}{2}mv^2 + 30mg = \tfrac{1}{2}m(100)^2 + 1.7mg$$

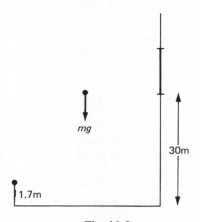

Fig. 10.2

so, using the approximation $g = 10$,

$$v^2 = 100^2 - (20 \times 28.3); \quad v = 97.1 \text{ m s}^{-1} \qquad 10\text{-}7$$

This is an upper limit for the velocity because air resistance would slow the bullet down by an amount which could be calculated precisely only if the path of the bullet was taken into account (the lost energy is dissipated as heat).

Use of equations of motion, neglecting air resistance, would derive result 10-7 less directly, since both horizontal and vertical components of motion would have to be considered.

10.232   Gravitational potential energy for an extended system
Let us model the extended system by a set of particle masses m_i whose position vectors are (at the chosen instant of time) $\mathbf{r}_i = (x_i, y_i, z_i)$ say, where the z_i's are measured vertically upwards. Taking $z = 0$ as the origin of potential energy, the total gravitational potential energy of the system is equal to

$$\sum m_i g z_i = g \sum m_i z_i = gM\bar{z}$$

where $M = \sum m_i$ = total mass of the system, and $\bar{z} = z$-coordinate of the system's centre of mass.

> The gravitational potential energy associated with a system of particles or bodies in the earth's local gravitational field is equal to
> $$Mg\bar{z} \qquad 10\text{-}8$$

In this context, the expression earth's 'field' means the region of space in which the (constant) gravitational force $mg\mathbf{k}$ acts on every mass m. The quantity $\mathbf{G} = g\mathbf{k}$ by which each mass has to be multiplied to give the force on it, communicates the essential mathematical characteristic of this region, and, in mathematical contexts, may be referred to as the **gravitational field**.

10.233   Potential energy due to tension in an elastic spring/string
Let a spring of natural length l and stiffness k have one end fixed at O, and let the particle which we are considering be attached at the other end, P, where $\overrightarrow{OP} = r\mathbf{I}$. Any elementary displacement $d\mathbf{r}$ from P may be written

$$d\mathbf{r} = \mathbf{I}\, dr + \mathbf{J}\, du + \mathbf{K}\, dv, \quad \text{(see FIG. 10.3)}$$

where $\mathbf{I}, \mathbf{J}, \mathbf{K}$ are three mutually perpendicular unit vectors; and force $\mathbf{F}$ on the particle is assumed to be

$$\mathbf{F} = -k(r-l)\mathbf{I}$$

by Hooke's law (§ 4.4). Thus

$$-\mathbf{F} \cdot d\mathbf{r} = +k(r-l)\, dr$$

$$\int -\mathbf{F} \cdot d\mathbf{r} = k(r-l)^2/2 + \text{constant}$$

292 An introduction to mechanics and modelling

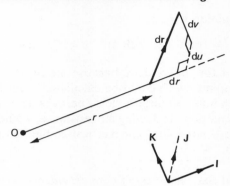

Fig. 10:3

If an unextended spring position, in which $r = l$, is taken as zero of potential energy, then the constant of integration is zero, and the potential energy function is

$$V = \tfrac{1}{2}k(r-l)^2$$
$$= \tfrac{1}{2}kx^2$$

where x is the extension $(r-l)$ of the spring. A compression of the spring is covered by the same analysis, although $(r-l)$ takes negative values. The potential energy for compression y is $\tfrac{1}{2}ky^2$.

For an elastic **string** on the other hand, the tension is zero when $x < 0$. Consequently $V = 0$ for $x < 0$.

Thus for an elastic spring,
$$V = \tfrac{1}{2}kx^2 \quad (x \gtreqless 0) \qquad\qquad 10\text{-}9$$

For an elastic string,
$$V = \tfrac{1}{2}kx^2 \quad (x > 0)$$
$$= 0 \quad\quad\;\; (x \leq 0)$$

The range of validity of these expressions is determined by the range of acceptability of the Hooke's law approximation (from which the V values derive).

Intuitive ideas about energy as a physical entity suggest that any actual energy resides in the spring/string itself. Nevertheless, so long as the particle is attached to the spring, $\tfrac{1}{2}kx^2$ is a **potential** energy of the particle: we have a model which obviates the need for any further reference to the spring's presence.

Energy modelling will later be discussed for extended systems (rather than for a single particle), and a different picture of the location of the potential energy $\tfrac{1}{2}kx^2$ then becomes appropriate (§ 10.44).

Example
The active part of a catapult is modelled (see FIGS. 10.4 and 10.5) as a pair of parallel elastic strings, each of which acquires a tension $225x$ (N) when extended

The energy model 293

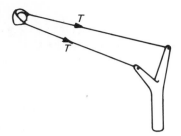

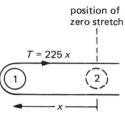

Fig. 10.4 Fig. 10.5

by distance x (m). Find how far a boy must extend the strings before releasing a stone of mass 0.02 kg, assuming that he wants the stone to be projected at a speed of 9 m s^{-1}.

The value of k is 225, and there are two stretched strings. We assume that the extension is horizontal or, if it is not, that it is acceptable to neglect the small gravitational potential energy (less than mgx, i.e. than $0.2x$) associated with the rise or fall of the stone before it leaves the catapult.

The required extension x is therefore given by equating the sums of kinetic energy and potential energy at the two positions marked 2 and 1 in FIG. 10.5;

$$\tfrac{1}{2}(0.02)9^2 + 0 = 0 + 2 \times \tfrac{1}{2}(225)x^2$$

i.e.

$$x^2 = 0.81/225$$
$$x = 0.9/15 = 6 \times 10^{-2} \text{ m}$$

The boy must extend the strings by 6 cm to get the required speed of projection.

10.234 Potential energy and work done when $\mathbf{F}$ is the resultant of forces of different types

If $\mathbf{F}$ is the vector sum of forces $\mathbf{F}_1, \mathbf{F}_2, \mathbf{F}_3, \ldots$, then

$$\int \mathbf{F} \cdot d\mathbf{r} = \int (\mathbf{F}_1 + \mathbf{F}_2 + \mathbf{F}_3 + \cdots) \cdot d\mathbf{r}$$
$$= \int \mathbf{F}_1 \cdot d\mathbf{r} + \int \mathbf{F}_2 \cdot d\mathbf{r} + \int \mathbf{F}_3 \cdot d\mathbf{r} + \cdots$$

| i.e. the work done by the resultant is the sum of the work done by the separate forces into which it may be analysed. | 10–10 |

It follows that if each of the forces $\mathbf{F}_1, \mathbf{F}_2, \mathbf{F}_3, \ldots$ is conservative, the corresponding potential energy functions being $V_1, V_2, V_3, \ldots$, then their resultant $\mathbf{F}$ is also conservative, with potential energy

$$V = V_1 + V_2 + V_3 + \cdots \qquad \qquad 10\text{–}11$$

294 An introduction to mechanics and modelling

If some, but not all, of the forces $\mathbf{F}_1, \mathbf{F}_2, \ldots$ are conservative, e.g. $\mathbf{F} = \mathbf{P}_1 + \mathbf{P}_2 + \cdots + \mathbf{Q}_1 + \mathbf{Q}_2 + \cdots$ where the $\mathbf{P}$'s are conservative but the $\mathbf{Q}$'s are not, we may still make use of any potential energy functions known for the $\mathbf{P}$'s. For equation 10–3 may be written

$$[\tfrac{1}{2}mv]_1^2 + \int_{\mathbf{r}_1}^{\mathbf{r}_2} -\mathbf{P} \cdot d\mathbf{r} = \oint_{\mathbf{r}_1}^{\mathbf{r}_2} \mathbf{Q} \cdot d\mathbf{r}$$

i.e. the increase in (kinetic energy + potential energy) = the work done by all forces not accounted for by potential energy terms.   10–12

The right-hand side can be evaluated only when the path C is known.

Example 1
The head of a jack-in-the-box (mass 0.1 kg) has become detached from its spring. Given that the spring has stiffness $k = 2 \times 10^3$ SI units, and that it is released, with the head on top of it, from an initial (vertical) compression of 3 cm, estimate the height risen by the jack-in-the-box's loose head.

Solution. Choosing the zero of gravitational potential energy at the initial position, and choosing the zero of energy due to the spring at the point at which the spring has zero extension, we have initially:

Kinetic energy = 0
Potential energy due to gravity = 0
Potential energy due to spring = $\tfrac{1}{2}k(0.03)^2$

Contact is lost when the end of the spring begins to move less rapidly than the head which it has been pushing (this occurs at the position of zero extension of the spring). At the position of greatest height, we therefore have:

Kinetic energy = 0
Potential energy due to gravity = $0.1gh \simeq h$ (taking $g = 10$)
Potential energy due to spring = 0

Apart from air resistance which we shall ignore, there are no other forces on the head; the equation of energy takes the form:

$$h = \tfrac{1}{2}k(0.03)^2 \quad \text{where } k = 2 \times 10^3$$
$$= 0.9 \text{ (m)}$$

Example 2
A pendulum consists of a heavy bob at one end of a light inextensible string, length l, whose other end is fixed (FIG. 10.6). At the lowest point of its swing, the bob is moving with speed V. Assuming that the string does not slacken and that the bob does not describe a complete circle, find the greatest height reached by the bob.

Solution 1. (Using the lowest point of the swing as zero of potential energy). The energy principle will be applied in the form of equation 10–12. The work

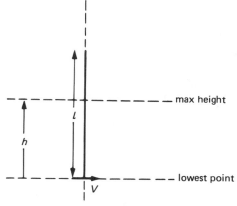

Fig. 10.6

done by the tension in the string is zero because **the elementary displacement dr of the bob is at all times perpendicular to the tension** T. The required height h above the lowest point of the swing (which we take as zero of gravitational potential energy) is achieved when the bob's velocity is instantaneously zero, and is therefore given by the equation of energy:

$$mgh = \tfrac{1}{2}mV^2;$$

so

$$h = V^2/2g.$$

Solution 2. (Using the point of suspension as zero of potential energy).

$$\text{Initial energy} = \tfrac{1}{2}mV^2 - mgl$$
$$\text{Final energy} = 0 - mg(l-h)$$
$$\tfrac{1}{2}mV^2 - mgl = -mgl + mgh$$
$$\tfrac{1}{2}mV^2 = mgh; \quad h = V^2/2g \quad \text{as before}$$

Exercise 1

1 Two particles, each of mass M, are moving in the same straight line with speeds V and v respectively, one overtaking the other. After collision they coalesce and move as one body. Find their common velocity from momentum considerations, and calculate the loss of kinetic energy. If they had originally been travelling in opposite directions, show that the loss of energy would have been greater by a factor $(V+v)^2/(V-v)^2$.

2 A mass of 3 kg is attached to one end of an elastic string of natural length 2 m whose other end is fastened to a fixed point O. The modulus of the string is such that the 3 kg mass, hanging vertically at rest, stretches it by 1 cm. The mass is held at O and allowed to fall vertically from rest. Apply the principle of conservation of energy to find the speed of the mass when the string is stretched by 3 cm, and find the maximum extension of the string.

Using the equation of energy, describe the main features of the subsequent motion. State some of the ways in which the model implied by the statement of the problem is likely to be an inadequate representation of reality.

3   A small ring, of mass m, is threaded on to a smooth wire bent into the form of a circle of radius a. The wire is fixed in a vertical plane and the ring is projected along the wire, starting from the lowest point, with speed $\sqrt{(kag)}$. Assuming that the ring goes completely round the wire, find the reaction between the ring and the wire when (a) the ring is at the lowest point, (b) the ring is at the highest point.

If $k=4$ find the position of the ring on the wire when the reaction is zero.

(**Hint:** use the equation of energy to relate position and speed; use the radial equation of motion, equation 3–17 to find the normal reaction.)

10.235   Potential energy of a particle acted on by an inverse square law attraction or repulsion

In FIG. 10.7 we postulate:

(a)   for attraction

$$\mathbf{F} = \frac{-K}{r^2}\mathbf{I} \quad (K>0)$$

(b)   for repulsion

$$\mathbf{F} = \frac{k}{r^2}\mathbf{I} \quad (k>0)$$

where $\mathbf{I}$ is the unit vector parallel to $\mathbf{r}$, $(\mathbf{I} = \mathbf{r}/r)$. Consider the law of attraction.

Write $d\mathbf{r} = \mathbf{I}\,dr + \mathbf{J}\,du + \mathbf{K}\,dv$ as in § 10.233 (FIG. 10.3)

$$-\mathbf{F}\cdot d\mathbf{r} = \frac{K}{r^2}\,dr$$

$$\int_c^r -\mathbf{F}\cdot d\mathbf{r} = \int_c^r \frac{K}{r^2}\,dr$$

$$= \frac{-K}{r} + \frac{K}{c}$$

which depends only on the end values r and c and is independent of the route, etc. taken between the points $\mathbf{c}$ and $\mathbf{r}$. The constant K/c may be made equal to zero by choosing for $\mathbf{c}$ a vector of infinite length, i.e. writing $c \to \infty$. With such a choice of zero of potential energy we have

> $$V = -K/r \qquad \qquad \text{10-13}$$
>
> for the attractive inverse square law force

and $V = +k/r$ for the inverse square law repulsion.

Example

At height 1000 km above the earth's surface, a rocket's speed is 20 000 km h^{-1}. Show that, whatever its direction of travel, it will not (without the supply of further power) escape from the earth's gravitational field.

Since the direction of travel is unspecified, the energy equation is a much more appropriate tool for this question than would be provided by equations of motion.

The energy model   297

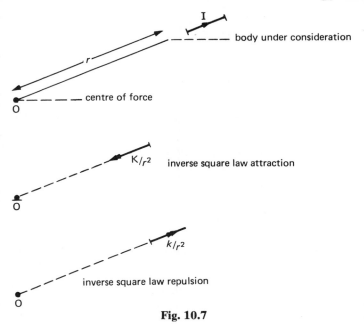

Fig. 10.7

Taking account of the gravitational potential energy due to the earth, and ignoring all effects due to the sun, planets, atmospheric drag, etc. the equation of conservation of energy takes the form

$$\tfrac{1}{2}mv^2 - K/r = \text{initial value} = \tfrac{1}{2}mv_0^2 - K/r_0 \quad \text{say} \qquad 10\text{--}14$$

where $K = GMm$ and v_0, r_0 are obtainable from the given information. Because m is a factor of both sides of this equation, its value need not be known. The problem-solver does need to know the values of GM and the radius of the earth. These are not given in the statement of the question, so recourse must be made to other sources of information (e.g. by reference to the index of this book).

Working in SI units we have

$$v_0 = \frac{20\,000\,000}{60 \times 60} = \frac{1 \times 10^5}{18} \text{ m s}^{-1}$$

$$GM = 6.7 \times 10^{-11} \times 6 \times 10^{24} = 4.02 \times 10^{14} \text{ SI units}$$

$$r_0 = (1000 + 6400) \times 1000 = 7.4 \times 10^6 \text{ m}$$

Inserting these values, equation 10–14 gives

$$\tfrac{1}{2}v^2 - GM/r = -3.889 \times 10^7 \qquad 10\text{--}15$$

Any increase in r, i.e. diminution in GM/r, requires a corresponding diminution in $\tfrac{1}{2}v^2$. However, since $v^2 \geqslant 0$, there is a limit beyond which $\tfrac{1}{2}v^2$ cannot further reduce. If and when r reaches the value for which $v = 0$, there can be no further increase in r. An upper limit on r, for all possible initial directions of motion is

298 An introduction to mechanics and modelling

therefore given by
$$GM/r_{max} = 3.889 \times 10^7$$
i.e. the rocket's path will be confined to the region
$$r \leq (4.02 \times 10^{14})/(3.889 \times 10^7) \simeq 1 \times 10^7 \text{ m}$$
from the earth's centre.

Confinement within some finite radius is inevitable whenever the right-hand side of an energy equation of form 10–15 is negative. For an object starting from radius r_0, the least speed which would give a non-negative right-hand side is that V_0 for which $\frac{1}{2}V_0^2 - GM/r_0 = 0$; the speed $V_0 = \sqrt{(2GM/r_0)}$ is called the escape velocity from radius r_0.

Exercise 2

1   Prove that the escape velocity of a rocket from the surface of the earth is about 40 000 km h^{-1}.
2   Estimate the potential energy of a rocket in the vicinity of the earth due to the gravitational attraction of the sun. Do you conclude that neglect of the sun's attraction in the example of § 10.235 invalidates the answer?
3   What reservations would you make about the estimate of 40 000 km h^{-1} in question 1? Attempt a very rough estimate of the work done against air resistance (you will have to assume something, or ascertain something about the diminution of air density with height).
4   A spacecraft orbits the earth, acted on only by the earth's gravitational force. When the craft is at distance $r = R_1$ from the earth's centre, its speed is exactly half of the escape speed from that position, but its direction of motion is at right angles to the radius vector **r**. Assuming that the orbit lies in a plane which passes through the earth's centre, write down the equation of energy of the spacecraft (introducing any necessary symbols).
   (i)   From the equation of energy, deduce that the spacecraft's speed is greatest when the craft is nearest to the earth, and least when the craft is at its greatest radial distance.
   (ii) Show that at the positions of least and greatest speed, $\dot{r} = 0$ and $v = r\dot{\theta}$ where θ has its usual significance as a polar coordinate. These two positions are called **apses** of the orbit.
   (iii) Use the transverse equation of motion in polar coordinates to show that $r^2\dot{\theta} = $ constant throughout the orbiting motion. Deduce that at the apses the coordinate r satisfies a relation of the form
$$A\left(\frac{1}{r}\right)^2 - \frac{1}{r} + \frac{3}{4R_1} = 0$$
   (iv) Given that the constant $A = \frac{1}{4}R_1$, find the distance of the second apse from the centre of the earth.

10.236   *Potential energy of a particle acted on by a radial force of magnitude $F(r)$*
Let **I** be unit vector in the radial direction.
$$\mathbf{F} = \mathbf{I}F(r); \quad d\mathbf{r} = \mathbf{I}\,dr + \mathbf{J}\,du + \mathbf{K}\,dv$$
$$\int -\mathbf{F} \cdot d\mathbf{r} = \int F(r)\,dr = G(r) \quad \text{say}$$
$$\int_c^r -\mathbf{F} \cdot d\mathbf{r} = G(r) - G(c)$$

whatever the path between the end-points **c** and **r**; i.e. **the force is conservative, with potential energy function** $V = G(r)$.

10.24 Non-conservative forces

10.241 Resisting force of constant magnitude
The work done by a resistive force of constant magnitude F_0 when the particle on which it acts travels from **c** to **r** along the curve CP of length s in FIG. 10.8 is equal to

$$-\int F_0 \, ds = -F_0 s$$

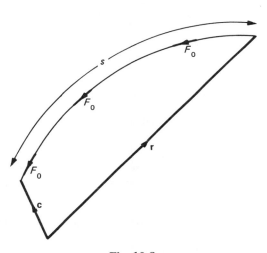

Fig. 10.8

Since the magnitude of s is not deducible from the values of **c** and **r** alone, the force F_0 is non-conservative.

10.242 Any resisting force
For any resisting force, $\int \mathbf{F} \cdot d\mathbf{r}$ is negative along any path. Along no path is

$$\int_{\mathbf{r}}^{\mathbf{c}} \mathbf{F} \cdot d\mathbf{r} = -\int_{\mathbf{c}}^{\mathbf{r}} \mathbf{F} \cdot d\mathbf{r}$$

Any resisting force is non-conservative. The deficit in (potential energy + kinetic energy) is usually attributable to production of heat.

10.243 Calculation of work done for non-conservative forces
The special nature of conservative forces which enables the work integral to be expressed in terms of a single, suitably chosen variable, and hence evaluated as an ordinary integral in one variable, is absent for non-conservative forces. The work integral cannot be evaluated unless the path is known. Only use of the

Example

Find the work done by the force $\mathbf{F} = -y\mathbf{i} + 3\mathbf{j}$ acting on a particle which moves in the x-y plane from the point $(1, 1)$ to the point $(2, 4)$ as in FIG. 10.9; (a) along the direct path AB; (b) along the path AQB consisting of segments AQ and QB parallel respectively to the axes of x and y; (c) along a curve of the form $(y-1) = k(x-1)^2$.

In each case, path element

$$\mathbf{dr} = \mathbf{i}\, dx + \mathbf{j}\, dy$$

$$\mathbf{F} \cdot \mathbf{dr} = -y\, dx + 3\, dy$$

$$\int \mathbf{F} \cdot \mathbf{dr} = -\int y\, dx + \int 3\, dy$$

In each case

$$\int_{(1,1)}^{(2,4)} 3\, dy = [3y]_1^4 = 9$$

the integral $\int y\, dx$ can be integrated only if y is expressed in terms of x (or dx in terms of y and dy) by means of an equation between x and y which is available because the particle is constrained to move along a specified path.

On path (a),

$$y - 1 = 3(x - 1); \quad dy = 3\, dx, \quad \text{i.e.} \quad dx = \tfrac{1}{3} dy.$$

$$\int_{AB} \mathbf{F} \cdot \mathbf{dr} = -\tfrac{1}{3} \int y\, dy + 9 = -(16 - 1)/6 + 9$$

$$= 6\tfrac{1}{2}$$

On path (b),

$$\int_{\text{path AQB}} \mathbf{F} \cdot \mathbf{dr} = \int_{\text{path AQ}} \mathbf{F} \cdot \mathbf{dr} + \int_{\text{path QB}} \mathbf{F} \cdot \mathbf{dr}$$

Moreover on AQ,

$$y = 1, \quad dy = 0, \quad \int_{AQ} \mathbf{F} \cdot \mathbf{dr} = -1$$

on QB,

$$dx = 0, \quad \int_{QB} 3\, dy = 9,$$

so in all

$$\int \mathbf{F} \cdot \mathbf{dr} = 8.$$

On path (c),
$$y - 1 = k(x-1)^2,$$
and, since the path must pass through the point B (2, 4),
$$4 - 1 = k(2-1)^2, \quad \text{i.e.} \quad k = 3.$$
$$y - 1 = 3(x-1)^2; \quad y = 1 + 3(x-1)^2$$
$$\int_{\text{curve AB}} \mathbf{F} \cdot d\mathbf{r} = -\int [1 + 3(x-1)^2] \, dx + 9$$
$$= -[x + (x-1)^3]_{x=1}^{x=2} + 9$$
$$= 7.$$

The work done is different on each of the three paths, the values being (a) $6\frac{1}{2}$; (b) 8; and (c) 7 units.

Exercise 3

1 Using the same force and end-points as those of the preceding example, evaluate the work done when the particle follows the path ARB in FIG. 10.9.

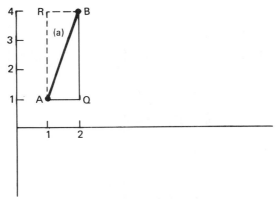

Fig. 10.9

2 ABCD is a unit square. Three point masses move from A (0, 0) to C (1, 1), the first (m) via B, the second (m) via D, and the third (m) directly along the diagonal AC. Each of the particles is acted on by the force $\mathbf{F}_1 = 2\mathbf{i} - 5xy\mathbf{j}$ as well as other forces which are not necessarily the same for the three particles. Find the work done by the force $\mathbf{F}_1$ on each of the three particles.

3 The polar equation $2/r = 1 + e \cos \theta$, with $e < 1$, represents an ellipse passing through the two points $r = 2$, $\theta = \pm \pi/2$ (i.e. $x = 0$, $y = \pm 2$), with focus at the origin, and of an elongation (parallel to the x-axis) which depends on the magnitude of e. (i) A particle describes the complete ellipse from $\theta = 0$ to 2π under the action of a force of magnitude $5/r^2$. Find the work done by the force if that force is always directed towards the origin, i.e. the focus. (**Hint:** the components of the force in the radial and transverse directions are $-5/r^2$, 0 respectively; the components of the path element d$\mathbf{r}$ are dr, r dθ respectively, whence $\mathbf{F} \cdot d\mathbf{r} = ?$) (ii) What is the work done in the half-ellipse from $\theta = 0$ to $\theta = \pi$?

4 A mass describes the same ellipse as in question 3 under the action of several forces, one of which, F_3 say, is a transverse force of magnitude $5/r^2$ acting always in the direction of θ decreasing. Find the work done by F_3 in one complete revolution in the positive sense of θ.

10.3 Power

10.31 Definitions

The rate of achieving a specified speed (and hence a specified kinetic energy) is often almost as important as the eventual speed/energy itself. Similarly, we are interested not only in the total amount of energy consumed by an electric light bulb, but in the rate of consumption – which determines the brightness.

If $E(t)$ denotes the energy of a body at time t, then

> the net 'power' supplied to the body is equal to its rate of increase of energy, dE/dt.   10–16

This is often used in the form: the energy increment δE provided in time δt by a source of power P is $\delta E = P\delta t$. In mechanical applications

$$\delta E \simeq \mathbf{F} \cdot \delta \mathbf{r}$$

and

> $$\text{mechanical power} = \lim_{\delta t \to 0}\left(\frac{1}{\delta t}\mathbf{F} \cdot \delta \mathbf{r}\right) = \lim\left(\mathbf{F} \cdot \frac{\delta \mathbf{r}}{\delta t}\right)$$
> $$= \mathbf{F} \cdot \mathbf{v} \qquad \qquad 10\text{–}17$$

(**Note** that $1/\delta t$ behaves like the scalar factor k in the scalar product law $k(\mathbf{a} \cdot \mathbf{b}) = \mathbf{a} \cdot k\mathbf{b}$.)

10.32 Unit of power

> The SI unit of power is the watt (W). 1 W is an energy transfer of 1 joule per second.
>
> watts = joules per second = newtons × metres per second

Note that, in calculating energy transfers, it would be misguided to think in terms of the sign of $\mathbf{F} \cdot \mathbf{v}$ when the direction of transfer is much clearer by direct energy considerations.

10.33 Example

What power must the engine of a crane supply in order to be able to lift $\frac{1}{2}$ tonne girders at 80 metres per minute?

The force required for steady motion with no acceleration is $\frac{1}{2}$ tonne-force = 4900 N. The power required during the unaccelerated motion is therefore $80 \times 4900/60$ W $\simeq 6.53$ kW. Note that equation 10–17, which applies to the force, gives this result more readily than equation 10–16 whose application requires the use of § 10.232.

Exercise 4

1. What power must the engine of a car produce in order that it may sustain unaccelerated motion along a flat road at (a) 40 km h^{-1}, (b) 120 km h^{-1}, assuming that the resistances to motion are equivalent to a force of 20 kgf at all speeds? (c) If the mass of the car is 1000 kg, and it accelerates uniformly from 40 km h^{-1} at time $t=0$ to 120 km h^{-1} at time $t=30$ seconds, find the power used at any time t in the interval $0 < t \leq 30$.
2. A ship whose maximum power output is 30 000 kW has a maximum speed of 10 m s^{-1}. Given that the resistance to the motion of the ship is proportional to the square of the speed, find what power would be required to enable the ship to move at 12 m s^{-1}.
3. Estimate the power required to pump 1000 gallons of water per minute from a depth of 20 m and deliver it through a pipe of cross-section 30 cm^2. (1 gallon of water has a mass of approximately 4 kg and a volume of approximately 4000 cm^3. The speed of the water may be calculated from the cross-section of the pipe and the volume delivered per second. Friction effects may be neglected for the purposes of the estimate.)
4. In the accelerated motion of the car in question 1(c) calculate the amount of energy supplied by the engine between $t=0$ and $t=30$ seconds. Compare this with the increase in kinetic energy of the car and reconcile your results with the principle of conservation of energy.
5. A meteorite of mass 2×10^4 kg falls directly towards a planet of mass 10^{26} kg and radius 5×10^7 m, which may, for the purposes of discussion, be considered to be stationary. Show that in falling from infinity to the surface of the planet, an amount of energy of the order of 3×10^{12} J could be radiated in the form of heat and light.
   State what assumption is implied by this estimate and suggest why the energy radiated might be (a) greater than, (b) less than this amount.

10.4 Work done by forces acting on extended systems, and consequences for the use of potential energy

10.41 Work done by a moving force applied to the surface of a stationary body

If a rigid body is stationary, each of its parts has zero kinetic energy and constant potential energy. This remains true even if a force **F**, of magnitude F, acting on its surface, as in FIG. 10.10, moves its point of contact from A to B, a displacement **d** of magnitude d say.

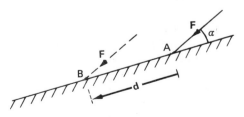

Fig. 10.10

As already stated, the work done by force **F** is independent of the mass of the particle on which it acts. In the proposed circumstances, we may envisage the force **F** as acting on a particle of negligible mass which moves across the face of the rigid body. The work done by the force is equal to $Fd \cos \alpha$ ($= \mathbf{F} \cdot \mathbf{d}$). The change in (kinetic energy + potential energy) for the system of motionless particles which constitutes our model of a rigid body is zero; it follows that the energy $Fd \cos \alpha$ if conveyed from the moving particle to the rigid body must be interpreted as a different kind of energy; the physicist tells us that it is mainly heat.

> When two bodies instantaneously have a point of contact, P, and the point $P = P_1$ of the first body has a different velocity from that of the point $P = P_2$ of the second body, then part of the work done by the interaction forces will be unaccounted for by the change in (potential energy + kinetic energy) for the composite system. The lost energy manifests itself as heat or in other forms such as light, sound, electricity, magnetism, chemical change, etc. Friction forces have this nature. What the physicist sees as a situation in which heat, etc. are produced, the mathematician describes as a situation in which non-conservative forces are involved, and for which the equation kinetic energy + potential energy = constant may therefore not be used.

Exercise
In FIG. 10.11, a rough box slides a distance l down the slope. Given that the coefficient of sliding friction between box and slope is μ', estimate the heat generated. (Identify and name any datum quantities needed in addition to l, α, μ'.)

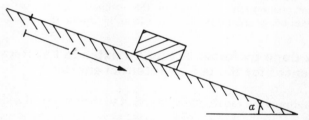

Fig. 10.11

10.42 Work done by internal forces of a rigid body

10.421 Contact forces between particles
$d\mathbf{r}$ is the same for each of two particles in contact; the forces are $\mathbf{F}$, $-\mathbf{F}$ say (Newton's third law); in particular the work done by internal contact forces in any displacement of any rigid body is equal to

$$\sum \int (\mathbf{F} \cdot d\mathbf{r} - \mathbf{F} \cdot d\mathbf{r}) = 0$$

10.422 Forces at a distance between particles of a rigid body
According to the model of Chapter 6, forces at a distance (within the body) can only be attractions or repulsions, e.g. the forces between the two particles A and B of FIG. 10.12 act along the line AB. In any motion of the rigid body of

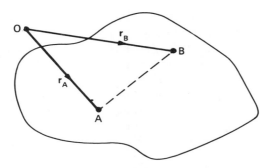

Fig. 10.12

which A and B are points, AB is constant, so the displacement ds of B relative to A can have no component parallel to AB. Consequently $\mathbf{F} \cdot \mathbf{ds} = 0$. If A's displacement is $d\mathbf{r}_A$, the work done by the pair of equal and opposite tension or thrust forces $\mathbf{F}$ is

$$\mathbf{F} \cdot d\mathbf{r}_A - \mathbf{F} \cdot (d\mathbf{r}_A + \mathbf{ds}) = -\mathbf{F} \cdot \mathbf{ds} = 0$$

Like the contact forces, internal attractions/repulsions in total do no work.

| In any displacement of a rigid body, internal forces do no work | 10–18 |

Actual bodies are not perfectly rigid; but in many circumstances deformations are so small that energy losses due to non-conservative internal forces are negligible.

10.43 Two common force-pairs which do no work
I   The pair of normal reactions between two rigid (undeformable) bodies (FIG. 10.13) in sum do zero work. For, either the normal displacements at any point of contact are the same, and the normal forces equal and opposite; or the normal displacements are unequal, in which case, since the bodies cannot interpenetrate, they separate, i.e. exert zero normal force on each other.

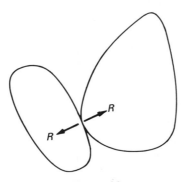

Fig. 10.13

306 An introduction to mechanics and modelling

In rolling, neither normal nor tangential component pairs do any net mechanical work (by the same argument as § 10.421).

When impulsive forces occur, the fiction of a perfectly rigid body is impossible to sustain. In a collision of two bodies, energy is usually 'lost' in the form of heat.

> We conclude that the principle of conservation of energy may be applied globally to a system of rigid bodies, without any need to consider work done by forces between them at rolling contacts or at smooth contacts, provided there are no impulsive collisions.   10–19

Results 10–18 and 10–19 justify the application of the conservation law 'kinetic energy + potential energy = constant' to a single rigid body with rolling or smooth contacts with other fixed bodies, provided any additional applied forces are conservative. (**Exercise:** in one sentence, state why.)

II The pair of tensions at the ends of an inextensible string which passes round one or more small or light pulleys (FIG. 10.14) do no work. For, the tensions at the two ends have equal magnitudes, and the displacements of the ends have equal components parallel to the length of the string (see FIG. 10.14); if positive work $T\,ds$ is done at one end, work $-T\,ds$ is done by the tension at the other end; the net work done by the pair is zero.

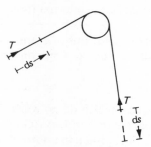

Fig. 10.14

> The principle of conservation of energy may be applied globally to a system of bodies connected by inextensible strings (passing perhaps over light pulleys whose axes are fixed in space) without any need to consider work done by the tensions in those strings.   10–20

Example

A light string passes over a light frictionless pulley whose axis is fixed. Masses m and $2m$ are attached at the two ends, and hang freely. Given that the two masses are level at the moment of release from rest, find the velocity of each when the $2m$ mass is 1 m lower than the mass m.

Consider the system consisting of the two particles. Taking the initial level of the masses as zero of (gravitational) potential energy, and using the nomenclature

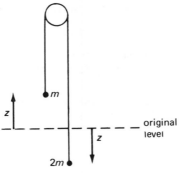

Fig. 10.15

indicated in FIG. 10.15, in the final position,

$$z + z = 1 \text{ (metre)}; \quad z = \tfrac{1}{2}$$

Now
$$\text{initial KE} + \text{PE} = 0 + 0$$
$$\text{final KE} + \text{PE} = \tfrac{1}{2}mv^2 + \tfrac{1}{2}(2m)v^2 + mgz - 2mgz$$

whence
$$3mv^2/2 - mg/2 = 0; \quad v = \sqrt{(\tfrac{1}{3}g)}$$

The velocity of the mass m is thus $\simeq 1.8 \text{ m s}^{-1}$ upwards. The velocity of the mass $2m$ is $\simeq 1.8 \text{ m s}^{-1}$ downwards.

10.44 Potential energy of a system in motion including a spring

In FIG. 10.16 the pair of equal and opposite tensions $\mathbf{T}$, $-\mathbf{T}$ say, acting at A and B, do no net work so long as the distance AB remains unchanged. When length AB changes we may write

$$d\mathbf{r}_B = d\mathbf{r}_A + d\mathbf{u} + \mathbf{I}\, dx$$

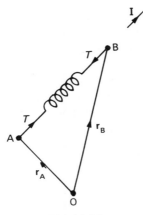

Fig. 10.16

where **I** is unit vector parallel to $\overrightarrow{AB}$ and d**u** is perpendicular to **I**, and

$$\mathbf{T} = +kx\mathbf{I}$$

Then the total work which the pair of tensions acting on the end-points A and B are capable of doing while the spring returns to its natural length is

$$\int_{\text{extended}}^{\text{unextended}} \mathbf{T} \cdot d\mathbf{r}_A - \mathbf{T} \cdot (d\mathbf{r}_A + d\mathbf{u} + \mathbf{I}\, dx)] = \int_{\text{extended}}^{\text{unextended}} -\mathbf{T} \cdot \mathbf{I}\, dx$$

$$= \int_x^0 -kx\, dx$$

$$= \tfrac{1}{2}kx^2$$

This potentiality for work is independent of the position of A as well as the route whereby B reverts to a position of zero extension. The force-pair is conservative, and the system is said to have potential energy equal to $\tfrac{1}{2}kx^2$.

The reader may better appreciate the relation of this calculation to the original definition of potential energy for a particle, viz.

$$\int_c^r -\mathbf{F} \cdot d\mathbf{r}$$

if that original definition is rewritten as

$$\int_r^c \mathbf{F} \cdot d\mathbf{r}$$

(the two forms being equivalent for conservative **F**).

Potential energy $\tfrac{1}{2}kx^2$ is associated with the extension x of the spring and is independent of the location of A or the orientation of AB.   10–21

10.45 Potential energy of a pair of bodies with an inverse square law force acting between them

By analogy with § 10.44 above, the reader may confirm the existence of a potential energy $V = -K/r$ (or $+k/r$) for the moving system consisting of the **pair** of bodies between which the inverse square law attraction K/r^2 (or repulsion k/r^2) acts.

Enough has been said to allow the idea of potential energy of **a system** to take off in its own right as an element in energy modelling.

We remind our readers once again that we have been discussing idealized situations. In real-life problems, such models provide a useful framework for a first approach, but corrections and modifications will usually be necessary before the investigation is concluded.

Assignment or discussion
In § 4.21 it was stated that measures of force alone could not distinguish between active and passive partners, or between the winner and the loser of a contest.

The energy model 309

Discuss, in the light of §§ 10.241, 10.242 and 10.41 how the concept of energy could be used to provide the sought-after discrimination.

10.5 Kinetic energy of a system of particles, or rigid body; applications of energy modelling

Using the particle-model of any system (e.g. a rigid body), and the same notation as that used in Chapter 6, the reader will be able to fill out the argument which shows that – according to this model –

---

the kinetic energy of any system of particles is

$$\sum_i \tfrac{1}{2} m_i v_i^2 = \sum_i \tfrac{1}{2} m_i (\dot{\bar{r}} + \dot{s}_i) \cdot (\dot{\bar{r}} + \dot{s}_i) = \tfrac{1}{2} M \bar{v}^2 + \sum_i \tfrac{1}{2} m_i \dot{s}_i^2 \qquad 10\text{--}22$$

which equals the kinetic energy of the whole mass moving with the centre of mass ($\tfrac{1}{2} M \bar{v}^2$) **plus** the kinetic energy of the motion relative to the centre of mass ($\sum \tfrac{1}{2} m_i \dot{s}_i^2$).

This is a general result. For the particular case of a rigid body moving 'in a plane' (cf. § 6.73) the awkward summation $\sum \tfrac{1}{2} m_i \dot{s}_i^2$ may be replaced by a single, easily calculated term. For, every $\dot{s} = \omega \rho_i$, so

---

the kinetic energy (KE) of a rigid body moving in a plane equals

$$\tfrac{1}{2} M \bar{v}^2 + \tfrac{1}{2} I_G \omega^2 \qquad 10\text{--}23$$

where $I_G = \sum_i m_i \rho_i^2$ (cf. equation 6–26).

When a rigid body rotates about a fixed axis, equation 10–23 may be replaced by the simpler formula,

$$KE = \tfrac{1}{2} I_0 \omega^2 \qquad 10\text{--}24$$

where the suffix 0 locates the axis of rotation, e.g. by signifying a point on the axis of rotation. For, if the body has angular velocity ω, the velocity of point-particle m_i of the body is $R_i \omega$ where R_i is the perpendicular distance of m_i from the axis of rotation. Hence

$$KE = \tfrac{1}{2} m_i v_i^2 = \tfrac{1}{2} \sum_i m_i (R_i \omega)^2 = \tfrac{1}{2} \dot{\theta}^2 \sum_i m_i R_i^2$$
$$= \tfrac{1}{2} I_0 \omega^2$$

where $I_0 = \sum_i m_i R_i^2$ is a geometric characteristic of the body.

10.51 Example

Two uniform rods AB and BC, each of mass m, length $2l$ and moment of inertia $I_G = ml^2/3$, are freely jointed at B. The end C moves along a smooth horizontal wire, and the rod AB can rotate freely in a vertical plane about the end A which is fixed on the wire. Show that if the system is released from rest with the rods horizontal,

$$2l(4 - 3\cos^2 \theta)\dot{\theta}^2 = 3g \sin \theta$$

where θ is the angle turned through by either rod at time t.

The first activity is to draw a figure to clarify the geometry (FIG. 10.17).

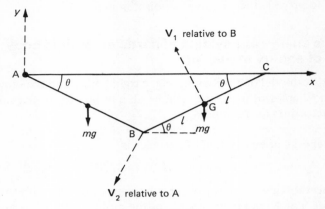

Fig. 10.17

Plan of action: all the joints and contacts are smooth, and the effect of gravity can be accounted for by potential terms; consequently the appropriate method is to express potential energy + kinetic energy at any time t, for which $\angle BAC = \theta$, in terms of θ and $\dot\theta$. When this expression is equated to the initial value of potential energy + kinetic energy we shall obtain a relation between θ, $\dot\theta$ which we expect to be equivalent to the stated relation.

Taking the horizontal level of AC as zero of potential energy (PE), we use the formula $V = mg\bar{z}$ (equation 10–8) for each of the two rods:

$$\text{PE} = -mgl \sin\theta - mgl \sin\theta = -2mgl \sin\theta \qquad 10\text{–}25$$

The kinetic energy of rod AB, of which end A is fixed, is most readily evaluated as $\tfrac{1}{2}I_A\dot\theta^2$, provided that the value of I_A is determinable. By reference to any tables of moments of inertia of bodies of standard shape (e.g. that given in § 6.95 of this book), we find $I_A = 4ml^2/3$. Alternatively I_A may be calculated from $I_G = \tfrac{1}{3}ml^2$ by use of the parallel axis theorem, equation 6–32; $I_A = \tfrac{1}{3}ml^2 + ml^2 = 4ml^2/3$. Hence

$$\text{KE of rod AB} = \tfrac{1}{2}(4ml^2/3)\dot\theta^2$$

Rod BC has no fixed point, so its KE must be evaluated by use of formula 10–23, with $I_G = \tfrac{1}{3}ml^2$. Since angle BCA $= \theta$, and AC is a fixed direction, the magnitude of the angular velocity of rod BC at any time is $\dot\theta$. The second term in 10–23 is thus of magnitude $\tfrac{1}{2}(ml^2/3)\dot\theta^2$.

The velocity $\bar v$ of the centre of mass G must be calculated with care. The easiest way of proceeding is to note that the coordinates of G in FIG. 10.17 are

$$\bar x = 3l \cos\theta; \quad \bar y = -l \sin\theta$$

The components of G's velocity are thus

$$\dot{\bar x} = \frac{d}{dt}(3l\cos\theta); \quad \dot{\bar y} = \frac{d}{dt}(-l\sin\theta)$$
$$= -3l\dot\theta \sin\theta \qquad\qquad = -l\dot\theta \cos\theta$$

whence

$$\bar{v}^2 = l^2\dot{\theta}^2(9\sin^2\theta + \cos^2\theta) = l^2\dot{\theta}^2(9 - 8\cos^2\theta)$$

The same result may be obtained by calculating the vector sum of $\mathbf{V}_1$ and $\mathbf{V}_2$ from FIGS. 10.17 and 10.18. $\mathbf{V}_1$ is the velocity of G relative to B and $\mathbf{V}_2$ is the velocity of point B, the senses in both cases corresponding to θ increasing.

$$v^2 = V_1^2 + V_2^2 + 2V_1V_2\cos(\pi - 2\theta)$$
$$= (l\dot{\theta})^2 + (2l\dot{\theta})^2 - 4(l\dot{\theta})^2\cos 2\theta$$
$$= l^2\dot{\theta}^2(5 - 4\cos 2\theta) = l^2\dot{\theta}^2(9 - 8\cos^2\theta)$$

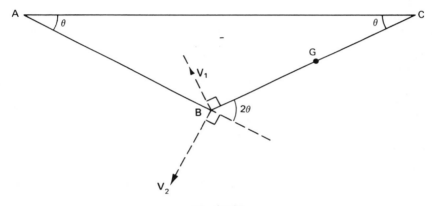

Fig. 10.18

We have now written down sufficient information about each of the four terms needed to evaluate potential energy + kinetic energy for the moving system in terms of θ and $\dot{\theta}$. The initial value of kinetic energy + potential energy (with our choice of zero of gravitational potential energy) is equal to 0, so the equation of energy takes the form

$$-2mgl\sin\theta + \tfrac{2}{3}ml^2\dot{\theta}^2 + \tfrac{1}{6}ml^2\dot{\theta}^2 + \tfrac{1}{2}ml^2\dot{\theta}^2(9 - 8\cos^2\theta) = 0$$

Dividing through by $2ml$ reduces this equation to

$$-g\sin\theta + l\dot{\theta}^2(8/3 - 2\cos^2\theta) = 0$$

whence

$$2l\dot{\theta}^2(4 - 3\cos^2\theta) = 3g\sin\theta$$

10.52 Supplementation of the energy equation

The energy equation supplies only one relationship. This is usually enough to determine the motion of a single particle moving along a known path, but when two or more coordinates are required to specify the position of a particle or rigid body, then two or more relations are required to determine the motion.

312 An introduction to mechanics and modelling

The first possible type of supplementation to look for is a **geometrical connection** between the chosen coordinates, such as the rolling condition $x = a\theta$ which is available for use in question 2 of Exercise 6 below.

If the chosen coordinates are geometrically independent, a second relation can often be obtained by use of the principle of **conservation of linear momentum,** or of **angular momentum.** The modelling example of § 10.53 below will show the energy equation in use in conjunction with a variety of supplementary techniques. The reader is already equipped to try the following examples.

Exercise 5

1 A wheel of mass M, radius R, and moment of inertia about the centre of mass $= I_G$, is set rolling with angular velocity ω up a rough plane inclined at angle α to the horizontal. How far up the plane will it go?

2 Write down the equation of energy for a cylinder of mass M, radius a and moment of inertia I_G about its axis when it is rolling down a plane inclined at angle α to the horizontal. Express this equation in terms of the distance $x = a\theta$ of the point of contact at any time from the initial point of contact, and $\dot{x}$. By differentiating this equation with respect to x, find the acceleration of the cylinder.

3 A flywheel is movable in smooth bearings about a horizontal axis and is set in motion by a descending body of mass m which hangs from one end of a light string coiled round the axle of radius a. The mass m is found to descend a distance h in time t. Using energy relations, or otherwise, show that the moment of inertia of the wheel and axle about its axis is

$$\left(\frac{gt^2}{2h} - 1\right) ma^2$$

4 A uniform rod AB of mass m_1 and length $2a$ is smoothly hinged at A to a fixed point. A second uniform rod BC of mass m_2 and length $2b$ is smoothly hinged at B to the first rod. The rods swing in the vertical plane. Find their kinetic energy in terms of $\dot{\theta}$ and $\dot{\phi}$ at the instant when AB and BC make angles θ, ϕ respectively with the downward vertical. (**Note:** many of the techniques used in the example of § 10.51 are again applicable here; § 2.28 is also relevant.)

10.53 Example: investigative model of an automatic closing device

A design engineer is considering the possibility of making an automatic closing device for a sliding door, using a mechanism like that indicated in FIG. 10.19. The ends of the metal bar AB are connected by smooth joints (a modelling assumption) to blocks which are free to slide in horizontal and vertical channels as indicated. The speed of arrival of end B at C must be sufficient to operate a spring-loaded catch at C. What calculations might the engineer usefully make before constructing a mock-up of the device?

Step A

Much of the modelling has already been done in drawing the diagram and describing the proposed device. We assume that the engineer can test the catch to find a value for the impulse K required to operate it. (It will be an assumption of the model that the impulse K provides a valid criterion of whether or not the catch will be activated.) K will be regarded as a known datum.

If friction forces were everywhere negligible, it would be possible, by application of 'potential energy + kinetic energy = constant' to deduce the velocity with

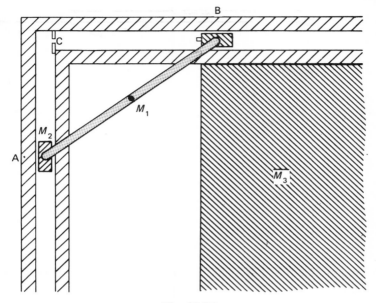

Fig. 10.19

which B meets C for any given starting conditions. This calculation looks so manageable compared with a full treatment including friction, that it is recommended as a 'first round' of inquiry.

Step B (i)
Let the bar AB be modelled as a thin rod of length $2l$ and mass M_1. Let the widths of the channels be ignored so CA is regarded as vertical. The block at end A is sufficiently modelled as a mass M_2. The block at B is attached to the door, block and door moving horizontally with the same velocity; let M_3 be their combined mass. Presumably the value of M_3 will be predominantly determined by the mass of the door itself: the designer is free to choose the size and mass of the block B which will most economically provide enough strength.

Step C (i)
Let θ denote the angle between AB and the horizontal in the general position (FIG. 10.20). To find the kinetic energy of the system we note that

$$\text{velocity of B} = \frac{d}{dt}(2l\cos\theta) = (-)2l\sin\theta \cdot \dot\theta$$

$$\text{velocity of A} = -\frac{d}{dt}(2l\sin\theta) = (-)2l\cos\theta \cdot \dot\theta$$

and velocity $\bar v$ of mid-point G of rod AB has components $(-l\sin\theta \cdot \dot\theta, -l\cos\theta \cdot \dot\theta)$; so,

$$\tfrac{1}{2}M_1\bar v^2 = \tfrac{1}{2}M_1(l^2\dot\theta^2\sin^2\theta + l^2\dot\theta^2\cos^2\theta) = \tfrac{1}{2}M_1 l^2\dot\theta^2$$

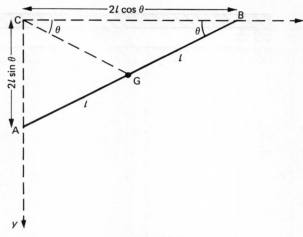

Fig. 10.20

(That the magnitude of $\bar{v}$ is $l\dot{\theta}$ could also have been deduced from the facts that CG = l and that angle BCG = θ, FIG. 10.20.)

$$\text{KE of rod AB} = \tfrac{1}{2}M_1\bar{v}^2 + \tfrac{1}{2}I_G\dot{\theta}^2$$
$$= \tfrac{1}{2}M_1l^2\dot{\theta}^2 + \tfrac{1}{2}\cdot\tfrac{1}{3}M_1l^2\dot{\theta}^2$$
$$= \tfrac{2}{3}M_1l^2\dot{\theta}^2$$

Total KE of moving system $= \tfrac{2}{3}M_1l^2\dot{\theta}^2 + \tfrac{1}{2}M_2(4l^2\cos^2\theta)\dot{\theta}^2 + \tfrac{1}{2}M_3(4l^2\sin^2\theta)\dot{\theta}^2$
$$= l^2\dot{\theta}^2(\tfrac{2}{3}M_1 + 2M_2\cos^2\theta + 2M_3\sin^2\theta)$$

PE of the moving system $= -M_2g\cdot 2l\sin\theta - M_1gl\sin\theta$

If $\theta = \dot{\theta} = 0$ is chosen as the initial situation, the value of $\dot{\theta}$ when $\theta = \pi/2$ is given by

$$l^2\dot{\theta}^2(\tfrac{2}{3}M_1 + 2M_3) - lg(2M_2 + M_1) = 0$$

i.e.

$$\dot{\theta}^2 = \frac{3}{2}\cdot\frac{M_1 + 2M_2}{M_1 + 3M_3}\cdot\frac{g}{l} \qquad 10\text{-}26$$

Step B (ii) (to find the corresponding impulse delivered)
The magnitude of any impulse directly relates to the change of momentum which it produces. An assumption about the motion after impact is therefore required as well as information about velocities before impact. We shall assume (i.e. make the modelling assertion) that the impulses I and J at B and A when B arrives at the catch (FIG. 10.21) are such that both the velocity of G and the angular velocity of bar AB are at once reduced to zero. (Note that the impulse I at C required to bring B to rest will not by itself bring end A to rest. That end will collide with the walls of the channel, an impulse J just sufficient to keep end A at rest being thereby delivered.)

The energy model 315

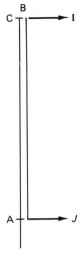

Fig. 10.21

Step C (ii)
The equations of impact may be written: (horizontal momentum)

$$[M_1(l\sin\theta\cdot\dot\theta)+M_3(2l\sin\theta\cdot\dot\theta)]_0^{\pi/2}=I+J$$

i.e., since $\sin\pi/2=1$ and $\sin 0=0$,

$$l\dot\theta(M_1+2M_3)=I+J \qquad 10\text{--}27$$

where $\dot\theta$ has the value given by equation 10–26.
  (Angular momentum change of bar AB about G)

$$M_1l^2/3\,\dot\theta=lI-lJ$$

i.e.

$$\tfrac{1}{3}M_1l\dot\theta=I-J \qquad 10\text{--}28$$

From 10–27 and 10–28,

$$I=l\dot\theta(M_3+\tfrac{2}{3}M_1)$$

By Newton's third law, the impulse delivered by B on the catch is of equal magnitude: it is this quantity which we require to be greater than K,

$$l\dot\theta(M_3+\tfrac{2}{3}M_1)>K \qquad 10\text{--}29$$

From 10–26 and 10–29 we deduce that the masses M_1, M_2 and M_3 must be such that

$$\frac{3}{2}\frac{M_1+2M_2}{3M_3+M_1}\cdot l>\frac{K^2}{g(M_3+\tfrac{2}{3}M_1)^2} \qquad 10\text{--}30$$

The mass M_3 of the door will be chosen mainly on grounds unconnected with the automatic latching device: we may regard the value of M_3 to be given, whereas the quantities $m_1 = M_1/M_3$ and $m_2 = M_2/M_3$ are variables to be chosen by the designer of that mechanism. Economy will probably dictate that m_1, $m_2 \ll 1$. Condition 10–30 may be rewritten

$$\frac{(m_1+2m_2)}{1+\tfrac{1}{3}m}(1+\tfrac{2}{3}m_1)^2 l > \frac{2K^2}{gM_3^2}$$

i.e., retaining only first-order terms in m_1 and m_2, and noting that to this order $(1+\tfrac{1}{3}m_1)^{-1} = 1-\tfrac{1}{3}m_1$,

$$(m_1+2m_2)l > 2K^2/(gM_3^2) \qquad 10\text{--}31$$

Step D_1
For chosen $2l$ = desired width of the open doorway, inequality 10–31 tells the designer that increasing the mass M_2 is more effective than increasing M_1. The mass M_1 will therefore be chosen to give sufficient strength, whereafter that sliding mass M_2 will be prescribed which adequately satisfies 10–31. For very smooth horizontal and vertical channels the designer might first try out his design allowing a safety margin of, say, 50% in the value of K (by writing $3K/2$ instead of K on the right-hand side of the equality derived from equation 10–31), and might consequently consider for his first trial an M_2 value of

$$M_2(\equiv m_2 M_3) = 9K^2/(lgM_3) - \tfrac{1}{2}M_1 \qquad 10\text{--}32$$

Should 10–32 give a negative value, the designer will not worry – his choice of M_1 has already ensured the satisfaction of condition 10–31, and M_2 may take any value. If he wishes, he may use a stronger spring catch (greater K) without requiring an excessively large M_2.

Step E_1
The measurement of most of the quantities in inequality 10–31 presents no difficulty: the practicability of assigning a value of K is less apparent, and without a means of deducing numerical conclusions, our work would be a frivolous exercise rather than a practical investigation. The writer performed a crude experiment on the spring catch of the door of a kitchen cabinet. The experiment is represented in FIG. 10.22. The impulse due to a mass of $\tfrac{1}{4}$ kg dropped from a height of $\tfrac{1}{2}$ m (conveyed through an inextensible string passing over a pulley) just opened the door. Using the equation of energy, $v = \sqrt{(2gh)}$; the estimated value of K was $mv = \tfrac{1}{4}\sqrt{10}$, whence $K \approx 1$. The writer is satisfied that a useful order-of-magnitude value of K can quite easily be attributed to any chosen spring catch.

Three further investigations seem highly desirable, namely:

(i) a consideration of friction effects;
(ii) a consideration of whether it will be easy to open the door again;
(iii) a consideration of closing from a part-opened position, or fixing the bar AB so that its final and/or initial positions are inclined.

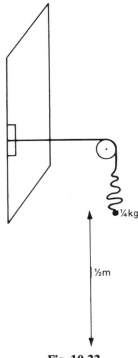

Fig. 10.22

Of these, (iii) involves no new principles, and will not be carried out here. For (ii), the reader may like to provide an argument showing that – irrespective of the mechanism used to operate the catch – the impulse required to jerk the door into the fully open position will normally be at least $2K$. Matter (i) will be the subject of the next round of inquiry which we now undertake.

Step A_2
We shall first find estimates of the normal reactions R_1, R_2 and R_3 (FIG. 10.23) between the blocks A, B, the door, and the channels in which they run, on the assumption that the channels are smooth. From those estimates of the R's, estimates will be deduced of the energy loss caused by friction when the channels are **not** considered to be 'perfectly smooth', so an improved equation of energy may be written down, including a correction term for friction. The plan is manageable even though, like most successive approximation methods, it lacks elegance.

Step B_2
It will be postulated that the door must 'slide parallel' i.e. that the torque on it must not be so great that the top right-hand corner of either the door or of block B presses upwards on the channel in which it runs. (This postulate ensures that only the sum $R_2 + R_3$ will be needed in subsequent calculations of the friction force.)

318   An introduction to mechanics and modelling

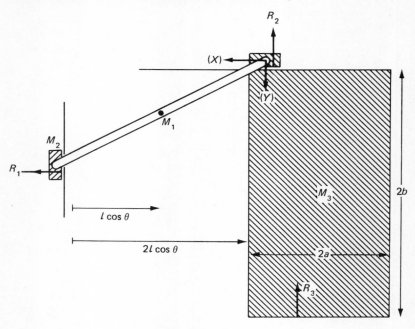

Fig. 10.23

Step C_2
The equations of motion of the mass centre of the frictionless moving system (derived from $\mathbf{F} = M\ddot{\mathbf{r}} \equiv M_1\ddot{\mathbf{r}}_1 + M_2\ddot{\mathbf{r}}_2 + M_3\ddot{\mathbf{r}}_3$) are: (horizontal)

$$R_1 = -M_1 \frac{d^2}{dt^2}(l\cos\theta) - M_3 \frac{d^2}{dt^2}(2l\cos\theta) \qquad 10\text{-}33$$

(vertical)

$$R_2 + R_3 - (M_1 + M_2 + M_3)g = -M_1 \frac{d^2}{dt^2}(l\sin\theta) - M_2 \frac{d^2}{dt^2}(2l\sin\theta)$$

$$10\text{-}34$$

We note that $R_1 > 0$ for $\ddot{\theta} > 0$, and $(R_2 + R_3) > 0$.

The energy losses when friction is taken into account may therefore be estimated during closing as

$$\int \mu R_1 \,|dy| + \int \mu(R_2+R_3)\,|dx|, \quad \text{where } y = 2l\sin\theta,\ x = 2l\cos\theta$$

$$= -\mu(\tfrac{1}{2}M_1 + M_3)\int \ddot{x}\,dy + \mu Mg \cdot 2l + \mu \int (\tfrac{1}{2}M_1 + M_2)\ddot{y}\,dx$$

where M has been written for $(M_1 + M_2 + M_3)$, and we have used the fact that, during closing, $dy > 0$, $dx < 0$ (as well as $\ddot{x} < 0$). (The consideration of signs is tricky, and the reader will do well to write out his own version of the steps leading to equation 10–35.) Since

$$\int \ddot{x}\,dy + \int \ddot{y}\,dx = \int (\ddot{x}\dot{y} - \ddot{y}\dot{x})\,dt$$

$$= \int \frac{d}{dt}(\dot{x}\dot{y})\,dt$$

$$= [\dot{x}\dot{y}]_{\theta=0}^{\theta=\pi/2}$$

$$= 0$$

(since when $\theta = 0$, $\dot{x} = -2l \sin\theta \cdot \dot{\theta} = 0$ and when $\theta = \pi/2$, $\dot{y} = 2l \cos\theta \cdot \dot{\theta} = 0$).

$$\int \ddot{x}\,dy = -\int \ddot{y}\,dx$$

The energy loss due to friction thus takes the form,

$$\text{Energy loss} = 2\mu Mgl - \mu M \int_{x=2l}^{x=0} \ddot{y}\,dx \qquad 10\text{–}35$$

That $\ddot{y} < g$ may be assumed (or justified by an energy argument), so

$$-\int_{x=2l}^{x=0} \ddot{y}\,dx \leq 2gl$$

It follows that

$$2\mu Mgl \leq \text{energy loss} \leq 4\mu Mgl$$

We may write energy loss $= n\mu Mgl$ where $2 < n < 4$. With this correction term, the energy equation (evaluated at $\theta = \pi/2$) has the form

$$l^2 \dot{\theta}^2 (\tfrac{2}{3}M_1 + 2M_3) - lg(2M_2 + M_1) = -n\mu Mgl$$

whence

$$l^2 \dot{\theta}^2 = lg(2M_2 + M_1 - n\mu M)/(\tfrac{2}{3}M_1 + 2M_2) \qquad 10\text{–}36$$

It is apparent from this that the device will not work at all unless $2M_2 + M_1 > \mu nM$ (n is not known precisely, but $2 < n < 4$).

The new form of the condition for operating the catch is

$$\frac{(2M_2 + M_1 - n\mu M)}{(\tfrac{2}{3}M_1 + 2M_2)} (M_3 + \tfrac{2}{3}M_1)^2 > \frac{K^2}{lg} \qquad 10\text{–}37$$

The 'parallel sliding' condition referred to in step B₂ is satisfied if

$$(R_2 + R_3)a \geq bX + \mu bR_3 + aY$$

where the significance of X and Y can be seen from FIG. 10.23. To obtain a bold estimate, put $R_2 \approx Y$, $R_3 \approx M_3 g$, and $X \ll M_3 g$. This yields

$$\mu \leq (a-b)/b \qquad 10\text{–}38$$

as a condition which it would be wise to satisfy.

320 An introduction to mechanics and modelling

Step D_2
The design engineer should find it useful to check whether his proposed values of K, l, M_1, M_2, M_3, μ, a and b satisfy the inequalities 10–37 and 10–38. The former, in particular, imposes severe restrictions on the value of μ which may be tolerated (we recall that $n > 2$ and may be nearer to 4). It must be recognized that when surfaces rub together over a long period, the value of μ may undergo progressive change; the designer must satisfy himself that such an increase in friction will not unduly reduce the lifetime of the closing device. He might wish to consider the possibility of incorporating roller bearings.

Step E_2
The foregoing calculations were approximate; in particular the friction forces were estimated from values of R_1, R_2 and R_3 appropriate to perfectly smooth channels. This will lead to overestimates of friction effects. Further mathematical work is unlikely to be cost-effective: to spend one day on a calculation which will save the equivalent of many days on construction trials and materials is a good use of time; it would not be well-judged to spend a week on slight improvements which in the end would provide less reliable information (and little other illumination) than one day's practical work. The writer decides to terminate the mathematical investigation at this point.

10.6 Discussion of the equilibrium of bodies acted upon only by conservative forces

10.61 Equilibrium of a particle

Suppose that all the forces acting on a particle are conservative, that their vector sum is **F**, and that the sum of their potentials, i.e. the potential of **F**, is $V(\mathbf{r})$.

Using rectangular cartesian axes we may write

$$\mathbf{F} = (X, Y, Z) \quad \text{and} \quad d\mathbf{r} = (dx, dy, dz)$$

The change in potential in the small displacement d**r** satisfies

$$-dV = X\,dx + Y\,dy + Z\,dz \qquad \text{10–39(a)}$$

(this is a consequence of the definition

$$V(\mathbf{r}) = -\int_c^{\mathbf{r}} \mathbf{F} \cdot d\mathbf{r}).$$

Equation 10–39(a) may be re-expressed, using the notation of partial differentiation, in the three equations,

$$-\partial V/\partial x = X; \quad -\partial V/\partial y = Y; \quad -\partial V/\partial z = Z \qquad \text{10–39(b)}$$

The condition for the equilibrium of the particle, starting from rest, is therefore

$$\partial V/\partial x = 0; \quad \partial V/\partial y = 0; \quad \partial V/\partial z = 0 \qquad \text{10–40}$$

The energy model

In the partial derivatives of equations 10–39(b) and 10–40, it is assumed that V is expressed as a function of x, y and z only. No-one has found a wholly satisfactory and neat way of overcoming the shortcomings of the partial derivative notation. In interpreting partial derivatives, the reader must be guided by the context.

Example

To find the equilibrium position of a mass M hanging at the end of a light elastic string of modulus λ and natural length l_0.

Taking the zero of gravitational potential energy at the level of the point of suspension, when the mass is vertical distance x below this point (where $x > l_0$), and using equations 10–8 and 10–9, the potential energy is

$$V = -Mgx + \frac{\tfrac{1}{2}\lambda}{l_0}(x - l_0)^2$$

$\partial V/\partial x = \mathrm{d}V/\mathrm{d}x$ when V involves only the variable x

$$= -Mg + \frac{\lambda}{l_0}(x - l_0)$$

For equilibrium

$$\lambda(x - l_0)/l_0 = Mg, \quad \text{i.e.} \quad x = l_0 + Mgl_0/\lambda \qquad 10\text{–}41$$

Completeness might also require us to consider the possibility $x < l_0$, in which case the second term on the right-hand side of the equation for V would be absent: the requirement $-Mg = 0$ which then arises, cannot be satisfied by any choice of x, so 10–41 is the only solution. [As usual, checking the 'obvious' gives a satisfying reassurance, but only at the expense of (often considerable) extra work.]

10.611 Use of generalized orthogonal coordinates

Instead of cartesian coordinates, any orthogonal system of coordinates may be used. Adhering to a common notation, an element $\mathrm{d}\mathbf{r}$ is written

$$\mathrm{d}\mathbf{r} = (h_1\,\mathrm{d}u, h_2\,\mathrm{d}v, h_3\,\mathrm{d}w)$$

(e.g. if, in a plane, r, θ coordinates are used, $\mathrm{d}\mathbf{r} = (\mathrm{d}r, r\,\mathrm{d}\theta)$, so with $u = r$ and $v = \theta$, $h_1 = 1$ and $h_2 = r$). Using the same general u, v, w coordinates, let F_u, F_v, F_w denote the components of $\mathbf{F}$ in the directions of $h_1\,\mathrm{d}u$, $h_2\,\mathrm{d}v$, $h_3\,\mathrm{d}w$ respectively; then

$$-\mathrm{d}V(u, v, w) = F_u h_1\,\mathrm{d}u + F_v h_2\,\mathrm{d}v + F_w h_3\,\mathrm{d}w$$

whence $-\partial V/\partial u = h_1 F_u$ etc.

i.e. $$F_u = -\frac{1}{h_1}\frac{\partial V}{\partial u}; \quad F_v = \frac{-1}{h_2}\frac{\partial V}{\partial v}; \quad F_w = \frac{-1}{h_3}\frac{\partial V}{\partial w} \qquad 10\text{–}42$$

and the condition for equilibrium is

$$\partial V/\partial u = \partial V/\partial v = \partial V/\partial w = 0 \qquad 10\text{–}43$$

10.62 Graph of potential energy, and its uses

Suppose we know the initial energy E_0 of a particle which moves under conservative forces, and suppose that we plot the sum ϕ ($= V$) of the potentials of all those forces. In the simplest case ϕ is a function of one coordinate (which we shall call u), and the graph of ϕ may then be drawn, for example as in FIG. 10.24, for a hypothetical set of forces.

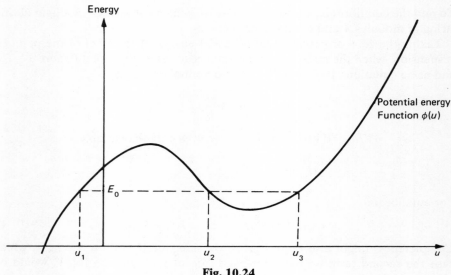

Fig. 10.24

We know that at all times

$$\text{PE} + \text{KE} = E_0 \quad \text{where } \text{KE} = \sum \tfrac{1}{2} m_i v_i^2 \geq 0$$

It follows that

$$\text{PE} = E_0 - \sum \tfrac{1}{2} m_i v_i^2$$

i.e.

$$\phi \leq E_0$$

at all times in the actual motion.

Now look at FIG. 10.24 in which the shape of a particular potential energy function ϕ is shown. In any motion with energy E_0, the coordinate u must (see FIG. 10.24) be restricted to one of the ranges

$$-\infty \leq u \leq u_1 \quad \text{or} \quad u_2 \leq u \leq u_3$$

in which $\phi \leq E_0$.

If we know the initial values of u and du/dt, we can make further deductions about the nature of the subsequent motion. Thus if, **initially**, $u_2 < u < u_3$ and du/dt is > 0, then u continues to increase until du/dt changes from $+$ to $-$, i.e. until du/dt passes through the value zero.

When $\dot{u} = 0$, kinetic energy is zero and $E_0 = \phi$. By consulting the graph we see that this first occurs at $u = u_3$.

What happens next? The point $u = u_3$ is not a point of equilibrium as the requirement $d\phi/du = 0$ (see 10–43) is satisfied only at maxima and minima of the graph of ϕ. So, after reaching $u = u_3$ we cannot have $\dot{u} = 0$ maintained; nor can u increase beyond u_3; we deduce that $\dot{u}$ must change sign, and motion continues with u decreasing until $u = u_2$.

The graph has enabled us to conclude that under the stated initial conditions, the ensuing motion is oscillatory, with coordinate u taking all values in $u_2 \leq u \leq u_3$. This is sometimes called being 'trapped in a potential well'. The sections of the graph which are higher than E_0 are called potential barriers.

If initially $u < u_1$ and $\dot{u} > 0$, then by similar reasoning, the coordinate increases to $u = u_1$ after which u diminishes without limit.

If initially $\dot{u} < 0$, the reader should have no difficulty in drawing his own conclusions about the subsequent course of the motion in each of the separate cases $u_2 < u < u_3$ and $u < u_1$ initially.

The case of $\phi =$ function of two independent coordinates u and v may be considered in the same way. If axes of u and v are taken at right angles to one another in a horizontal plane (directions which have no necessary connection with the geometrical significance of u and v), then $\phi(u, v)$ may be plotted as a surface whose height above the horizontal u-v plane is a measure of ϕ (FIG. 10.25). For any given value of E_0 certain regions on the ϕ-surface are forbidden, others allowed, as in the two-dimensional curve of FIG. 10.24. Potential wells and potential barriers may exist, but in addition to straightforward humps and dips, a surface may have horizontal elements of area where the ground rises to right and to left, but descends before and behind like a mountain pass – such a point is called a saddle-point of the surface (e.g. point S of FIG. 10.25).

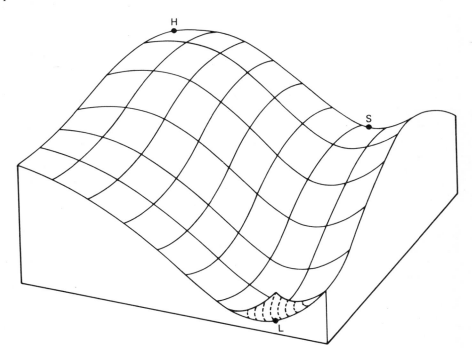

Fig. 10.25

Exercise 6

1. A particle of mass m is acted on by a force whose potential energy is $V = ax^2 - bx^3$, where a and b are positive constants.
   (i) Sketch the graph of V.
   (ii) Determine in terms of a and b the value V_0 of V at the (local) maximum of the potential.
   (iii) The particle starts at the origin $x = 0$ with positive velocity u_0 in the x-direction. Show that, if $u_0 < u_c$ where u_c is a certain critical velocity, the particle will remain confined to a region near the origin.
   (iv) State the value of u_c in terms of m, a and b.
   (v) Prove that the greatest negative value of x which can be achieved with a positive projection velocity u_0 is $-\tfrac{1}{3}a/b$.
   (vi) Explain why, when $u_0 < u_c$, the end-points of the region to which the particle is confined cannot be positions of equilibrium.

10.63 Stability of equilibrium, and small oscillations

If, from a position of equilibrium, a particle is given a small velocity, its initial energy E_0 will be slightly greater (viz. by $\tfrac{1}{2}mv^2$) than the value of ϕ at the equilibrium position. Initially, the coordinate u can only move away from the stationary value of ϕ. From the argument of § 10.32 above, applied to FIG. 10.25, we infer that, if the stationary value of ϕ is a minimum (e.g. point L) the ensuing motion will consist of small oscillations about the equilibrium position; while if the stationary value is a maximum (e.g. point H) the particle will not return to the initial 'equilibrium' position unless some other mechanism intervenes to reverse the direction of motion.

We describe the above behaviour by saying that a maximum of ϕ corresponds to a position of **unstable** equilibrium, while at a minimum, the equilibrium is stable.

Another way of looking at the situation is to consider not a small velocity, but a zero starting velocity from a position slightly displaced from the equilibrium position. The reader should study the implications of such a displacement in the context of FIG. 10.24, and should come to the same conclusions about the stability/instability of minima/maxima as before; indeed, the impossibility of regaining the maximum is even clearer in this case.

> A position of equilibrium is stable or unstable according as $d^2\phi/du^2$ is positive or negative. If $d^2\phi/du^2$ is zero, the equilibrium is called 'neutral', and the behaviour of higher derivatives must be considered.

10–44

10.64 Example

A uniform heavy rod has two smooth, and comparatively light, rings fixed at its ends A and B. The rings may slide on two fixed smooth wires OP and OQ. The plane OPQ is vertical and P and Q lie below O and both OP and OQ > AB. The inclinations of the wires to the horizontal are α and β ($\beta > \alpha$), (see FIG. 10.26). Show that one position of equilibrium of the rod is given by $2 \tan \theta = \cot \alpha - \cot \beta$, where θ is the inclination of the rod to the horizontal. Prove that the equilibrium is stable.

The energy model

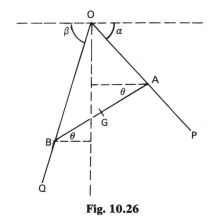

Fig. 10.26

Let AB represent the rod whose length we write as l, and whose centre of mass (mid-point) is G. Let $OA = a$, $OB = b$, and let $\bar{y}$ denote the depth of G vertically below O.

$$\bar{y} = \tfrac{1}{2}(a \sin \alpha + b \sin \beta)$$

$$\frac{a}{l} = \frac{\sin (\beta - \theta)}{\sin (\alpha + \beta)} \; ; \qquad \frac{b}{l} = \frac{\sin (\alpha + \theta)}{\sin (\alpha + \beta)}$$

so

$$\bar{y} = l[\sin \alpha \sin (\beta - \theta) + \sin \beta \sin (\alpha + \theta)]/2 \sin (\alpha + \beta)$$

and

$$d\bar{y}/d\theta = l[\sin \beta \cos (\alpha + \theta) - \sin \alpha \cos (\beta - \theta)]/2 \sin (\alpha + \beta)$$

$$\phi = -Mg\bar{y}$$

Positions of equilibrium are given by $d\bar{y}/d\theta = 0$ and the equilibrium will be stable if, also $d^2\phi/d\theta^2 > 0$, i.e. $d^2\bar{y}/d\theta^2 < 0$. The rod is in equilibrium when

$$\sin \beta \cos (\alpha + \theta) - \sin \alpha \cos (\beta - \theta) = 0$$

i.e.

$$\cos \theta (\sin \beta \cos \alpha - \sin \alpha \cos \beta) - 2 \sin \alpha \sin \beta \sin \theta = 0$$

i.e.

$$2 \tan \theta = \cot \alpha - \cot \beta \qquad \qquad 10\text{--}45$$

Now

$$d^2\bar{y}/d\theta^2 = [-\sin \beta \sin (\alpha + \theta) - \sin \alpha \sin (\beta - \theta)]/2 \sin (\alpha + \beta)$$

The angles α, β and θ all lie between 0 and $\pi/2$. Consequently

$$\sin \beta \sin (\alpha + \theta) \text{ is positive}$$

and, as $\beta > \theta$ (A being below O),

$$\sin \alpha \sin (\beta - \theta) \text{ is positive}$$

Thus $d^2 \bar{y}/d\theta^2$ is negative and the equilibrium is stable.

Exercise 7

1. Using the expression $\frac{1}{2}k(r-l_0)^2$ for the potential energy of a spring (in the absence of gravity), and using r as one of the three coordinates in applying the method of § 10.611, verify that the natural length l_0 is the equilibrium value of r irrespective of the spring's orientation.

2. Using the same expression for the potential energy as in question 1, but writing $r = \sqrt{(x^2 + y^2 + z^2)}$, show that equations 10–40 lead to the same solution as in question 1, but also suggest there may be a second solution. Is the second solution valid? Discuss.

3. A particle of mass m slides on a smooth plane which is inclined at angle α to the horizontal. The particle is attached to one end of a spring of stiffness k whose other end is fixed to a point on the plane. Find the extension of the spring in the equilibrium position, and verify that the equilibrium is stable.

4. A rough plank of thickness $2b$ is laid across a fixed cylinder of radius a, its length being perpendicular to the axis of the cylinder, and rests in equilibrium at an angle α with the horizontal. Find an expression for the potential energy when the plank is rolled through an angle θ from the equilibrium position, and deduce that the equilibrium is stable provided $b < a \cos^2 \alpha$.

5. An open cylindrical can whose height is 12 cm and diameter 12 cm is poised on the top of a rough sphere. Show that the least diameter of the sphere consistent with stability is 9.6 cm. (Use the composite body theorem to find the centre of mass of the can.)

Answers to exercises

(The last digit given is not necessarily significant: most answers have been calculated using the approximation $g \simeq 10$ m s^{-2}.)

Chapter 1

Exercise 1 (p. 21)
1 (i) Yes; addition, subtraction; (ii) Yes; substraction; (iii) No; (iv) Yes; multiplication, division.
2 No.
3 No.
4 No.

Exercise 2 (p. 24)
1 $7\mathbf{v} - 5\mathbf{u}$.
4 (i) $\frac{1}{8}(\mathbf{a} + 7\mathbf{b})$; (ii) $(11\mathbf{a} + 3\mathbf{b})/14$.
5 Draw $\overrightarrow{CP}$ parallel to $\overrightarrow{AB}$; $\angle PCD$ (with appropriate sign) is the required angle.

Exercise 3 (p. 27)
1 (i) $\frac{1}{2}(\mathbf{a} + \mathbf{b})$; (ii) $\frac{1}{2}(\mathbf{a} + \mathbf{b})$. Diagonals bisect one another.
3 (i) $\frac{1}{3}(\mathbf{a} - 2\mathbf{b})$; (ii) $7(\mathbf{a} + \mathbf{b})/9$; (iii) $7:2$.
6 $-9:1$ ($\overrightarrow{AR} = \frac{1}{8}\overrightarrow{BA}$).

Exercise 4 (p. 28)
1 (i) 53.1 below x-axis (or $-53.1°$ with x-direction); 5; (ii) 26.5° above x-axis; $2\sqrt{5} = 4.47$; (iii) 53.1 above negative direction of x-axis (or 126.9° with x-direction); 5; (iv) 78.7° below negative direction of x (or 258.7° with x-direction); $\sqrt{26} = 5.1$.
2 $\sqrt{34}$ ($=5.83$); 31° above negative direction of x-axis.
3 Same as question 2.
4 $\mathbf{a}$: 18.4° below x-axis; $\sqrt{10}$. $\mathbf{b}$: 63.5° above negative x-direction; $\sqrt{5}$. $\mathbf{a}+\mathbf{b}$: 26.5° above x-axis; $\sqrt{5}$. $(2k-1)/(3-k) = \tan 30°$; $k = (3+\sqrt{3})/(1+2\sqrt{3}) = 1.06$.
5 Equation 1-5.

Exercise 5 (pp. 33–4)
1 4.07; 10.4° above the x-axis.
2 $(-4, 1)$.
3 18; 60° above AB, i.e. along AD.
4 4.66; 25.1° below x-axis.

5 $X = F_1 \cos \alpha_1 - F_2 \cos \alpha_2 - F_3 \cos \alpha_3 + F_4 \cos \alpha_4$; $Y = F_1 \sin \alpha_1 - F_2 \sin \alpha_2 + F_3 \sin \alpha_3 - F_4 \sin \alpha_4$;
6 $F = \frac{1}{2}W$; $T = W\sqrt{3}/2$.
7 $T_1 = 2W/(1+\sqrt{3}) = 0.732W$; $T_2 = W\sqrt{6}/(1+\sqrt{3}) = 0.897W$.

Exercise 6 (pp. 36–7)
1 $-5\mathbf{i}+\mathbf{j}$; $-(4\mathbf{i}+13\mathbf{j})/3$; $\sqrt{26}$; $(-4/3, -13/3)$.
2 (i) $\mathbf{J} = -\sin \alpha \mathbf{i} + \cos \alpha \mathbf{j}$; (ii) $\overrightarrow{OB} = \cos \alpha \cos \beta - \sin \alpha \sin \beta)\mathbf{i} + (\sin \alpha \cos \beta + \cos \alpha \sin \beta)\mathbf{j}$.
3 $\overrightarrow{OP} = \mathbf{a} + \lambda \mathbf{b} = (a_1 + \lambda b_1)\mathbf{i} + (a_2 + \lambda b_2)\mathbf{j}$; $x = a_1 + \lambda b_1$; $y = a_2 + \lambda b_2$.
4 $-2\mathbf{i}+4\mathbf{j}-4\mathbf{k}$; $(0, -3, 6)$.
5 $3\sqrt{5}$; $(-\mathbf{j}+2\mathbf{k})/\sqrt{5}$; direction cosines $0, -1/\sqrt{5}, 2/\sqrt{5}$.
6 $2\sqrt{6}/3$; $4/3$; $\sqrt{5}/3$.
7 $65.9°, 144.7°, 65.9°$; $24.1°, 54.7°, 24.1°$.

Chapter 2

Exercise 1 (p. 40)
4 $6t\mathbf{a}$; $\mathbf{a}-\mathbf{b}$; $-\mathbf{a}/t + 2\mathbf{b}/t^3$.

Exercise 2 (pp. 44–5)
Imagine yourself in a position of responsibility: are you prepared to stand by your answer?

Exercise 3 (pp. 51–2)
2 $(6, 0)$.
3 $\dot{r} \sin \theta + r\dot{\theta} \cos \theta$.
4 $(\ddot{r} - r\dot{\theta}^2) \sin \theta + (2\dot{r}\dot{\theta} + r\ddot{\theta}) \cos \theta$; $\ddot{r} - r\dot{\theta}^2$; two functions whose graphs intersect do not generally have the same slope/curvature/derivatives at the point of intersection.

Exercise 4 (p. 54)
2 $35.3°$.
3 $100.9°$.

Exercise 5 (pp. 64–5)
1 $\sqrt{10}$; $\sqrt{5}$.
2 (i) $(3, 1, -1)$; $(1, -2, 1)$; $(4, -1, 0)$; $(2, 3, -2)$; (ii) $\sqrt{11}$; $\sqrt{6}$; $\sqrt{17}$; $\sqrt{17}$.
5 $82.2°$.
6 $3/\sqrt{29}$ (≈ 0.557).
7 (i) $5\mathbf{k}$; (ii) $-\mathbf{i}-4\mathbf{j}-7\mathbf{k}$; (iii) -4; (iv) -4; (v) -4.
8 (i) $\mathbf{r} \wedge \mathbf{s}$; (ii) $(\mathbf{r} \wedge \mathbf{s})/|\mathbf{r} \wedge \mathbf{s}|$; (iii) $(\mathbf{r} \wedge \mathbf{s}) \cdot \mathbf{t}/|\mathbf{r} \wedge \mathbf{s}|$.
10 $2\mathbf{i}+2\mathbf{j}-\mathbf{k}$; $-\mathbf{i}-3\mathbf{j}+2\mathbf{k}$; $-\mathbf{i}+\mathbf{j}-\mathbf{k}$; $101.1°$; $p = \frac{1}{3}\sqrt{26} \approx 1.70$.
11 $\mathbf{b} \wedge \mathbf{a} = \mathbf{c} \wedge \mathbf{a}$.
12 $36.1°$; $90°$; $53.9°$; $\pm(16\mathbf{i}-17\mathbf{j}-27\mathbf{k})/35.1$.

16 $m(\ddot{r}-r\dot{\theta}^2) = R$; $\dfrac{m}{r}\dfrac{d}{dt}(r^2\dot{\theta}) = S$.

17 $m\ddot{s} = T$; $m\dot{s}^2/\rho = N$.

Exercise 6 (pp. 68–9)

1
```
         0       a
  ───────┼───────▶───────
  −∞<λ<0   0<λ<1    1<λ<∞
```

2 (i) $\mathbf{r} = \lambda(\mathbf{i}+\mathbf{j}+\mathbf{k})$ i.e. $x = y = z$ $(=\lambda)$;
(ii) $\mathbf{r} = \mu(\mathbf{i}+\mathbf{j})$ i.e. $x = y$ $(=\mu)$, $z = 0$; and 2 similar;
(iii) $\frac{1}{3}\sqrt{3}, \frac{1}{3}\sqrt{3}, \frac{1}{3}\sqrt{3}; \frac{1}{2}\sqrt{2}, \frac{1}{2}\sqrt{2}, 0; \frac{1}{2}\sqrt{2}, 0, \frac{1}{2}\sqrt{2}; 0, \frac{1}{2}\sqrt{2}, \frac{1}{2}\sqrt{2}$;
(iv) 54.7°.

3 (i) $x/4 = y/3 = z/2$; (ii) $x/4 = y/3$, $z = 0$; $y/3 = z/2$, $x = 0$; $z = \frac{1}{2}x$, $y = 0$.
(iii) $4/\sqrt{29}, 3/\sqrt{29}, 2/\sqrt{29}; 4/5, 3/5, 0; 0, 3/\sqrt{13}, 2/\sqrt{13}; 2/\sqrt{5}, 0, 1/\sqrt{5}$.
(iv) 42°, 56.1°, 68.2° (v) 60.1°, 75.6°, 44.3°.

4 (i) $\mathbf{r} = 6\mathbf{j}+2\mathbf{k}+\lambda(\mathbf{i}-7\mathbf{j}-9\mathbf{k})$; (ii) $x = (y-6)/(-7) = (z-2)/(-9)$.

5 $\mathbf{r} = \mathbf{i}+\mathbf{j}+\mathbf{k}+\lambda(\mathbf{i}+\mathbf{j})$: $x = y$, $z = 1$;
$\mathbf{r} = \mathbf{i}+\mathbf{j}+\mathbf{k}+\mu(\mathbf{j}+\mathbf{k})$; $y = z$, $x = 1$;
$\mathbf{r} = \mathbf{i}+\mathbf{j}+\mathbf{k}+\nu(\mathbf{i}+\mathbf{k})$; $z = x$, $y = 1$.

6 $10\mathbf{i}+5\mathbf{j}+6\mathbf{k}$ (i) $\mathbf{r} = \lambda(10\mathbf{i}+5\mathbf{j}+6\mathbf{k})$; (ii) $\mathbf{r} = 4\mathbf{i}+2\mathbf{j}+5\mathbf{k}+\mu(2\mathbf{i}+\mathbf{j}+\mathbf{k})$.

7 $2\mathbf{i}-6\mathbf{j}+3\mathbf{k}$; 7; $\mathbf{r} = 2\mathbf{i}+3\mathbf{j}-\mathbf{k}+\lambda(2\mathbf{i}-6\mathbf{j}+3\mathbf{k})$.

8 $(21, 8, 2)$, $(-15, -16, -6)$.

Exercise 7 (pp. 70–1)

1 (i) $\mathbf{r} = 3\mathbf{i}+\mathbf{j}+2\mathbf{k}+\lambda(\mathbf{i}-\mathbf{j}-\mathbf{k})$; (ii) $\mathbf{r} = 3\mathbf{i}+\mathbf{j}+2\mathbf{k}+\lambda(\mathbf{i}-\mathbf{j}-\mathbf{k})+\mu(\mathbf{i}+2\mathbf{j}+\mathbf{k})$
or, better, $\mathbf{r} \cdot (\mathbf{i}-2\mathbf{j}-3\mathbf{k}) = 7$, i.e. $x - 2y - 3z = 7$.

2 Line $\mathbf{r} = (1+\lambda)\mathbf{i}-2\mathbf{j}+(1-2\lambda)\mathbf{k}$; plane $\mathbf{r} \cdot (\mathbf{i}-2\mathbf{k}) = 0$; point $(1.2, -2, 0.6)$.

4 $\mathbf{r} = \mathbf{i}-\mathbf{j}+2\mathbf{k}+\lambda(5\mathbf{i}+5\mathbf{j}+\mathbf{k})$; units of length ÷ units of force; $5/\sqrt{51}, 5/\sqrt{51}, 1/\sqrt{51}$.

5 No. Equivalent neither to a single force nor to a couple.

6 $(0, 0, 0)$, $(1, 1, 0)$, $(2, 0, -3)$ satisfy requirements, but many others.

Exercise 8 (p. 76)

3 2.19, 90°, 50.8°, 39.2°.   **4** $\frac{1}{3}\sqrt{3} = 0.577$.

5 $\frac{2}{3}\sqrt{6} = 1.63$ **6** $\pm\frac{1}{2}(\mathbf{a}\wedge\mathbf{b}+\mathbf{b}\wedge\mathbf{c}+\mathbf{c}\wedge\mathbf{a})$.

Exercise 9 (p. 78)

1 (i) independent; (ii) independent; (iii) dependent.

Chapter 3 Assignment

Congratulations on any serious attempt!

Chapter 4

Exercise 2 (pp. 96–100)

1 (a) $\alpha = 53.1°$, $T = 5$; (b) $T_A = 2.06$, $T_B = 1.60$; (c) $\alpha = 0$, $T = 3$;

330 An introduction to mechanics and modelling

  (d) $\alpha = 90°$, $\beta = 53.1°$; (e) $\alpha = 8.7°$, $T = 1.21$; (f) $T_A = 9\sqrt{2}/2$, $T_B = \sqrt{2}/2$; (g) $T_A = 3 - 4\sqrt{3}/3 \simeq 0.69, T_B = 8\sqrt{3}/3 \simeq 4.62$: (h) $T = 0.984$, $\theta = 31.4°$; (i) $T = 4.11$, $\alpha = 66.2°$.

3 Probably flexible; probably not light.
4 $T = R = 1/2\sqrt{3} \simeq 0.289$ kgf.
5 $\frac{1}{4}\sqrt{3}$, $\frac{1}{4}$ kgf
6 $F_1 = F_2$ if equilibrium, or, if pulley is small or light, in motion.
7 (a) 5; (b) 8; (c) $T_1 = \frac{1}{2}W$; $T_2 = \frac{3}{4}W$; $T_3 = F = \frac{7}{8}W$; (d) $\frac{1}{8}P$.
8 Not 'free-running' in contexts where a fly's weight is significant.
9 Vector sum of four $= W$ upwards.
10 Each $= \frac{1}{4}W$ upwards.
11 (i) Simplicity; (ii) Strains in legs and asymmetric loading ignored, so a real table might bear less than suggested by answer to 10.
12 (i) 3/5, 4/5 kgf; (ii) 23/25, 11/25 kgf; (iii) 12.8, −15.4; equilibrium cannot be sustained – string BC will slacken and B move to a new position.

Exercise 3 (pp. 114–16)

1 $2g/9 \simeq 2.2$ m.
3 1 s; 6 m s^{-1} downwards.
4 Other forces besides mg act after reaching the ground, so the equation $\ddot{x} = -g$ no longer holds good.
5 Velocity of centre is constant.
6 kt/m; $k^2t^2/ms = 2k^2t^2/(kt^2 + 2ma)$.
7 $2ma\omega^2 \sinh \omega t \equiv ma\omega^2(e^{\omega t} - e^{-\omega t})$.
8 2.02, 2.85; (a) and (b) all points of body have same velocity/acceleration at all times.
9 (i) $\sqrt{(g/21)}$; (ii) 60°.
10 mc^2/s.
12 If θ is angle with downward vertical, $(-R) = Mg \cos\theta + \frac{1}{2}Ml\dot{\theta}^2$; $S = Mg \sin\theta + \frac{1}{2}Ml\ddot{\theta}$.
13 (i) $\frac{1}{2}V$; (ii) $0 + a\ddot{\theta}$, $0 + a\dot{\theta}^2$; (iii) 0, $V/2a$, $\frac{1}{4}mV^2/a$; (iv) $2\pi a$.
14 $a\dot{\theta}^2 + g \sin\theta$, $a\ddot{\theta} - g \cos\theta$.
15 $mg \sin\alpha \cos\alpha/200(M + m \sin^2\alpha)$ (after start from rest).
16 3960 m; 58.5 m s^{-1}.

Chapter 5

Exercise 1 (pp. 130–2)

1 $5:8$.
2 $m_1\ddot{x}_1 = -m_2\ddot{x}_2$; $m_1\ddot{y}_1 = -m_2\ddot{y}_2$; $[m_1\dot{x}_1] = -[m_2\dot{x}_2]$; $[m_1\dot{y}_1] = -[m_2\dot{y}_2]$.
3 $0.84:1$; horizontal component of impulse negligible for period of time involved.
4 $38/9$ m s^{-1}; as question 3.
5 (a) $\frac{1}{2}$ m s^{-1}; (b) $4/7$ m s^{-1}.
6 1.15×10^4 kg; air resistance; diminish; SI.

Answers to exercises 331

7 $(44+4/9)$N; no, the earth left behind also exerts force.
8 (i) 115.2 N; (ii) 144 N.
9 $(7, -9)$; $\sqrt{(130)}$; 52.1° 'below' **i**.
10 Constant velocity.
11 True.
12 46.9°.
13 $0.446 V$.
14 Ignoring friction, 389 N, 944 N.
15 10900 N.
16 (ii) The assumption that no unnamed factor is involved; (iii) £ of wages per cup of tea drunk.
17 61°.

Exercise 2 (pp. 133–5)
1 $m = m_0 F/F_0$.
2 $m = m_0 l_0/l$.
3 6×10^{24} kg.

Exercise 3 (pp. 143–6)
1 0.826.
2 1.174.
3 On the bridge.
4 1.73 s.
5 (i) $6.22\,m$ N downhill; 3.05 m, $6.1\,\text{m s}^{-1}$ uphill; (ii) $1.93\,m$ N downhill; 0, 0; (iii) $2.40\,m$ N uphill; 0, 0; (iv) $3.14\,m$ N uphill; 0.28 m, $0.56\,\text{m s}^{-1}$ downhill.
6 $mg(\sin \alpha - \mu \cos \alpha)$; $mg(\sin \alpha - \mu \cos \alpha)/(\cos \alpha + \mu \sin \alpha)$; $mg(\sin \alpha - \mu \cos \alpha)/(\cos \beta + \mu \sin \beta)$ where $\beta = \tan^{-1} \mu$.
7 $Mmg/(m+M)$; $mg/(m+M)$ horizontally, $mg/(m+M)$ vertically.
10 $\tfrac{2}{3}mg$; $7g\sqrt{3}/9$.
11 Between $1.0g$ and $3.5g$ i.e. $\simeq 10\,\text{m s}^{-2}$ and $35\,\text{m s}^{-2}$ in the sense which gives support.
12 (i) 10 N; (ii) 5.6 N; (iii) 0.
13 $\simeq 1$ radian, or $\simeq 58°$.
14 $20\,\text{s}^{-1}$.
15 $5.8\,\text{s}^{-1}$.
16 $27m \cos 3t = 27m(2-x)$; constant velocity; $m\ddot{x} = 27m(2-x)$.
19 $M(M+m)g \cos \alpha /(M + m \sin^2 \alpha)$.

Chapter 6

Exercise 1 (pp. 158–9)
1 $4b/9$; $13b/18$.
2 $2/7(m)$ along line joining O to position of 5 kg mass.
3 Point $(0, 0)$.

332 An introduction to mechanics and modelling

5 (i) on that axis; (ii) in that plane; (iii) on the line of intersection of the planes; (iv) at the point of intersection of the three planes.
8 $M\ddot{x} = 0$; $M\ddot{y} = -Mg$.
10 $2\ddot{x} = 4$; $\ddot{x} = 2$; $x = 9$ m.
11 $M\ddot{\mathbf{r}} = \mathbf{F}$ ($= \sum \mathbf{F}_i$); if couplings slacken, different parts may have different accelerations; $\simeq 1.5$ m s^{-1}.

Exercise 3 (pp. 172–4)
1 (a) $x\ddot{y} - y\ddot{x}$; (b) $xY - yX$.
4 $-r_i\dot{\theta}^2$, $r_i\ddot{\theta}$.
5 (i) $a\Omega$; (ii) $a\Omega$; (iii) $4a\Omega$ from starting point; (iv) 0, Ω;
(v) coordinates $a(1-\cos 4\Omega)$, $a(4\Omega - \sin 4\Omega)$ measured from original position parallel and perpendicular to original diameter, and velocity components $a\Omega \sin 4\Omega$, $a\Omega (1 - \cos 4\Omega)$.
6 $x = l \cos \Omega t$, $y = l\Omega t + t^3/2M + l \sin \Omega t$ (or equivalent for different choice of axes).
7 (i) $4 \times m(a/2)^2 = ma^2$; (ii) mb^2; (iii) $m(a^2 + b^2)$.
8 20.
9 (i) 119; (ii) 111.

Exercise 4 (pp. 181–2)
1 (i) $5/3$ s^{-1}; (ii) $16/3$ m s^{-1}; (iii) $\frac{5}{6} \sin 2 \simeq 0.76$; (iv) $-(89 + \cos 10)/30 \simeq -2.94$.
2 (a) 1; 1; 0; (b) $\sqrt{5}$; $\sqrt{2}$; 3.
6 $\frac{2}{3}ma^2$.
8 0.045; 0.09; 0.14.
9 0.2 N m.
10 Reduction by 8 m s^{-1}.

Chapter 7

Exercise 1 (p. 188)
1 (a) $\ddot{x}$; (b) $\dot{x}\ddot{x}$.
2 (i) $\frac{1}{2}\dot{s}^2 = s^2 + 3s + C_1$; (ii) $\dot{s} = t^2 + 3t + C_2$; (iii) $\frac{1}{2}\dot{\theta}^2 = -2\theta^2 + C_3$; (iv) $\frac{1}{2}\dot{p}^2 = -3 \cos p + C_4$; (v) $\dot{y}^2 = C_5 e^y$.
3 $3 \ln 3 - 2 \ln 2 - 1 \simeq 0.91$.

Exercise 2 (p. 189)
1 e^t, e^{2t}.
2 e^{2t}, e^{3t}.
3 e^{-2t}, e^{3t}.
4 e^{2t}, $t e^{2t}$.

Exercise 3 (p. 190)
1 $A e^t + B e^{2t}$.
2 $A e^{2t} + B e^{3t}$.
3 $A e^{-2t} + B e^{3t}$.
4 $e^{2t}(A + Bt)$.

Exercise 4 (p. 192)
1 $s = A e^{3t} + B e^{-3t}$.
2 $s = A \cos 2t + B \sin 2t$.
3 $x = A e^{2t} + B e^{3t}$.
4 $x = A e^{-2t} + B e^{-3t}$.
5 $s = A e^{2t} + B e^{-t}$.
6 $\theta = A e^t + B e^{2t}$.
7 $y = A e^x + B e^{-4x}$.
8 $z = e^t(A \cos 2t + B \sin 2t)$.
9 $p = A \cos \sqrt{5} t + B \sin \sqrt{5} t$.

Answers to exercises 333

Exercise 5 (pp. 202–3)
1. (i) 5; (ii) $4\sqrt{5}$; (iii) 2; (iv) $\sqrt{10}$; (v) $\frac{1}{2}\sqrt{34}$ (Note $x = \frac{1}{2}(3\sin 2t - 5\cos 2t)$; (vi) 26.
2. (i) 1.066; (ii) -8.035; (iii) 3.181; (iv) -6.835.
3. Writing $n^2 = \Omega^2 - (1/2t_0)^2$, $x = x_0 e^{-(t/2t_0)}[\cos nt + (1/2nt_0)\sin nt]$, $\dot{x} = -\Omega^2(x_0/n) e^{-t/2t_0} \sin nt$.
4. $\frac{1}{8}e^{-t}(4\cos 2\sqrt{2}t + \sqrt{2}\sin 2\sqrt{2}t)$; $-(9/8)\sqrt{2}\, e^{-t} \sin 2\sqrt{2}t$.
5. (a) $x = (9e^{-t} - e^{-9t})/16$; (b) $x = (e^{-t} - e^{-9t})/16$.
6. $e^{(-\frac{1}{2}\pi\sqrt{2})}$; $\frac{1}{2}\pi\sqrt{2}$.
8. $k = 2.68$, $\omega = 1.84$.
9. Yes (to expected accuracy), $0.32I$; $1.61I$.

Exercise 6 (pp. 206–7)
1. (i) $A + B e^{-t} - (\cos 4t + 4 \sin 4t)/68$; (ii) $A + B e^{2t} - (2t^3 + 3t^2 + 3t)/12$; (iii) $A \cos 3t + B \sin 3t - (\sin 5t)/16$; (iv) $A e^t + B e^{5t} + (\cos t - 8 \sin t)/26$; (v) $A e^t + B e^{-5t} - \frac{1}{2}t e^t$; (vi) $A \cos 3t + B \sin 3t - \frac{1}{6}t \cos 3t$.
2. $x = \dfrac{F_0}{k_1 + k_2 - p^2}\left[\sin pt - \dfrac{p}{\sqrt{k_1+k_2}} \sin\sqrt{(k_1+k_2)}\,t\right]$; $\dot{x} = \dfrac{F_0 p(\cos pt - \cos\sqrt{(k_1+k_2)}t)}{k_1 + k_2 - p^2}$; $p = \sqrt{(k_1+k_2)}$; x initially proportional to t^3.

Exercise 7 (p. 210)
1. (i) $q = 5E_0[1 - e^{-4t}(\cos 2t + 2 \sin 2t)]$; (ii) $100 \sin 2t$.

Exercise 8 (pp. 212–13)
2. $(1/k) \ln \frac{3}{2}$; $(g/k^2)(1 - \ln \frac{3}{2})$.
4. 86 m.
5. (i) $28.2\,\text{m s}^{-1}$, 0.66 s (ii) $29.5\,\text{m s}^{-1}$, 0.64 s.
6. $uc/\sqrt{(u^2+c^2)}$.

Exercise 9 (p. 217)
1. (i) $y = (t+C)\exp(-t)$; (ii) $y = \frac{1}{4}x^3 + C/x$; (iii) $y = x^4/6 + C/x^2$; (iv) $z = C\, \text{cosec}\, t + \frac{2}{3}\sin^2 t$; (v) $y = 1/(t + 1 + C \exp t)$; (vi) $y = \cos x(C - 3 \ln \cos x)^{1/3}$.
2. (i) $y = (3x + 4 + 5x^{-3})/12$; $-\infty < x < 0$ or $0 < x < \infty$; (ii) $y = (2t - t^2)^{-1/2}$; $0 < t < 2$; (iii) $v^2/(4 + v^2) = (1/5)\exp 8t$; $-\infty < t < \infty$; (iv) $y = \ln x + (2 - \ln 4) \ln 4/\ln x$; $0 < x < \infty$.
3. $v = \exp -\kappa t(1 - 0.005t)[V + g\int_0^t \exp \kappa t(1 + 0.005t)\, dt]$.
4. $v^2 = \exp[(2\kappa/a) \exp(-ay)][v_0^2 \exp(-2\kappa/a) - 2g \int_0^y \exp\{-2\kappa/a \exp(-ay)\}\, dy]$.

Exercise 10 (pp. 223–4)
No answers – you are on your own.

Chapter 8

Exercise 1 (pp. 228–9)
1. 31.1 units to right of O.
2. 1.77 units below O.

334 An introduction to mechanics and modelling

4 12 g acting vertically down at $x = 7/12$.
5 $(m_1 + m_2 + \cdots)g$ at $x = (m_1 x_1 + m_2 x_2 + \cdots)/(m_1 + m_2 + \cdots)$.
6 $\sqrt{3}$; $x = 6$.
7 Magnitude $\sqrt{3}$; with D as origin and DC as x-axis, coordinates $(-a, 0)$.
8 $x = -5/4$.

Exercise 2 (pp. 241–3)
4 $F_1 - R_2 = 0$; $R_1 + F_2 - 3W = 0$; $6(F_2 + R_2 \tan \alpha) - 7W = 0$.
5 (a) $R_2 = F_1 = 0$; (b) $F_2 = 0$; (c) $R_1 = 3W$; (d) $R_2 = F_1$.
6 No; Yes; $\mu_1 \tan \alpha \geq 7/18$. A right-hand side larger than 7/18 will be necessary.
13 $X_1 = 0$; $Y_1 = W$; $N_1 = LW$.
14 When man is at distance x, $X_1 = 0$; $Y_1 = 5W$; $N_1 = W(L + 4x)$.
15 $N_2 = 0$; $\tfrac{1}{2}W$.
17 (i) $\tfrac{1}{2}W$; $\tfrac{1}{2}W$. (ii) $X_3 = 0$; $Y_3 = W(L - x)/2L$; $N_3 = (-)Wx(2L - x)/4L$.

Exercise 3 (pp. 248–9)
1 Upper hinge, horizontal pull $Mgb/2d$; lower hinge, horizontal push $Mgb/2d$ and vertical force Mg.
2 $Y_1 = \tfrac{1}{2}Mg + Mah/d$; $Y_2 = \tfrac{1}{2}Mg - Mah/d$.
4 (vi) $M\ddot{x} = Mg \sin \alpha - X$; $Y = Mg \cos \alpha$; $Ma^2 \ddot{\theta} = aX$; (vii) $\ddot{x} = \tfrac{1}{2}g \sin \alpha$.
5 $\mu \geq \tfrac{1}{2} \tan \alpha$.
6 $Ma^2 \ddot{\theta} = \mu M g a \cos \alpha$.
10 $V_0^2 / 2\mu g$.
12 Writing $AB = a$, $BC = b$, tipping or sliding will occur according as $\tfrac{1}{2}Mga/b$ (the tension required for tipping) is less than or greater than $\tfrac{1}{2}Mg(\mu + \mu')a/\{a + (\mu - \mu')b\}$, the tension needed for sliding, i.e. according as $a \lessgtr 2b\mu'$.

Chapter 9

Exercise 2 (pp. 269–70)
3 (a) 3/7 s, 1.1 m; (b) 1/7 s, 0.12 m.
4 $h = 7a/5$.
6 (i) $2V/7$; (ii) zero angular velocity.
7 $9V/14$, cue ball $3V/14$.

Exercise 3 (p. 274)
2 ω_1 does not matter; $0.2\omega_2/\sqrt{(\omega_2^2 + \omega_3^2)}$ determines initial rotation about axis parallel to that of ω_2; $0.2\omega_3/\sqrt{(\omega_2^2 + \omega_3^2)}$ determines angle which initial velocity makes with line of centres. (Factor 0.2 applies only if slipping occurs throughout period of impulse.)

Exercise 4 (pp. 278–81)
1 (i) $20/p$; (ii) $10/p$ in opposite sense; (iii) $N_0 = Pd$.
3 (i) Safer; (ii) None.

Answers to exercises   335

4 (i) $\ddot{x} = \ddot{x} + c\ddot{\theta}\sin\theta + c\dot{\theta}^2\cos\theta$; $\ddot{y} = c\ddot{\theta}\cos\theta - c\dot{\theta}^2\sin\theta$; (iii) $I_G\ddot{\theta} = Mc(\mu'\sin\theta - \cos\theta)(g + c\ddot{\theta}\cos\theta - c\dot{\theta}^2\sin\theta)$.
8 Be careful to match the signs of θ, $\dot{\theta}$.

Assignment
Congratulations on any progress you have made.

Chapter 10

Exercise 1 (pp. 295–6)
1 $\frac{1}{2}(V+v)$; $\frac{1}{4}M(V-v)^2$.
2 6.2 m s^{-1}; 21 cm; alternate simple harmonic motion (while string is stretched) and free rise and fall under gravity while string is slack, rising to point of initial release once in each cycle. Weight of string, air resistance, departure from Hooke's law, disturbance from vertical (by slack part of string), yield of support.
3 (a) $mg(k+1)$ inwards; (b) $mg(k-5)$ inwards; when radius makes $\sin^{-1}(\frac{2}{3}) = 41.8°$ above horizontal.

Exercise 2 (p. 298)
2 $10^9 \times$ rocket mass; no: it is the **change** of potential energy which matters, $\delta(\gamma M/r) = -\gamma M/r^2 \delta r$; and **change** in potential energy due to sun $\ll$ that due to earth.
4 (iv) $\frac{1}{3}R_1$.

Exercise 3 (pp. 301–2)
1 5.
2 $-\frac{1}{2}$; 2; $+\frac{1}{3}$ (AB parallel to x-axis).
3 (i) 0; (ii) 5e.
4 -5π.

Exercise 4 (p. 303)
1 (a) 2.2 kW; (b) 6.7 kW; (c) $508(15+t)/729$.
2 43 200 kW.
3 ≥ 30 kW (because of what is ignored, this can be regarded only as an order of magnitude estimate).
4 627 kJ; 494 kJ; energy loss due to friction mainly dispersed as heat.
5 (a) Assumptions include no potential contribution from other planets and sun;
   (b) assumption that all available energy is converted to heat/light.

Exercise 5 (p. 312)
1 $\omega^2(Ma^2+I_G)/2Mg\sin\alpha$.
2 $Mga^2\sin\alpha/(Ma^2+I_G)$.
4 $\frac{2}{3}m_1a^2\dot{\theta}^2 + \frac{1}{6}m_2b^2\dot{\phi}^2 + \frac{1}{2}m_2[4a^2\dot{\theta}^2 + b^2\dot{\phi}^2 + 4ab\dot{\theta}\dot{\phi}\cos(\phi-\theta)]$.

Exercise 6 (p. 324)
1 (ii) $4a^3/27b^2$; (iv) $[8a^3/27b^2m]^{1/2}$.

Exercise 7 (p. 326)
3 $(mg\sin\alpha)/k$.

Abbreviations and symbols

$a, b, \ldots p, q, \ldots$	constant and ... variable scalars.
$\mathbf{a}, \mathbf{b}, \ldots \mathbf{p}, \mathbf{q}, \ldots$	constant and ... variable vectors.
CF	complementary function (§ 7.3).
c.g.s.	system of units based on centimetre, gram and second.
CG	centre of gravity (§ 8.3).
CM	centre of mass (§ 6.2).
2D	two-dimensional.
3D	three-dimensional.
e	constant for which $\frac{d}{dx}(e^x) = e^x$; ($e \simeq 2.718$).
e	(i) eccentricity of ellipse (§ 5.312).
	(ii) coefficient of restitution (§ 5.34).
$\exp x$	same as e^x.
$f(x)$	function of x.
$\mathbf{F}$	force; often vector sum of all forces on a body.
g	constant representing acceleration due to gravity (near earth).
G	label for position of CM.
G	gravitational constant 6.7×10^{-11} SI units (§ 5.31)
$\mathbf{G}$	gravitational field (§ 5.31).
$\mathbf{i}, \mathbf{j}, \mathbf{k}$	unit vectors respectively parallel to x-, y-, z-axes.
I_0, I_G or I	moment of inertia about axis through O, G, or an unstated point.
$\mathbf{I}, \mathbf{J}, \mathbf{K}$	mutually perpendicular unit vectors, whose directions may or may not be constant with time.
$\mathbf{J}, J$	impulse, magnitude of impulse.
J	Joule (unit of energy).
k	(i) general constant (scalar).
	(ii) stiffness of spring (§ 4.4).
KE	kinetic energy.
kg	kilogram.
ln	logarithm to base e; $\ln x \equiv \log_e x$.
L, M, N	components of moment $\mathbf{M}$.
m	metre
$m, m_1, m_2 \ldots, M$	masses
mi h^{-1}	miles per hour; formerly written m.p.h.
M_0	(2D) magnitude of moment of force about 0; ($M_0 = pF = xY - yX$).
M_G	(2D) magnitude of moment of force about G; ($M_G = (x - \bar{x})Y - (y - \bar{y})X$.].
$\mathbf{M}_0$	(3D) moment of force about O ($\mathbf{M}_0 = \mathbf{r} \wedge \mathbf{F}$).

Abbreviations and symbols 337

$\mathbf{M}_G$	(3D) moment of force about G $[\mathbf{M}_G = (\mathbf{r} - \bar{\mathbf{r}}) \wedge \mathbf{F}]$.
MI	moment of inertia (§ 6.733).
n, N	number of objects considered.
N	newton (unit of force).
N	see L, M, N above.
$\mathbf{N}$	$= N\mathbf{k} =$ component vector of angular momentum.
$0(\theta)$	same order as θ, i.e. limit $0(\theta)/\theta$ is finite.
p	(i) perpendicular distance.
	(ii) an angular velocity.
PE	potential energy (§ 10.22).
PI	particular integral (§ 7.3).
r	magnitude of $\mathbf{r}$; radius of sphere/orbit/earth, etc.
$\mathbf{r}$	position vector.
$\bar{\mathbf{r}}$	position vector of centre of mass.
R, S	radial and transverse components of force (§ 2.33).
$\mathbf{R}$	(i) position vector relative to a moving point,
	(ii) component vector in, say, horizontal plane, of position vector $\mathbf{r}$.
s	second.
s	distance or arc length.
$\mathbf{s}$	position vector relative to centre of mass; $\mathbf{s} = \mathbf{r} - \bar{\mathbf{r}}$.
SI	system of units based on metre, kilogram and second.
$t, t_1, t_2 \ldots$	times (usually measured in seconds).
u	value of velocity, often an initial value.
u, v, w	mutually perpendicular coordinates or components.
w, W	weight.
X, Y, Z	(i) components of $\mathbf{F}$ (see above) in $\mathbf{i}, \mathbf{j}, \mathbf{k}$ frame.
	(ii) cartesian coordinates in $\mathbf{I}, \mathbf{J}, \mathbf{K}$ frame.
x, y, z	components of $\mathbf{s}$ (see above).

INDEX

Acceleration, 41
  components, 47, 48, 49
  constant, 109
  due to gravity, 108, 132
Active/passive nature of forces, 88
Air resistance, 220
Algebraic laws for vectors, 19
  for scalar and vector products, 62, 63
Algebraic symbols, 6
Amplitude, 200
Angle between vectors, 22
Angle of a triangle, 55
Angles between lines and planes, 70
Angular equation of motion, 174
Angular momentum, 164
  fundamental theorem, 167
  summary of main results, 183, 184
Angular velocity, 43
Anticommutative law, 63
Antiparallel, 22
Antiparallelism, 33
Apparent weight, 135
Applied forces, 151
Approximation, 147
Apses of an orbit, 298
Arbitrary constants, 195
Associative law, 62
Attractive law, 163
Attributes of force, 86
Axis of a wrench, 234ff.

Base vectors, 27
Bead on wire, 114
Beats, 207
Billiard balls, 265

Calculus, important formal results, 186, 187
Capacitor, 208
Cartesian components of velocity and acceleration, 47
Categories of force, 178
Central axis of wrench, 234ff.
Centre of gravity, 235, 236
Centre of mass, 101, 129, 153, 158
  position, 159
  summary of formulae, 183
  for standard bodies, 184
Centrifugal force, 136
Choice of coordinates, 218
Coefficient of friction, 138
  values, 139
Coefficient of restitution, 146, 266
  values, 147
Collisions, 121
Collision experiments, 118, 123
Collision of billiard balls, 265ff.
Commutative law, 62
Complementary function, 204
Components, 27, 53, 55
Component forms of Newton's second law, 101
Components equivalent to resultant, 241

Components of force, 29
  of a moment, 62
  of a vector product, 61
Composite body theorem for centre of mass, 159
Compound pendulum, 237
Condenser, 208
Conical pendulum, 105
Conservation of energy, 294ff.
  for systems of bodies, 306
Conservation of linear momentum, 122, 128, 129
Conservative force, 287, 289ff
Constant acceleration, 109
Constants of proportionality, 8
Constraints, 176
Contact forces, 163
Conversion of units, 102
Coplanar vector, 77
Coriolis force, 136
Cosmological frame of reference, 263
Couples, 228, 232, 233

D'Alembert's principle, 164
Damage, 252, 257
Damped harmonic motion, 196, 201
Damping, 197
Dashpot, 196
Definitions, 7, 9
Derivatives of vectors, 38
  of unit vectors, 45
Derived units, 9
Determination of centre of mass, 159
Diagrams to model forces, 88
Differential equations, 186ff.
  linear, 189ff.
  homogeneous, 190
Differentiation of vectors, 38
  unit vectors, 45
Dimensional analysis, 9
Direction cosines, 34, 53, 54
Distance, 4
Distributive law, 62
Drag coefficient, 212, 219
Dynamically equivalent bodies, 240

Earth, mass and radius, 134, 261
Elastic modulus, 95, 96
Electric circuits, 208
Electromotive force, 208
Ellipse, 134
Empirical laws, 8, 147
Energy, 285ff.
Energy equation, 287, 289, 295
Equating coefficients of vectors, 77
Equation of angular motion, 177
Equation of energy, 287, 289, 295
Equation of planes, 69
Equation of translation, 174
Equations of a straight line, 36, 66
Equations of motion, 124

61
88
118
235

Edinburgh Mathematical Society

HUNDREDTH SESSION 1981-82

The fifth Ordinary Meeting of the session will be held in Lecture Theatre 3 of the Appleton Tower, University of Edinburgh, at 4.30 p.m. on Friday, 12th February 1982.

Professor F. J. Ursell

"Uniform Asymptotics of Integrals"

Tea will be served from 4.00 p.m. in the concourse area of the Appleton Tower.

Department of Mathematics,
James Clerk Maxwell Building,
The King's Buildings,
Edinburgh, EH9 3JZ.

T. A. GILLESPIE,
J. MARTIN,
Honorary Secretaries.

Index

Equilibrium, conditions for, 91, 232
   using energy, 320ff.
Equivalent bodies, 240
Equivalent single force, 228
Equivalent systems of force, 225ff.
Escape velocity, 296ff.
Exponential function, significance of, 188ff.
External forces, 150

Fall of a raindrop, 217ff.
Flexible, 94, 95
Force, 29, 86, 123, 124
Forced oscillations, 203
Forces at an axis, 239
Forces of contact, 243ff.
Forces, how to enumerate, 113
Foucault pendulum, 261
Frames of reference, 1, 258ff.
Frenet-Serret formulae, 69
Friction, 138
Frictionless hinge, 244

General solutions, 195
Generalized orthogonal coordinates, 321
Geometrical constraint, 247
Geometry in 3D, 65
Geometry of the ellipse, 134
Graphical methods, 143
Gravitation, 93, 132
Gravitational forces, 236
Gravitational potential energy, 290, 291
Gyroscopic effects, 176

Harmonic Oscillator, 192, 199, 200
Hinges, 244
Homogeneous differential equation, 190
Hooke's Law, 94, 95, 197
Hubble's Law, 263

Impulse, 126, 129, 131, 180, 252
Inductance, 209
Inextensible, 95
Inhomogeneous differential equation, 190, 203
Integrating factor, 216
Internal forces, 150
Inverse square law (gravitation), 132
   potential, 296

Jointed rods, 43
Joule, 302

Kepler, 134
Kinetic energy, 252, 286, 287
   of a system, 309
   of a rigid body, 309
Know-how, 131

Laboratory frame of reference, 260
Ladders, 279
Law of association, 62, 63
Law of commutation, 62
Law of distribution, 62, 63
Laws of nature, 5, 8
Light bodies, 94, 125
Light pulleys, 113
Limit of sequence of vectors, 39

Line of action of resultant, 227
Linear dependence, 76
Linear differential equations, 189ff.
   with variable coefficients, 214
Linear first-order differential equations, 214ff.
Linear independence, 76
Linear momentum of a system, 155
Logarithmic decrement, see common ratio exponent, 207

Mass, 100, 106, 117, 121
   additive property, 130
   centre, 154
   of the earth/sun, 133, 134
Mathematical model, 1
Matter, 1
Measurement of force, 90
Medians of a triangle, 24
Modelling, 1, 16, 79, 84, 217, 267
   examples, 79, 217, 251, 267, 312
Modelling of objects and forces, 252, 253
Modulus of elasticity, 95, 96
Moment of a force, 62
   sum of moments, 164
Moment of inertia, 170, 173
   formulae for standard bodies, 185
Moment of momentum, see angular momentum
Moments in 3D, 233
Moments of external forces, basic result, 167
Moments, theorems, 230, 232
Momentum, 117, 126
Motion in a circle, 50, 103–5
   on a sphere, 141

Newton (unit of force), 29, 101, 124
Newton's definitions, 116
Newton's law of gravitation, 132
Newton's laws of motion, 117
Newton's second law, 52, 93, 100
Newton's third law, 90, 92
Non-conservative forces, 299
Normal reaction, 136
Normal to a curve, 47
Numerical methods, 143

Orbits, 105, 134, 298
Oscillations, 192ff.

Parallel axis theorem, 177
Parallelism, 33
Parametric equations, 67
Particles, 123
Particular integrals, 204
   trial forms, 206
Pendulum, 237
Period of s.h.m., 201
   of orbit, 106
Perpendicular distances, 74
Pitch of wrench, 235
Point particle modelliing, 1, 150
Points, 15
Polar coordinates, 104
Polar equation of ellipse, 134
Position vector, 29
Potential energy, 287ff.
   barrier/well, 323

340 An introduction to mechanics and modelling

Potential energy of elastic spring/string, 291, 292, 308
Potential for inverse square law attraction/repulsion, 296
Potentials for standard force-types, 290ff.
Power, 57, 302
Predictions, 2
Principle of superposition, 189
Product of vectors, 52
Projectile, 107
   example, 112
Pulleys, 98, 113, 249

Radial and transverse components, 49
Radian measure, importance of, 187
Radius of curvature, 48
Radius of gyration, 185
Rain, 217
Rational mechanics, 3
Reaction, 88, 92, 117
   at an axis, 239
Real-life problems, 79
Reference tables, 139, 147, 148, 184, 185
Relative acceleration, 42
Relative velocity, 41, 44, 130
   examples, 311, 313
Repulsive forces, 163
Resistance (electrical), 209
Resistance proportional to velocity, 210, 222
   proportional to square of velocity, 211, 222
Resistances and use of energy, 299
Resisting medium, 210ff.
Resonance, 205
Resultant, 18, 19, 30
Reynold's number, 219
Rigid, non-rigid, 90
Rolling and/or slipping, 268
Rotating tube, 115
Rotation about a fixed axis, 175, 176
   about a moving axis, 182
   of rigid body, 163
Rough, 138
Rules of algebra, 1

Satellite, 105, 134
Scalar product, 55
Scalars, 20, 21
Scale dependence, 8
Second order differential equation, 190
Sets of forces, 88, 156
Shortest distance, 75
Sidereal reference frame, 263
Sign conventions, 110
Simple harmonic motion, 192, 199, 200
Simple pendulum, 237
Skew lines, 21, 75
Sliding condition, 256
Sliding friction, 139
Small oscillations, 324
Small pulleys, 113
Smooth, 138
Smooth hinge, 244
Space, 4
Speed, 43
Sphere, motion on, 141
Spring, oscillation of, 192
Stability of equilibrium, 324

Standard bodies, formulae for centres of mass and moments of inertia, 184, 185
Stationary frame of reference, 1
Stiffness, 197
Stokes' law, 220
Strings, 94
Structure of mechanics, 264
Subtraction of vectors, 22
Sum of vectors, 18, 19
Sun, mass and radius, 134, 262
Superposition, 189, 191
Swerving of a ball, 270
Symmetry properties and determination of centre of mass, 159
System of particles model, 150

Tables of physical data, 148
Tangential and normal components, 47
Tension, 94
Three dimensional space, 19, 20
   motion in, 175
   systems of forces in, 233, 234
Time, 3
Transients, 205
Translation (motion of), 122, 174
Triangle law, 18
Triple products of vectors, 63
Two-dimensional motion of a rigid body, 163ff.
   definition, 167

Uniqueness theorem, 194, 195
Unit of power 302
Unit vectors, 22
Units, 4, 8, 102
Unknown forces, 176
Unstable equilibrium, 324

Vectors, 17
Vector addition, 18
Vector algebra, 15, 19, 23
Vector area, 60, 71
Vector equality, 17
Vector law of addition for forces, 92
   of composition, 18
Vector product, 58
Vector products of $\mathbf{I, J, K}$, 60
Vector sum, 18
Vector sum of forces, 156
   significance, 152
Velocity, 39
   components, 47, 48, 49
   instantaneous change of, see impulse
Velocity and acceleration, 47
Volume of tetrahedron or parallelopiped, 73

Water resistance, 223
Watt (definition), 302
Wedge problem, 116
Weight, 93, 106, 135
Weightlessness, 93, 135
Work, 287
Work done at contact points, 303
   smooth or rolling contacts, 305
Work done by the resultant, 293
   by internal forces, 304
Wrench, 234

Zero of potential energy, 289
Zero vector, 22